Allergy and Asthma

NEW TRENDS AND APPROACHES
TO THERAPY

Allergy and Asthma

NEW TRENDS AND APPROACHES
TO THERAPY

EDITED BY

A.B. KAY

Professor,
Department of Allergy
and Clinical Immunology,
The National Heart and Lung Institute,
London

BLACKWELL SCIENTIFIC PUBLICATIONS

OXFORD LONDON

EDINBURGH BOSTON MELBOURNE

Editorial offices:
Osney Mead, Oxford OX2 0EL
(*Orders*: Tel: 0865 240201)
8 John Street, London WC1N 2ES
23 Ainslie Place, Edinburgh EH3 6AJ
3 Cambridge Center, Suite 208
Cambridge, Massachusetts 02142, USA
107 Barry Street, Carlton
Victoria 3053, Australia

First published 1989

Set by Macmillan India Ltd, Bangalore 25; printed and bound in Great Britain by Billing and Sons Ltd, Worcester

DISTRIBUTORS

USA
Year Book Medical Publishers
200 North LaSalle Street
Chicago, Illinois 60601
(*Orders*: Tel: (312) 726-9733)

Canada
The C. V. Mosby Company
5240 Finch Avenue East
Scarborough, Ontario
(*Orders*: Tel: 416 298-1588)

Australia
Blackwell Scientific Publications
(Australia) Pty Ltd
107 Barry Street
Carlton, Victoria 3053
(*Orders*: Tel: (03) 347-0300)

British Library
Cataloguing in Publication Data

Allergy and asthma
1. Man. Allergies. Therapy—serials
2. Man. Bronchi. Asthma. Therapy
I. Kay, A.B. (A. Barry)
616.97′06

ISBN 0–632–02239–6

Contents

List of contributors

J.P. Arm *Department of Respiratory Medicine, Guy's Hospital, London SE1 9RT*

R.M. Auty *Fisons plc—Pharmaceutical Division, Research and Development Laboratories, Loughborough, Leicestershire LE11 0RH*

M.K. Bach *Hypersensitivity Diseases Research, The Upjohn Company, Kalamazoo, Michigan 49001, USA*

P.J. Barnes *Department of Thoracic Medicine, National Heart and Lung Institute, Brompton Hospital, Dovehouse Street, London SW3 6LY*

P. Braquet *Institut Henri Beaufour, Research Laboratories, 17 Avenue Descartes, F-92350 LePlessis-Robinson, France*

B. Björkstén *Department of Pediatrics, Faculty of Health Sciences, Linköping University, Sweden*

N. Chand *Wallace Laboratories, Cranbury, New Jersey 08512-0181, USA*

C. De Vos *UCB Pharmaceutical Sector, Braine-l'Alleud, Belgium*

J.M. Dewdney *Beecham Pharmaceuticals, Research Division, Great Burgh, Yew Tree Bottom Road, Epsom, Surrey KT18 5XQ*

W. Diamantis *Wallace Laboratories, Cranbury, New Jersey 08512-0181, USA*

S. Dreborg *Department of Pediatrics, Faculty of Health Sciences, University Hospital, S-581 85 Linköping, Sweden*

R.J. Flower *School of Pharmacy and Pharmacology, University of Bath, Claverton Down, Bath BA2 7AY*

A.W. Ford-Hutchinson *Department of Pharmacology, Merck Frosst Canada Inc., PO Box 1005, Pointe Claire-Dorval, Quebec, Canada H9R 4P8*

S.T. Holgate *Immunopharmacology Group, Medicine I, D Level, Centre Block, Southampton General Hospital, Southampton, Hants SO9 4XY*

D. Hosford *Institut Henri Beaufour, Research Laboratories, 17 Avenue Descartes, F-92350 LePlessis-Robinson, France*

D.H. Katz *Division of Immunology, Medical Biology Institute, and Quidel, 11077 North Torrey Pines Road, La Jolla, California 92037, USA*

A.B. Kay *Department of Allergy and Clinical Immunology, National Heart and Lung Institute, Brompton Hospital, Dovehouse Street, London SW3 6LY*

T.H. Lee *Department of Allergy and Allied Respiratory Disorders, Guy's Hospital, London SE1 9RT*

J.M. Mencia-Huerta *Institut Henri Beaufour, 1 avenue des Tropiques, 91952 Les Ulis, France*

J.L. Perhach *Department of Clinical Pharmacology and Pharmacokinetics, Wallace Laboratories, PO Box 1001, Cranbury, New Jersey 08512-0181, USA*

G.D. Phillips *Immunopharmacology Group, Medicine I, D Level, Centre Block, Southampton General Hospital, Southampton, Hants SO9 4XY*

A. Rosenberg *Wallace Laboratories, Cranbury, New Jersey 08512-0181, USA*

R.D. Sofia *Wallace Laboratories, Cranbury, New Jersey 08512-0181, USA*

S.I. Wasserman *Department of Medicine, School of Medicine, UCSD Medical Center, University of California at San Diego, San Diego, California 92103, USA*

Preface

The development of new forms of therapy for bronchial asthma and other diseases which have a substantial allergic component is one of the most active areas of medical research today. Traditionally the approach to treatment has been allergen avoidance, antiallergic drugs and, where indicated, immunotherapy. Some recent advances in these broad areas of asthma and allergy treatment are covered in the present volume. New ideas on the prevention of allergy are discussed with particular reference to neonatal sensitization. There are many new and novel approaches to pharmacotherapy. Historically, lipid mediators, particularly 5-lipoxygenase products and platelet-activating factor, are thought to play an important role in the asthma process. The future promise of agents which inhibit the formation and action of leukotrienes and platelet-activating factor is addressed and a number of new and novel agents which inhibit the interaction of agonists with specific receptors or prevent their formation are described. The prospect of altering membrane phospholipids by dietary manipulation with fish oil is another intriguing approach. It is recognized that in bronchial asthma and a number of allergic diseases, a specialized inflammatory component is prominent and therefore, not surprisingly, compounds have been developed which affect, for instance, eosinophil accumulation, the mucosal mast cell and infiltration of other cell types. Compounds such as nedocromil sodium, an antiasthma drug with steroid-sparing effects; cetirizine, a selective H_1-antagonist which affects infiltrating inflammatory cells, particularly eosinophils; azelastine, a novel oral anti-asthma compound with several modes of action, are all described. Intriguingly, the prospect of modulating the allergic response by biotechnology is closely becoming a reality. This is discussed in a general sense and more specifically in the chapter which deals with factors identified and characterized which suppress the IgE response. Immunotherapy by traditional hyposensitization using improved and safer vaccines is also considered. These contributions are based on a meeting held at the Royal Society of Medicine in November 1987. The presentations provoked a lively discussion and an edited version of these interchanges is included to reflect, apart from anything else, the enthusiasm and buoyancy of the field. This volume would not have been possible without the skill and dedication of Miss Jennifer Mitchell, the editorial assistant and technical editor.

A.B. KAY

I

Lipid Mediators and their Antagonists

1

The future promise of agents which inhibit the formation and action of lipid mediators

M.K. BACH

Summary

A number of inhibitors of 5-lipoxygenase (5-LO) and antagonists of leukotriene (LT) D_4 have been tested for efficacy in various provocation-testing models of asthma. While some beneficial activity was seen in some of these trials, on the whole the tests have failed to reveal the profound beneficial effects which had been hoped for. These findings raise questions about the direction which should be followed in future trials. On the one hand, it may be premature to conclude that, in fact, the products of the 5-LO pathway of arachidonate metabolism are not the key aetiological agents in asthma which they have been postulated to be. While it may be true that LTD_4 is not responsible for causing these symptoms by itself, the combined activity of LTC_4 and LTD_4 or the combination of all the sulphidopeptide leukotrienes and LTB_4 may still be responsible for causing a major part of the symptoms. To definitively exclude this possibility it will be necessary to combine carefully designed clinical challenge protocols of a potent and selective 5-LO inhibitor with objective and quantitative measurement of the effects of this inhibitor on the levels of the products of the 5-LO pathway in the target organ (presumably employing bronchoalveolar lavage) at time points which are critically related to the times for the development of symptoms. In this context, it may be more meaningful to consider the development of the so-called 'late-phase' response than to concentrate primarily on the prevention of acute bronchospasm. On the other hand, recent findings of the pro-inflammatory properties of eicosanoids from other branches of the cascade, such as the products of the 15-LO or some of the metabolites of prostaglandin (PG) H_2 (9α, 11β-PGF_2), suggest that it may be appropriate to consider developing inhibitors to modulate these pathways. Finally, it may prove advantageous to focus on antagonists of platelet-activating factor (PAF-acether) since some of the more recent data suggest that this mediator may play a more central role as the initiator of the inflammatory or chronic component of the asthmatic syndrome. It seems unlikely, however, that a single agent can be found which will effectively control all the lipid mediator-dependent symptoms. Thus it may become necessary to consider developing dual or multiple-specificity inhibitors, or, more practically, combinations of inhibitors each of which is selective for a certain pathway.

Introduction

Slow-reacting substance of anaphylaxis (SRS-A) has long been believed to play a central role in the elicitation of the symptoms of asthma. The elucidation of the structure of the leukotrienes, the demonstration that these materials are derived from arachidonic acid and the demonstration that SRS-A is in reality a mixture of the leukotrienes, LTC_4, LTD_4 and LTE_4, has resulted in great commitment of effort to the discovery of potent and selective antagonists of these substances or inhibitors of their formation. A number of potent inhibitors and antagonists have been found, and in the last year or two the results of initial clinical trials with some of these substances have been reported. On the whole, these trials have shown lack of activity or, at best, modest activity.

It is my task to present a framework within which we can view the results and, hopefully, also address future approaches. I propose to do this by first briefly reviewing the products of the 5-LO pathway in the context of the biochemistry of their formation and interconversions, pointing out where there might be further opportunities for looking for pharmacological inhibitors. I shall then expand my discussion to a consideration of other eicosanoids which have more recently become candidate aetiological agents for inflammation and asthma and, from there, to PAF-acether which, although a lipid, is not directly related to the eicosanoids. As I conclude the chapter, I hope that I shall have left you with the impression that the complex interactions firstly among the lipid mediators themselves and secondly among the many other inflammatory signals to which the lungs are exposed with the lipid mediators or the cells producing them make it highly unlikely that a single 'magic bullet' can be found which will control this condition. I would rather hope that attention in the future will shift to a more careful consideration of the rational design of drug trials in which combinations of drugs are used, each of which is capable of effectively blocking one or a limited number of mediators.

Products of 5-lipoxygenase as mediators of anaphylaxis and inflammation

I have listed the lipid mediators with which we shall be concerned in Table 1.1. The belief that the sulphidopeptide leukotrienes are responsible for some of the symptomatology of asthma goes back to the early 1960s [1]. More

Table 1.1 Lipid mediators of anaphylaxis and inflammation

	5-Lipoxygenase	15-Lipoxygenase	Cyclooxygenase
PAF	LTB_4	15-HETE	$PGF_{2\alpha}$
	LTC_4 } SRS-A	LXA (?)	PGD_2
	LTD_4 } SRS-A	LXB (?)	$9\alpha,11\beta$-PGF_2
	LTE_4 } SRS-A		TxA_2

recently it has been shown by several groups that inhalation of aerosols of these substances causes changes in lung function which are characteristic of the bronchoconstriction of asthma; furthermore, asthmatic volunteers appear to be somewhat more susceptible to these substances than are normal individuals [2–5]. To satisfy the third of Dale's criteria in the proof of the role of these mediators in asthma, efforts have been made to find receptor antagonists for the leukotrienes or inhibitors of their synthesis. The search for antagonists is complicated by the fact that, even though LTC_4 is converted to LTD_4 which in turn is metabolized to LTE_4, there are distinct receptors for at least two of these substances [6, 7] and there is indirect evidence suggesting the existence of a distinct receptor for the third [8]. Furthermore, the search for receptors for LTC_4 is complicated by the fact that glutathione-S-transferases, which are ubiquitous enzymes, mimic receptors, i.e. LTC_4 can bind to them in a saturable manner [9]. Thus, it may be disappointing but perhaps not completely unexpected that even potent receptor antagonists of LTD_4 such as LY-171 883 and L-649 923 only had marginal effects in allergen-challenge protocols in human asthmatic volunteers, even though it could be shown that at least one of these compounds was able to antagonize the effects induced by inhalation challenge with LTD_4 [10, 11].

Recognizing these difficulties and, further, the possible contributions to the syndrome by LTB_4, there may be more appeal to finding inhibitors of the biosynthesis of the 5-LO products than to relying on what would necessarily have to be a series of specific antagonists. A considerable number of inhibitors have been described [reviewed in 10] and some clinical trials have been reported [12–14]. One of these [14] was more in the nature of a safety trial than an efficacy trial in that only the effect on a cholinergic stimulus was monitored, and even though hyperreactivity to acetylcholine is a hallmark of asthma, one would not expect this symptom to abate over a 4-day treatment course. The other two trials also failed to reveal benefits in exercise [12] or antigen challenge protocols [12, 13]. In the case of piriprost [12], treatment was by inhalation which should have caused optimal inhibition in the lungs at the site where this would be the most desirable. Unfortunately, bronchoalveolar lavage was not included in the clinical protocols and we thus do not know if the administered dose of drug was sufficient to cause inhibition of the 5-LO at the target site. A possible substitute for such information, although somewhat less desirable, would have been to monitor inhibition of the 5-LO in blood *ex vivo* after drug administration. However, the known reversibility of the inhibition by piriprost [15] and the rapid clearance of this drug from blood precluded performing any meaningful *ex vivo* tests. In the second trial, the plasma levels of nafazatrom, which had been given orally, were reported as less than 10% of those which would have been required to achieve a 50% inhibition of the 5-LO [13]. Thus, the question of the possible therapeutic benefit to be derived from the inhibition of the 5-LO remains open.

Studies of the biosynthesis of the leukotrienes over the last few years have helped pinpoint a number of new approaches in the search for potential inhibitors (Fig. 1.1). It was felt at one time that inhibitors of the LTC synthase, the enzyme responsible for the conjugation of glutathione with LTA_4, might offer unique possibilities of selectivity since it was found that this enzyme differs in many critical aspects from the other known glutathione-S-transferases [16]. However, considerations such as the desirability of including the

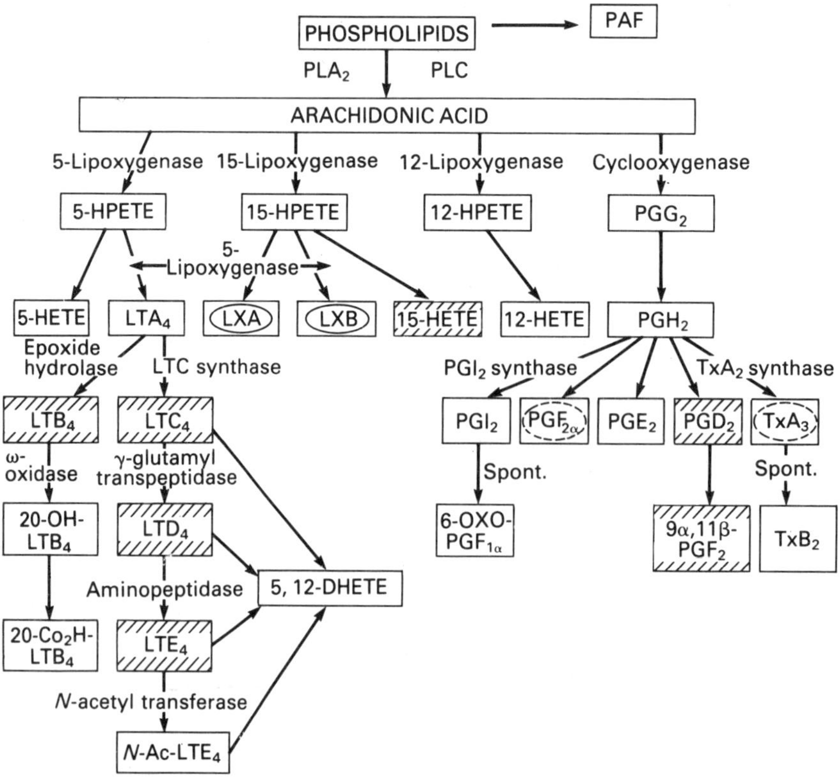

Fig. 1.1 Biosynthesis and metabolism of lipid mediators of anaphylaxis.

inhibition of the formation of LTB_4 in the spectrum of activities of a potential inhibitor make this enzyme a less desirable target than is the 5-LO or the mobilization of arachidonate from phospholipid pools.

Recent studies have shown that the 5-LO is an exquisitely regulated enzyme. In addition to the fact that the action of the enzyme appears to be 'suicidal' in that the enzyme appears to be destroyed in the course of the reaction, it is now clear that the 5-LO has requirements for calcium, ATP, a hydroperoxide and at least three larger cofactors. One of these cofactors, which is found in the high-speed pellet fraction from leucocyte homogenates, appears to serve as an anchor to which the enzyme can attach reversibly in a calcium-dependent reaction [17]. The other cofactors are present in the high-speed supernatant fraction from these cells and can be separated from the bulk of the 5-LO activity by ammonium sulphate precipitation and ion-exchange chromatography. Preliminary results have suggested [Bach, unpublished observations] that the action of one cytosolic cofactor is more strongly dependent on the ATP concentration than is the activity of the 5-LO itself. Furthermore, demonstration of the activity of this cofactor in *in vitro* enzyme incubations is only possible under conditions where the amount of enzyme used is limiting the reaction. Failure to find evidence for such a cofactor in purified enzyme preparations from other cells in reports from two other groups [18, 19] may be due to their use of larger amounts of enzyme in their incubations. It is clear from the above that there is at least a formal resemblance between the regulation of the 5-LO and our present understanding of the regulation of protein kinase C. It is interesting, therefore, that a recent paper [20] suggested that the antiallergic drug, sodium cromoglycate, may be acting by inhibiting protein kinase C.

It has become increasingly apparent in recent years that the 'mobilization' of arachidonate from phospholipid pools is a complex event (see Fig. 1.1). Early results [21, 22] had shown that the 5-LO prefers to utilize arachidonate from endogenous phospholipid pools over exogenously supplied, free arachidonate. Arachidonate can be mobilized by at least two quite distinct pathways in the cell: one depends on the phosphoinositide cycle in which arachidonate mobilization occurs via a phospholipase C which is coupled to a diacylglycerol lipase. In the other, which appears to predominate in more protracted arachidonate-dependent reactions, mobilization depends on a phospholipase A_2. It has also been recognized that the extent of arachidonate mobilization and the utilization of the mobilized arachidonate depend considerably on the stimulus employed even when the same cell population is studied [23–25]. Clark and his associates have described fascinating experiments which have led them to the conclusion that there is a positive-feedback loop which results in the mobilization of arachidonate in the presence of sulphidopeptide leukotrienes [26–31]. They have been able to show that:

1 This stimulation is a time-dependent reaction which requires both RNA and protein synthesis and involves phospholipase A_2 [26–28].
2 The effect is expressed preferentially on the arachidonate which is then converted to products of the cyclooxygenase pathway [27–29].
3 The effect can be preferentially inhibited by dexamethasone but not by aspirin [30].
4 It can be explained by the induced formation of a phospholipase A_2-stimulatory protein which this group have successfully isolated, characterized and cloned [31].

Finally, there is, of course, considerable evidence for the existence of multiple forms of phospholipase A_2 even in homogeneous cell populations [32]. These enzymes can differ in their substrate specificity (for head groups on the phospholipid or for the nature of the fatty acid to be cleaved), their requirement for calcium ion, their pH optimum and their soluble or membrane-bound character. I will not consider the voluminous literature dealing with the endogenous inhibitor(s) of the phospholipases which can be induced by corticosteroids, the macrocortins [33], although they too obviously offer opportunities for the regulation of this reaction.

Products of 15-lipoxygenase as mediators of anaphylaxis and inflammation

The first demonstration that 15-hydroxy-5,8,11,13-eicosatetraenoic acid (15-HETE) is the single major eicosanoid which is produced by specimens of asthmatic but not normal lung dates back to 1980 [34]. Several studies over the past few years have shown that granulocytes as well as primary isolates of epithelial cells from human and dog tracheas can generate large amounts of 15-HETE when they are incubated in the presence of relatively high concentrations of exogenous arachidonate or other toxic stimuli [35–40].

A variety of actions have been reported for 15-hydroperoxy-5,8,11,13-eicosatetraenoic acid (15-HPETE) and 15-HETE when these are used at μM concentrations but on the whole these actions are general reactions of lipid hydroperoxides and are not unique to the product of the 15-LO. More recently, several more specific actions have been reported, for which 15-HETE concentrations of only 0.01–1 μM were required. Among these are an inhibition of the expression of C3b complement receptors on B-cells [41], suppression of β-adrenergic receptor expression and potentiation of the response to histamine in guinea pig airways [42, 43] and a potent stimulation of mucus production in dog tracheas *in vivo* when only nanogram amounts of 15-HETE are instilled into the cranial thyroid artery [44]. Given the extremely low concentrations of 15-HETE which cause these responses and the fact that 15-HETE is apparently more active than is 15-HPETE in the dog model, it seems very unlikely that these responses require the further metabolism of the 15-LO products.

To the extent that the responses of 15-HETE which have just been described mimic or can explain the development of some of the symptoms of asthma, these observations satisfy the first two criteria of Dale's postulates in establishing a role for 15-HETE in the aetiology of asthma even though this role may not primarily involve bronchoconstriction. To test the possibility further, selective inhibitors of the 15-LO will be required. No such inhibitors have been described thus far, although some of the known 5-LO inhibitors do inhibit this enzyme as well.

Products of cyclooxygenase as mediators of anaphylaxis and inflammation

It was recognized in the early days of research with prostaglandins that $PGF_{2\alpha}$ was a potent bronchoconstrictor. In fact, it was proposed that the several thousand-fold increased susceptibility of asthmatics to the effects of this eicosanoid could explain the hyperreactive airways of asthmatics [45]. Thromboxane (Tx) A_2, which is also a smooth muscle contractant, was similarly considered as a possible contributor to bronchoconstriction. The bronchoconstrictive effect of synthetic TxA_2 in animals has been recently reported [46].

The potential contribution of PGD_2 (Fig. 1.2) to bronchoconstriction has been considered in the last few years. This has been encouraged by the observation that PGD_2 appears to be a major eicosanoid metabolite produced by human mast cells during anaphylactic challenge [47]. Indeed, careful analysis has shown that, of all the cyclooxygenase products which have been measured in nasal secretions, only PGD_2 levels appear to rise and fall *in vivo* in a manner which is temporally coincident with the rise and fall of allergic symptoms in volunteers who have been challenged with antigen

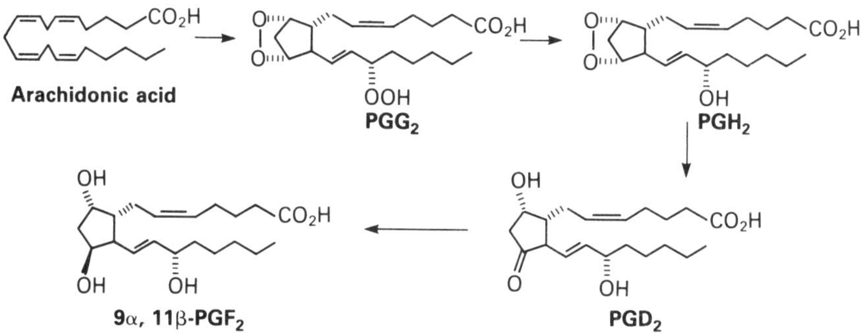

Fig. 1.2 Biosynthesis of PGD_2 and $9\alpha,11\beta$-PGF_2.

intranasally [48]. Administration of PGD_2 to human volunteers has been reported to cause profound changes in lung mechanics [49]. In addition, PGD_2 has been reported to potentiate airway responses to histamine and methacholine [50]. Recently, two groups have reported that PGD_2 can be metabolized to 9α,11β-PGF_2 [51, 52]. This material has now also been reported to be a potent bronchoconstrictor having four times the potency of PGD_2 in human airways [53].

One problem with the assignment of an aetiological role in asthma to a cyclooxygenase product is the failure of non-steroidal anti-inflammatory agents to control the symptoms of asthma even under circumstances where it can be readily demonstrated that TxB_2 synthesis in platelets is profoundly inhibited. Several explanations are possible: first, there is the possibility that the concentrations of the inhibitors are insufficient to inhibit cyclooxygenase in lung. Differences in susceptibility to these inhibitors between platelets and other cells have been recognized for a long time. More importantly, however, there is the recognition that, along with the inhibition of the formation of TxA_2 and possibly PGD_2 and $PGF_{2\alpha}$, these compounds will also inhibit the formation of prostacyclin which may well play a bronchodilatory role in the normal airways. Thus, there is justification for wishing to find selective inhibitors of the synthesis of PGD_2 or possibly of the conversion of PGD_2 to the 9α,11β-PGF_2 metabolite.

Platelet-activating factor as a mediator of anaphylaxis and inflammation

The existence of PAF-acether has been recognized for some 17 years [54], but it was not until the structure of this substance was established in 1979 [55, 56] that research with it really took off. In many ways, the developments with PAF-acether parallel the developments with the leukotrienes and by now it is clear that both these substances can be produced by a large variety of mammalian cells and, furthermore, can have profound effects on a truly staggering array of biological responses [57, 58].

The biosynthesis of PAF-acether involves the cleavage of a fatty acid residue from the hydroxyl radical on the middle carbon of the glycerol backbone of a plasmalogen (Fig. 1.3). This position in plasmalogens is often occupied by arachidonic acid so that arachidonate mobilization accompanies the biosynthesis of PAF-acether and, in that sense, the formation of this mediator is tied into the eicosanoid cascade as well.

Some of the evidence in support of the conclusion that PAF-acether may be an aetiological agent in asthma is as follows:

1 PAF-acether is released during anaphylactic challenge, e.g. of rabbit basophils [59], and the degradation product of PAF-acether, lyso-PAF, was

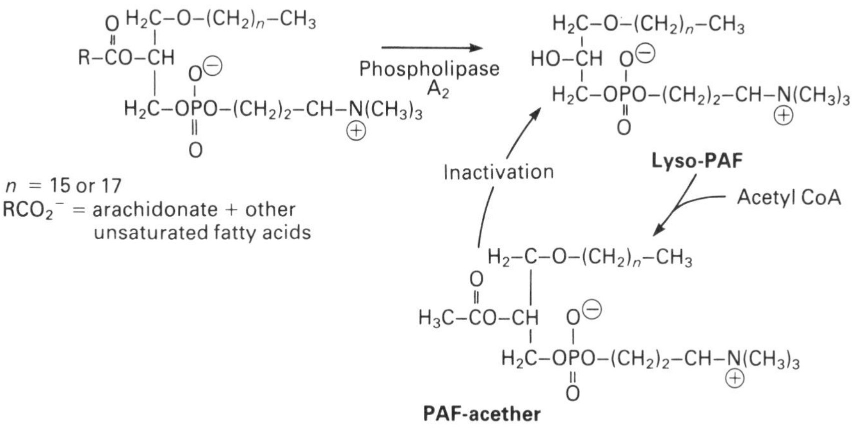

Fig. 1.3 Biosynthesis of PAF-acether.

found in the plasma of asthmatic patients with a time course which coincides with the development of the late-phase symptoms [60].

2 Passive administration of PAF-acether mimics many of the symptoms of asthma. Thus, intradermal injection of PAF-acether in humans caused the development of a weal and flare followed by an erythematous response 6 h later [61], while intratracheal administration to rabbits [62] and intratracheal or intranasal administration to humans [63, 64] caused acute changes in pulmonary function along with late-phase changes in the rabbits.

3 Minute concentrations of PAF-acether caused profound changes in protein transudation in guinea pig tracheas *in vivo* [65] and in tracheal mucus transport velocity in conscious sheep [66].

The potency of PAF-acether and its prolonged duration of action in the face of its rapid degradation in tissue suggest that there must be amplification mechanisms involved in its action. Indeed, it has been shown that PAF-acether markedly potentiates the activation of neutrophils by formyl-methionyl-leucyl-phenylalanine (FMLP) or LTB_4 [67] and at least some of its actions appear to involve products of the 5-LO as intermediates [68–71]. PAF-acether has also been reported to synergize the lethality of β_1-adrenergic inhibitors [71] and to cause a dose-dependent decrease in the expression of β_2-receptors in human lung [72]. All these observations have led to the suggestion that PAF-acether, and through it the platelet, may play a central role in the asthma syndrome [73]. On the other hand, there is considerable evidence for feedback regulation of PAF-acether formation by products of the 5-LO [74–76] and by neuropeptides [76].

Efforts to interfere with the PAF-acether-dependent events have centred on a search for receptor antagonists. The usual screen has involved tests for the inhibition of PAF-acether-induced platelet aggregation [57]; however, several lines of evidence [71, 77] have pointed to the existence of more than one type of receptor for PAF-acether, the receptors on platelets apparently being quite different from those on leucocytes. Some very potent antagonists have been found. They represent several diverse structural classes beginning with structural analogues of PAF-acether itself, ranging through various natural products and synthetic molecules which were designed on the basis of the activities of the natural products and ending with a variety of synthetic 'leads' [11]. The finding that some of the benzodiazepine tranquillizers can compete with PAF-acether and particularly that this activity can be dissociated from the central nervous system (CNS) activity of this class of molecules [78] is of particular interest. WEB-2086 is reportedly a potent benzodiazepine PAF-acether antagonist with no CNS activity. A number of the PAF-acether antagonists are already in various stages of clinical development but, to date, only one report concerning clinical efficacy in asthma has appeared [79]. In that study a total of 10 patients were treated; nine showed at least some improvement and six showed 'dramatic' improvement. If true and confirmed in larger studies, this would represent a much more optimistic rate of success than that encountered thus far with inhibitors of the 5-LO or antagonists of the leukotrienes.

Added complexities and promises for the future

It should be apparent from everything mentioned so far that, rather than considering the leukotrienes as the only, or even the central, aetiological agents in eliciting the symptoms of asthma, we have come to recognize that at best they are members of the 'pack'. It is clear that the list of interrelationships between the mediators, feedback loops or cross-regulations between mediators coming from different branches of the eicosanoid tree will continue to grow as we learn more about the intricacies of the system. Many of these interactions are actually not interactions between the mediators themselves but are rather a reflection of the fact that many cells in the body have receptors for these substances and, when these receptors are occupied, the respective cells go about carrying out the responses they happen to be programmed for. The result, obviously, is an enormously powerful amplification network.

To add complexity to what is already a multidimensional problem, we must introduce a consideration of the interactions of the lipid mediators, or the cells producing them, with non-lipid mediators and the effects of the lipid mediators on the responses of target cells to the non-lipid mediators. This is not the place to consider these interactions in any detail. I have already mentioned regulatory loops involving neuropeptides, PAF-acether and lipoxygenase products. I have also mentioned effects of the lipid mediators on the expression of adrenergic receptors on target cells, and the expression of receptors for complement on other cells. The list of such interactions is long and it would be foolhardy to believe that even the total abrogation of the effects of all lipid mediators would come close to inhibiting all the factors which are known to contribute to asthma or the inflammatory response. At the same time, it should be stressed that with the recognition of the ubiquitous distribution of the lipid mediators in the body and with the recognition that they play important roles in the regulation of numerous bodily functions aside from their role, if any, in the causation of asthma, one would have to predict severe toxic reactions if a complete inhibition or antagonism of these materials were ever achieved.

What, then, are the prospects for the chemical management of asthma with modulators of the lipid mediators? The verdict regarding the current strategies is not yet clear; a potent 5-LO inhibitor or a potent PAF-acether antagonist may be found which offers sufficient benefit that it will be possible to develop it as a single drug entity. However, one principle which emerges is that of selectivity. If we can find a modulator which is specific for only a narrow part of the whole cascade, and particularly if we can deliver the regulating drug in a 'focused' manner to the designated target tissue with minimal systemic distribution, we stand a much better chance of achieving desirable risk/benefit ratios. This, in fact, is the rationale behind developing antiasthma drugs to be delivered as aerosols rather than formulating them as oral preparations. A corollary to this wish to restrict activity to the target tissue is that, aside from demonstration of clinical efficacy in the target disease or in some suitable clinical model thereof, it may be difficult to document that such a drug, at the doses being administered, actually performed the intended task, e.g. it inhibited the formation of leukotrienes in the lung. No *ex vivo* studies with blood or leucocytes would be meaningful and even analysis of bronchoalveolar lavage may have limited utility given the constraints imposed on the timing of lavage and on the sectors of the lung which can be sampled.

But the more we develop selective inhibitors, the more there will be portions of the complex of mediators which will remain unaffected when we use a single inhibitor. Indeed, it is entirely possible to envision overproduction of some of these when the feedback controls exerted by one mediator on the cells producing another are manipulated. There are two potential ways of getting around this. One can attempt to design molecules which are endowed with dual or even triple specificities, e.g. 5-LO, 15-LO inhibitor and PAF-acether antagonist in the same molecule. It is obvious that designing such molecules and achieving the needed ratios of potency for the various end points will present major problems. In fact, recognizing that, in all likelihood, clinical asthma represents a multiplicity of underlying aetiologies, it may well be that one would not even want to have the same ratio of potencies for all patients. Thus, the more likely approach is to begin to think in terms of developing a series of highly selective drugs each of which is intended to regulate a single product and then to test these in combination without making *a priori* demands regarding their clinical efficacy as single entities. 'Objective' verification of the efficacies of these inhibitors taken singly by analysis of their effects on the production of the respective target mediators may be useful in place of the traditional demonstration of clinical efficacy.

In this context, I would like to briefly mention a line of experiments which my associate, John R. Brashler, and I have been carrying out over the past few years [80]. We have been interested in the combined effects of two drugs which are known to inhibit leukotriene synthesis when they are presented to a leukotriene-synthesizing system at the same time. We have found that, depending upon the detailed nature of the drugs involved, it is possible to achieve enormous potentiations (i.e. synergisms) when such combinations are used. A good example is the 100-fold increase in the potency of diethylcarbamazine as an inhibitor of leukotriene synthesis in the presence of concentrations of AA861 which, by themselves, are hardly inhibitory at all [80]. I would suggest that the same principle may apply when it comes to inhibitors affecting different pathways and it should be obvious that, as the drug doses are reduced, so is the danger of side effects.

I am well aware that the proposal to develop drugs to be used in combination without prior demonstration of the efficacy of the components of the combination taken singly is contrary to the prevailing regulatory climate. I would hold, however, that in the long run it is our responsibility as experts in the field to carry out the educational process which will be necessary to change the regulatory climate if we are convinced that by bringing about such changes, treatment of patients will improve.

References

1 Brocklehurst WE. Slow reacting substance and related compounds. *Prog. Allergy* 1962; **6**:539–558.

2 Holroyde MC, Altounyan REC, Cole M, Dixon M, Elliott EV. Selective inhibition of bronchoconstriction induced by leukotrienes C and D in man. *Adv. Prostaglandin Thromboxane Leukotriene Res.* 1982; **9**:237–242.

3 Weiss JW, Drazen JM, Coles N, McFadden ER Jr, Weller PF, Corey EJ, Lewis RA, Austen KF. Bronchoconstrictor effects of leukotriene C in humans. *Science* 1982; **216**:196–198.

4 Bisgaard H, Groth S, Tandorf E, Madsen F. The possible role of LTD_4 in asthma in humans investigated *in vivo*. *Biomed. Biochim. Acta* 1984; **43**:S327–330.

5 Smith LJ, Greenberger PA, Patterson R, Krell RD, Bernstein PR. The effect of inhaled leukotriene D_4 in humans. *Am. Rev. Respir. Dis.* 1985; **131**:368–372.

6 Hogaboom GK, Mong S, Wu HL, Crooke ST. Peptide leukotrienes: Distinct receptors for leukotriene C_4 and D_4 in the guinea pig lung. *Biochem. Biophys. Res. Commun.* 1983; **116**:1136–1143.

7 Mong S, Scott MD, Lewis MA, Wu HL, Hogaboom GK, Clark MA, Crooke ST. Leukotriene E_4 binds specifically to LTD_4 receptors in guinea pig lung membranes. *Eur. J. Pharmacol.* 1985; **109**:183–192.
8 Krilis S, Lewis RA, Drazen JM, Austen KF. Subclasses of receptors to the sulfidopeptide leukotrienes. In *Prostaglandins and Membrane Transport* (Eds Braquet P, Garay RP, Frohlich JC, Nicosia S), Raven Press, New York 1985, pp. 91–97.
9 Sun FF, Chan LY, Spur B, Corey EJ, Lewis RA, Austen KF. Identification of a high affinity leukotriene C_4-binding protein in rat liver cytosol as glutathione S-transferase. *J. Biol. Chem.* 1986; **261**:8540–8546.
10 Mathur PN, Callaghan JT, Farid NA, Sylvester AJ. The prevention of allergen induced late asthmatic response by LY 171883 a leukotriene antagonist. *Clin. Res.* 1986; **34**:580A.
11 Piwinski JJ, Kreutner W, Green MJ. Pulmonary and antiallergy agents. *Annu. Rep. Med. Chem.* 1987; **22**:73–84.
12 Mann JS, Robinson C, Sheridan AQ, Clement P, Bach MK, Holgate ST. Effect of inhaled piriprost (U-60, 257), a novel leukotriene inhibitor, on allergen and exercise induced bronchoconstriction in asthma. *Thorax* 1986; **41**:746–752.
13 Fuller RW, Maltby N, Richmond R, Dollery CT, Taylor GW, Ritter W, Philipp P. Oral nafazatrom in man: effect on inhaled antigen challenge. *Br. J. Clin. Pharmacol.* 1987; **23**:677–681.
14 Fujimura M, Sasaki F, Nakatsumi Y, Takahashi Y, Hifumi S, Taga K, Mifune JI, Tahaka T, Matsuda M. Effect of a thromboxane synthetase inhibitor (OKY-046) and a lipoxygenase inhibitor (AA 861) on bronchial responsiveness to acetylcholine in asthmatic subjects. *Thorax* 1986; **41**:955–959.
15 Bach MK, Bowman BJ, Brashler JR, Fitzpatrick FA, Griffin RL, Johnson HG, Major NJ, McGuire JC, McNee ML, Richards IM, Smith HW, Smith RJ, Speciale SC, Sun FF. Piriprost: a selective inhibitor of leukotriene synthesis. *Adv. Prostaglandin Thromboxane Leukotriene Res.* 1985; **15**:225–227.
16 Bach MK, Brashler JR, Peck RE, Morton DR Jr. Leukotriene C synthetase, a special glutathione S-transferase: properties of the enzyme and inhibitor studies with special reference to the mode of action of U-60,257, a selective inhibitor of leukotriene synthesis. *J. Allergy Clin. Immunol.* 1984; **74**:353–357.
17 Samuelsson B, Rouzer CA, Matsumoto T. Human leukocyte 5-lipoxygenase: an enzyme possessing dual enzymatic activities and a multicomponent regulatory system. *Adv. Prostaglandin Thromboxane Leukotriene Res.* 1987; **17**:1–14.
18 Ueda N, Kaneko S, Yoshimoto T, Yamamoto S. Purification of arachidonate 5-lipoxygenase from porcine leukocytes, and its reactivity with hydroperoxyeicosatetraenoic acids. *J. Biol. Chem.* 1986; **261**:7982–7988.
19 Hogaboom GK, Cook M, Newton JF, Varichio A, Shorr RGL, Saran HM, Crooke ST. Purification, characterization and structural properties of a single protein from rat basophilic leukemia (RBL-1) cells possessing 5-lipoxygenase and leukotriene A_4 synthetase activities. *Mol. Pharmacol.* 1986; **30**:510–519.
20 Lucas AM, Shuster S. Cromolyn inhibition of protein kinase C activity. *Biochem. Pharmacol.* 1987; **36**:562–565.
21 Kuehl FA Jr, Dougherty HW, Ham EA. Interactions between prostaglandins and leukotrienes. *Biochem. Pharmacol.* 1984; **33**:1–5.
22 Harvey J, Holgate ST, Peters BJ, Robinson C, Walker JR. Oxidative transformations of arachidonic acid in human dispersed lung cells: disparity between utilization of endogenous and exogenous substrate. *Br. J. Pharmacol.* 1985; **86**:417–426.
23 Haines KA, Giedd KN, Rich AM, Korchak G, Weissmann G. The leukotriene B_4 paradox: neutrophils can, but will not, respond to ligand–receptor interactions by forming leukotriene B_4 and its omega metabolites. *Biochem. J.* 1987; **241**:55–62.
24 Sellmayer A, Strasser T, Weber PC. Differences in arachidonic acid release, metabolism, and leukotriene B_4 synthesis in human polymorphonuclear leukocytes activated by different stimuli. *Biochim. Biophys. Acta* 1987; **927**:417–422.

25 Godfrey RW, Manzi RM, Clark MA, Hoffstein ST. Stimulus-specific induction of phospholipid and arachidonic acid metabolism in human neutrophils. *J. Cell. Biol.* 1987; **104**:925–932.
26 Clark MA, Littlejohn D, Conway TM, Mong S, Steiner S, Crooke ST. Leukotriene D_4 treatment of bovine aortic endothelial cells and murine smooth muscle cells in culture results in an increase in phospholipase A_2 activity. *J. Biol. Chem.* 1986; **261**:10713–10718.
27 Clark MA, Littlejohn D, Mong S, Crooke ST. Effect of leukotrienes, bradykinin, and calcium ionophore (A23187) on bovine endothelial cells: release of prostacyclin. *Prostaglandins* 1986; **31**:157–166.
28 Clark MA, Cook M, Mong S, Crooke ST. The binding of leukotriene C_4 and leukotriene D_4 to membranes of a smooth muscle cell line ($BC3H_1$) and evidence that leukotriene induced contraction in these cells is mediated by thromboxane, protein and RNA synthesis. *Eur. J. Pharmacol.* 1985; **116**:207–220.
29 Mong S, Wu HL, Clark MA, Gleason JG, Crooke ST. Leukotriene D_4 receptor-mediated synthesis and release of arachidonic acid metabolites in guinea pig lung: induction of thromboxane and prostacyclin biosynthesis by leukotriene D_4. *J. Pharmacol. Exp. Ther.* 1986; **239**:63–70.
30 Clark MA, Bomalaski JS, Conway TM, Wartell J, Crooke ST. Differential effects of aspirin and dexamethasone on phospholipase A_2 and C activities and arachidonic acid release from endothelial cells in response to bradykinin and leukotriene D_4. *Prostaglandins* 1986; **32**:703–708.
31 Clark MA, Conway TM, Shorr R, Crooke ST. Identification, isolation, characterization, and cloning of a phospholipase A_2 stimulatory protein rapidly induced by leukotriene D_4 treatment. *J. Cell. Biol.* 1986; **103**:224a.
32 Loeb LA, Gross RW. Identification and purification of sheep platelet phospholipase A_2 isoforms. Activation by physiologic concentrations of calcium ions. *J. Biol. Chem.* 1986; **261**: 10467–10470.
33 Flower RJ, Blackwell GJ. Antiinflammatory steroids induce biosynthesis of a phospholipase A_2 inhibitor which prevents prostaglandin generation. *Nature* 1979; **278**:456–459.
34 Hamberg M, Hedqvist P, Rådegran K. Identification of 15-hydroxy-5,8,11,13-eicosatetraenoic acid (15-HETE) as a major metabolite of arachidonic acid in human lung. *Acta Physiol. Scand.* 1980; **110**:219–221.
35 Vanderhoek JY, Bailey JM. Activation of a 15-lipoxygenase/leukotriene pathway in human polymorphonuclear leukocytes by the antiinflammatory agent ibuprofen. *J. Biol. Chem.* 1984; **259**:6752–6756.
36 McGuire J, McGee J, Crittenden N, Fitzpatrick F. Cell damage unmasks 15-lipoxygenase activity in human neutrophils. *J. Biol. Chem.* 1985; **260**:8316–8319.
37 Narumiya S, Salmon JA, Cottee FH, Weatherley BC, Flower RJ. Arachidonic acid 15-lipoxygenase from rabbit peritoneal polymorphonuclear leukocytes. Partial purification and properties. *J. Biol. Chem.* 1981; **256**:9583–9592.
38 Turk J, Maas RL, Brash AR, Roberts LJ II, Oates JA. Arachidonic acid 15-lipoxygenase products from human eosinophils. *J. Biol. Chem.* 1982; **257**:7068–7076.
39 Hunter JA, Finkbeiner WE, Nadel JA, Goetzl EJ, Holtzman MJ. Predominant generation of 15-lipoxygenase metabolites of arachidonic acid by epithelial cells from human trachea. *Clin. Res.* 1985; **33**:78A.
40 Holtzman MJ, Aizawa H, Nadel JA, Goetzl EJ. Selective generation of leukotriene B_4 by tracheal epithelial cells from dogs. *Biochem. Biophys. Res. Commun.* 1983; **114**:1071–1076.
41 Cook J, Delebassée S, Aldigier JC, Gualde N, Kazatchkine M. 15-HETE 'modulates' expression of C_{3b} receptor (CR1) antigen on peripheral blood B-lymphocytes. *Prostaglandins Leukotrienes Med.* 1986; **23**:201–206.
42 Nijkamp FP, Van Oosterhout AJM. Decreased lung β-adrenoreceptor function induced by 15-hydroxyarachidonic acid. *Eur. J. Respir. Dis.* 1984; **65** (suppl. 135): 221–225.
43 Folkers G, Nijkamp FP, Van Oosterhout AJM. Induction in guinea pigs of airway hyperreactivity and decreased lung β-adrenreceptor number by 15-hydroperoxyarachidonic acid. *Br. J. Pharmacol.* 1983; **80**:597–599.

44 Johnson HG, McNee ML, Sun FF. 15-Hydroxyeicosatetraenoic acid is a potent inflammatory mediator and agonist of canine tracheal mucus secretion. *Am. Rev. Respir. Dis.* 1985; **131**:917–922.
45 Mathé AA, Hedqvist P, Holmgren A, Svanborg N. Bronchial hyperreactivity to prostaglandin $F_{2\alpha}$ and histamine in patients with asthma. *Br. Med. J.* 1973: 193–196.
46 Richards IM, Oostveen JA, Griffin RL, Bunting S. Pulmonary pharmacology of synthetic thromboxane A_2. *Adv. Prostaglandin Thromboxane Leukotriene Res.* 1987; **17**:1067–1072.
47 Schleimer RP, MacGlashan DW Jr, Peters SP, Pinckard RN, Adkinson NF Jr, Lichtenstein LM. Characterization of inflammatory mediator release from purified human lung mast cells. *Am. Rev. Respir. Dis.* 1986; **133**:614–617.
48 Naclerio RM, Proud D, Togias AG, Adkinson NF Jr, Meyers DA, Kagey-Sobotka A, Plaut M, Norman PS, Lichtenstein LM. Inflammatory mediators in late antigen-induced rhinitis. *N. Engl. J. Med.* 1985; **313**:65–70.
49 Beasley CRW, Varley J, Robinson C, Holgate ST. Direct and reflex bronchoconstrictor actions of prostaglandin $(PG)D_2$ and its initial metabolite $9\alpha,11\beta$-PGF_2 in asthma. *Br. J. Clin. Pharmacol.* 1987; **23**:606P–607P.
50 Fuller RW, Dixon CMS, Dollery CT, Barnes PJ. Prostaglandin D_2 potentiates airway responsiveness to histamine and methacholine. *Am. Rev. Respir. Dis.* 1986; **133**:252–254.
51 Liston TE, Robert LJ II. Transformation of prostaglandin D_2 to $9\alpha,11\beta$-(15S)-trihydroxyprosta-(5Z, 13E)-dien-1-oic acid ($9\alpha,11\beta$-prostaglandin F_2): a unique biologically active prostaglandin produced enzymatically *in vivo* in humans. *Proc. Natl. Acad. Sci. USA* 1985; **82**:6030–6034.
52 Pugliese G, Spokas EG, Marcinkiewicz E, Wong PYK. Hepatic transformation of prostaglandin D_2 to a new prostanoid, $9\alpha,11\beta$-prostaglandin F_2, that inhibits platelet aggregation and constricts blood vessels. *J. Biol. Chem.* 1985; **260**:14621–14625.
53 Beasley CRW, Robinson C, Featherstone RL, Varley JG, Hardy CC, Church MK, Holgate ST. $9\alpha,11\beta$-prostaglandin F_2, a novel metabolite of prostaglandin D_2, is a potent contractile agonist in human and guinea pig airways. *J. Clin. Invest.* 1987; **79**:978–983.
54 Benveniste J. The mast cell derived pharmacologic mediators of anaphylaxis: Platelet activating factor. In *Immediate Hypersensitivity: Modern Concepts and Developments* (Ed. Bach MK), Marcel Dekker, New York 1978, pp. 625–634.
55 Benveniste J, Tencé M, Varenne P, Bidault J, Boullet C, Polonsky J. Semisynthesis and proposed structure of platelet activating factor (PAF): PAF acetate, an alkylether analog of lysophosphatidyl choline. *Comp. Rend.* 1979; **289**:1037–1040.
56 Demopoulos CA, Pinckard RN, Hanahan DJ. Platelet activating factor. Evidence for 1-*O*-alkyl-2-acetyl-*sn*-glycerol-3-phosphorylcholine as the active component (a new class of lipid chemical mediators). *J. Biol. Chem.* 1979; **254**:9355–9358.
57 Vargaftig BB, Braquet PG. PAF-acether today—relevance for acute experimental anaphylaxis. *Br. Med. Bull.* 1987; **43**:312–335.
58 Raphael GD, Metcalfe DD. Mediators of airway inflammation. *Eur. J. Respir. Dis.* 1986; **69** (suppl. 147): 44–56.
59 Hanahan DJ, Demopoulos CA, Liehr J, Pinckard RN. Identification of platelet activating factor isolated from rabbit basophils as acetyl glyceryl ether phosphorylcholine. *J. Biol. Chem.* 1980; **255**:5514–5516.
60 Nakamura T, Morita Y, Kuriyama M, Ishihara K, Ito K, Miyamoto T. Platelet activating factor in late asthmatic response. *Int. Arch. Allergy. Appl. Immunol.* 1987; **82**:57–61.
61 Archer CB, Page CP, Paul W, Morley J, MacDonald DM. Inflammatory characteristics of PAF-acether in the skin of experimental animals and man. *Int. J. Tissue. React.* 1985; **7**:363–365.
62 Page CP, Guerreiro D, Morley J. Platelet activating factor (PAF-acether) may account for late-onset reactions to allergen inhalation. *Agent Actions* 1985; **16**:30–35.
63 Chung KF, Dixon CMS, Barnes PJ. Platelet activating factor (PAF) and asthmatic airways: effects on calibre, responsiveness and circulating cells. *Am. Rev. Respir. Dis.* 1987; **135(4II)**: 159A.

64 Pipkorn U, Karlsson G, Bake B. Effect of platelet activating factor on the human nasal mucosa. *Allergy* 1984; **39**:141–145.
65 Rogers DF, Aursudkij B, Evans TW, Belvisi MG, Chung KF, Barnes PJ. Platelet activating factor increases protein transudation but not mucous secretion in guinea pig trachea *in vivo*. *Am. Rev. Respir. Dis.* 1987; **135**(4II): 160A.
66 Homolka J, Abraham WM, Rubin E, Nieves L, Wanner A. Effect of platelet activating factor on tracheal mucus velocity in conscious sheep. *Am. Rev. Respir. Dis.* 1987; **135**(4II):160A.
67 Baggiolini M, Dewald B. Stimulus amplification by PAF and LTB_4 in human neutrophils. *Pharmacol. Res. Commun.* 1986; **18**(suppl.):51–59.
68 Moodley I, Stuttle A. Evidence for a dual pathway of platelet activating factor. Induced aggregation of rat polymorphonuclear leukocytes. *Prostaglandins* 1987; **33**:253–264.
69 Bruijnzeel PLB, Koenterman L, Kok PTM, Hameling ML, Verhagen J. Platelet activating factor (PAF-acether) induced leukotriene C_4 formation and luminol dependent chemiluminescence by human eosinophils. *Pharmacol. Res. Commun.* 1986; **18**(suppl.):61–69.
70 Piper PJ, Stewart AG. Coronary vasoconstriction in the rat isolated perfused heart induced by platelet activating factor is mediated by leukotriene C_4. *Br. J. Pharmacol.* 1986; **88**:595–605.
71 Criscuoli M, Subissi A. PAF-acether induced death in mice. Involvement of arachidonate metabolites and β-adrenoreceptors. *Br. J. Pharmacol.* 1987; **90**:203–209.
72 Agrawal DK, Townley RG. Effect of platelet activating factor on beta-adrenoreceptors in human lung. *Biochem. Biophys. Res. Commun.* 1987; **143**:1–6.
73 Morley J, Sanjar S, Page CP. The platelet in asthma. *Lancet* 1984; **II**:1142–1144.
74 Saito H, Hirai A, Tamura Y, Yoshida S. The 5-lipoxygenase products can modulate the synthesis of platelet activating factor (alkylacetyl-glycerophosphorylcholine) in calcium-ionophore A23187-stimulated rat peritoneal macrophages. *Prostaglandins Leukotrienes Med.* 1985; **18**:271–286.
75 Billah MM, Bryant RW, Siegel MI. Lipoxygenase products of arachidonic acid modulate biosynthesis of platelet activating factor (1-O-alkyl-2-acetyl-*sn*-glycero-3-phosphocholine) by human neutrophils via phospholipase A_2. *J. Biol. Chem.* 1985; **260**:6899–6906.
76 Di Marzo V, Tippins JR, Morris HR. Neuropeptides and leukotriene release: effect of peptide histidine isoleucine and secretin in platelet activating factor-stimulated rat lung. *Neuropeptides* 1987; **9**:51–58.
77 Lambrecht G, Parnham MJ. Kadsurenone distinguishes between different platelet activating factor receptor subtypes on macrophages and polymorphonuclear leukocytes. *Br. J. Pharmacol.* 1986; **87**:287–289.
78 Casals-Stenzel J, Weber KH. Triazolodiazepines: dissociation of the PAF (platelet activating factor) antagonistic and CNS activity. *Br. J. Pharmacol.* 1987; **90**:139–146.
79 Rense-Bourgain M. The effect of GBE 761, a standardized extract of *Ginko biloba*, on asthma in children. In *Abstracts of the VIth International Conference on Prostaglandins and Related Compounds*, Florence, Italy, June 1986, p. 297.
80 Bach MK, Brashler JR. Modes of action of, and synergism among, inhibitors of sulfidopeptide leukotriene synthesis in calcium-ionophore-challenged rat basophil leukemia cells. *Proc. XII Int. Cong. Allergol. Clin. Immunol.* 1986; 256–262.

2

Agents which inhibit the 5-lipoxygenase and cyclooxygenase pathways and their relationship to the treatment of allergic responses

A.W. FORD-HUTCHINSON

Cyclooxygenase products and the allergic response

Arachidonic acid may be metabolized to a variety of biologically active products including prostaglandin (PG) D_2, $PGF_{2\alpha}$, PGI_2, PGE_2 and thromboxane (Tx) A_2, and such products, especially PGD_2—the major cyclooxygenase product produced by the human pulmonary mast cell, are released from human pulmonary tissue following antigen (IgE) challenge [1, 2]. Prostaglandins and TxA_2 exert their biological reactions through interaction with specific receptor sites [3] and such actions may have opposite biological effects. Thus, PGI_2 inhibits platelet aggregation and is a vasodilator whereas TxA_2 is a potent inducer of platelet aggregation and vasoconstriction. Similar opposing actions may occur in the lung, and PGD_2, $PGF_{2\alpha}$ and TxA_2 may induce bronchoconstriction and prostaglandins PGI_2 and PGE_2, under appropriate conditions, inhibit bronchoconstriction. The production of all cyclooxygenase products may be inhibited by non-steroidal anti-inflammatory drugs such as aspirin and indomethacin. More recently, selective approaches to modulating the action of cyclooxygenase products have been developed to exploit the opposing biological actions of the released products, in particular thromboxane synthetase inhibitors and thromboxane/prostaglandin endoperoxide receptor antagonists.

Evidence for a role for cyclooxygenase products in human bronchial asthma through the use of non-steroidal anti-inflammatory drugs has been contradictory. Approximately 1% of asthmatic patients improve and approximately 4% worsen ('aspirin-induced asthma') following treatment with cyclooxygenase inhibitors [4]. A possible role for cyclooxygenase products in the late response has been indicated by two studies [5, 6], although other studies have shown no effect on the late response but have indicated an effect on the subsequent bronchial hyperresponsiveness [O'Byrne, unpublished observations]. A role for cyclooxygenase products, in particular TxA_2, in antigen-induced bronchial hyperresponsiveness has also been indicated from animal studies in the sheep and dog [7, 8].

Thromboxane A_2/prostaglandin endoperoxide receptor antagonists may be particularly useful as they will selectively antagonize the contractile activities of TxA_2, prostaglandin endoperoxides and other prostanoids such as PGD_2 when these compounds interact with a thromboxane/prostaglandin (TP) receptor [3], but will allow prostanoids to interact with inhibitory or dilator receptors. An example of such a compound is L-655 240 (3-[1-(4-chlorobenzyl)-5-fluoro-3-methyl-indol-2-yl]2,2-dimethylpropanoic acid) [9]. This compound is a potent and selective TP-receptor antagonist as evidenced by both functional and receptor binding studies. Thus, L-655 240 inhibited the binding of [^{125}I]-PTA-OH on washed human platelets with an IC_{50} of 12.7 nM. This was reflected in potent inhibition of the aggregation of washed human platelets induced by the prostaglandin endoperoxide mimetic, U-44 069 (IC_{50} value 7 nM), and inhibition of aggregation of human platelet-rich plasma induced by U-44 069, U-46 619, TxA_2 and collagen but not ADP or platelet-activating factor (PAF-acether) [9]. This was also observed *in vivo* where, following oral administration to rhesus monkeys, L-655 240 inhibited *ex vivo* platelet aggregation induced by U-44 069 but not ADP [9]. L-655 240 also showed activity on smooth muscle preparations being a potent and competitive antagonist of contractions of the guinea pig tracheal chain, thoracic aorta ring and pulmonary artery induced by U-44 069 *in vitro* (pA_2 values 8.0, 8.0 and 8.4 respectively) and bronchoconstriction-induced in guinea pigs *in vivo* by arachidonic acid and U-44 069 [9]. The selectivity of this compound in the bronchial tree was indicated *in vitro* by non-competitive antagonism of contractions of the guinea pig trachea induced by PGD_2 and minimal activity against contractions induced by leukotriene (LT) D_4, $PGF_{2\alpha}$, histamine, serotonin and acetylcholine and *in vivo* by the failure to inhibit bronchoconstriction in the guinea pig induced by histamine, serotonin or acetylcholine [9].

This potent thromboxane antagonist has shown activity in a number of models of allergic bronchoconstriction including allergen-induced bronchoconstriction in hyperreactive rats [10], *Ascaris*-induced immediate responses in squirrel monkeys [11] and *Ascaris*-induced immediate and late-phase bronchoconstriction in sheep [Abraham, unpublished observations]. The compound may also inhibit TxA_2-induced hyperreactivity as indicated by results in squirrel monkeys [12]. In these experiments, squirrel monkeys were exposed to ascending concentrations of aerosolized acetylcholine (0.1–30 mg/ml, 3-min exposure) and a dose–response curve was established. The animals were then exposed to aerosols (0.01–1 mg/ml, 5 min) of the thromboxane/prostaglandin endoperoxide mimetic, U-44 069. One hour later, after pulmonary variables had returned to baseline, a new dose–response curve to acetylcholine was established. The results indicated that after exposure to U-44 069, the concentration–response relationship to acetylcholine was shifted approximately 1000-fold to the left with an increase in the maximal change in pulmonary resistance. In five animals, the enhanced response to acetylcholine was totally reversed by an aerosol of L-655 240 (1 mg/ml, 5 min). In contrast, L-655 240 failed to antagonize the responses to acetylcholine in animals not treated with U-44 069. These above results indicate a role for contractile prostanoids on allergen-induced bronchoconstriction and bronchial hyperreactivity in animal models and suggest that compounds such as L-655 240 may be useful for investigating the role of prostanoids in human bronchial asthma.

Leukotrienes and the allergic response

Leukotrienes are products of the 5-lipoxygenase (5-LO) pathway of arachidonic acid metabolism. Leukotriene B_4 is a 5,12-dihydroxy fatty acid with high-affinity, structurally specific receptors on various leucocyte populations including polymorphonuclear leucocytes, eosinophils and subsets of T-cells [13]. Interaction of LTB_4 with these receptors stimulates a number of leucocyte functions including chemotaxis, chemokinesis, aggregation and augmentation of natural killer (NK) cell activity [13, 14]. Because of these activities, LTB_4 has been proposed as a potential mediator of inflammatory reactions [13] and as such could mediate inflammatory aspects of allergic diseases, e.g. eosinophil migration into antigen-challenged lung.

The peptidolipid leukotrienes, LTC_4, LTD_4 and LTE_4, have been proposed as important mediators of allergic conditions [15]. These compounds collectively account for the biological activity known as slow-reacting substance of anaphylaxis (SRS-A) [15]. In most systems, rapid conversion of LTC_4 to LTD_4 occurs through the action of γ-glutamyl transferase, and subsequent conversion to LTE_4 occurs through the action of dipeptidases. Leukotriene D_4 has high-affinity, structurally specific receptor sites on a number of smooth-muscle preparations, interaction with which results in smooth-muscle contraction (bronchoconstriction or vasoconstriction). In addition, leukotrienes mediate other aspects of defective lung function including mucus production, decreased mucociliary clearance and changes in vascular permeability, although the exact nature of the receptors involved in these responses is not clear [15]. In addition, in humans, leukotrienes are potent bronchoconstrictor agents and when administered to asthmatics they may induce a long-lasting bronchial hyperresponsiveness but not in normal individuals suggesting an important additional role for leukotrienes in human bronchial asthma [Lee, unpublished observations]. Leukotriene E_4 has similar properties to LTD_4 but is less potent and shows a somewhat lower affinity for the LTD_4-receptor site. Leukotriene C_4 has been reported to have recognition sites on a number of tissues. It is now clear that many of these studies simply reflect binding of LTC_4 to glutathione-S-transferase. However, on certain tissues, LTC_4 induces contractions not mediated through the

LTD_4-receptor. This appears not to be the case on human trachea where LTC_4 acts through the LTD_4-receptor site. Thus, the role, if any, of LTC_4-receptor-mediated responses in humans is not clear.

Two therapeutic approaches are being developed to inhibit the production or antagonize the action of leukotrienes. These are LTD_4-receptor antagonists and 5-LO inhibitors. Examples of LTD_4-receptor antagonists are the orally active compound, L-649 923 (sodium[βS*,γR*]-4-[3-(4-acetyl-3-hydroxy-2-propylphenoxy)-propylthio]-γ-hydroxy-β-methyl benzene butanoate) [16] and the aerosol-active compound, L-648 051 (sodium 4-[3-(4-acetyl-3-hydroxy-2-propylphenoxy)-propylsulphonyl]-γ-oxo-benzene butanoate) [17]. These compounds were shown to be selective and competitive LTD_4-receptor antagonists using receptor-binding studies on human and guinea pig lung membranes and functional studies on isolated tissues, in particular guinea pig trachea, guinea pig ileum and human trachealis [16, 17]. Selectivity was indicated by the failure to antagonize a number of other contractile antagonists either *in vitro* on guinea pig tracheal chains or *in vivo* in anaesthetized guinea pigs. Two examples of orally active 5-LO inhibitors are L-651 392 (4-bromo-2,7-dimethoxy-3H-phenothiazine-3-one) [18] and L-656 224 (7-chloro-2-[(4-methoxyphenyl)methyl]-3-methyl-5-propyl-4-benzofuranol) [19]. These compounds are potent inhibitors of LTB_4 production in either human or rat polymorphonuclear leucocytes stimulated with ionophore A-23 187 or highly purified or partially purified porcine or rat 5-LO [18, 19] (Table 2.1). In the case of L-651 392, the compound was only active in 5-LO preparations in the presence of NADH or NADPH indicating that

Table 2.1 Inhibition of eicosanoids by L-651 392 and L-656 224 in various preparations containing 5-LO

		Mean IC_{50}	
Enzyme source	Stimuli	L-651 392	L-656 224
Rat peritoneal polymorphonuclear leucocytes	A-23 187	0.6×10^{-7} M	2.4×10^{-7} M
Human polymorphonuclear leucocytes	A-23 187	2.6×10^{-7} M	1.8×10^{-8} M
Rat leucocyte 5-LO	Arachidonic acid	0.8×10^{-7} M*	0.4×10^{-7} M
Purified porcine 5-LO	Arachidonic acid	ND	4×10^{-7} M

Both compounds were inactive on ram seminal vesicle cyclooxygenase, human platelet or porcine leucocyte 12-LO and soyabean 15-LO (for L-651 392 in the presence or absence of NADH or NADPH).
* Active only in the presence of NADH or NADPH.
ND = not done.

reduction of L-651 392 must occur for inhibition of the enzyme to be observed. The selectivity of these compounds was indicated by the failure to inhibit the production of products of the cyclooxygenase, 12-LO or 15-LO pathways of arachidonic acid metabolism [18, 19].

Such compounds have been tested in various animal models of allergic bronchoconstriction and have been used to indicate a major role for LTD_4 in allergen-induced bronchoconstriction in hyperreactive rats [10], *Ascaris*-induced, immediate and late-phase bronchoconstriction in squirrel monkeys [11, 20] and *Ascaris*-induced, late-phase responses in sheep [7, 21]. In the hyperreactive rat, the induction of bronchoconstriction by allergen has been correlated with a large increase in the production of leukotrienes as assessed by excretion of biliary metabolites [Foster *et al.*, unpublished observations]. Leukotriene D_4 may also mediate permeability responses to antigen and a role for LTD_4 and histamine has been defined in an allergic conjunctivitis model in the guinea pig [22, 23]. In this model, the inhibitory effects of LTD_4-receptor antagonists on the antigen response were correlated with inhibitory effects on the LTD_4 response [22] and the inhibitory effects of topical 5-LO inhibitors were correlated with inhibition of LTB_4 production in the conjunctiva following *ex vivo* challenge with ionophore A-23 187 [23].

Clinical trials with LTD_4-receptor antagonists have tested such compounds against LTD_4-induced bronchoconstriction in normal volunteers and antigen challenge of asthmatics. Early compounds, such as L-649 923, produced small shifts in the dose–response curve for LTD_4 (3.8-fold) but not histamine [24] and small effects on the immediate antigen response [25], making it difficult to determine whether the antigen results reflect lack of potency of the compound or a minor role for leukotrienes in immediate antigen responses (leukotrienes could be more important in late-phase responses and bronchial hyperresponsiveness). Testing with aerosol compounds, e.g. L-648 051 [17], which may be more efficacious due to high local deposition in the lungs, second-generation compounds at doses that produce up to 50-fold shifts in the LTD_4 dose–response curve or leukotriene-biosynthesis inhibitors that can be shown to block increased leukotriene production following antigen challenge in man, should allow for a definitive evaluation of the role of leukotrienes in allergic diseases including human bronchial asthma.

References

1 Adkinson NF Jr, Newball HH, Findlay S, Adams K, Lichtenstein LM. Anaphylactic release of prostaglandins from human lung in vitro. *Am. Rev. Respir. Dis.* 1980;**121**: 911–920.

2 Schulman ES, Newball HH, Demers LM, Fitzpatrick FA, Adkinson NF Jr. Anaphylactic release of thromboxane A_2, prostaglandin D_2 and prostacyclin from human lung parenchyma. *Am. Rev. Respir. Dis.* 1981; **124**:402–406.

3 Kennedy I, Coleman RA, Humphrey PPA, Levy GP, Lumley P. Studies on the characterisation of prostanoid receptors: a proposed classification. *Prostaglandins* 1982; **24**:667–689.
4 Hanley SP. Prostaglandins and the lung. *Lung* 1986; **164**:65–77.
5 Fairfax AJ, Hanson JM, Morley J. The late reaction following bronchial provocation with house dust mite allergen. Dependence on arachidonic acid metabolism. *Clin. Exp. Immunol.* 1983; **52**:393–398.
6 Shephard EG, Malan L, Macfarlane CM, Mouton W, Joubert JR. Lung function and plasma levels of thromboxane B_2, 6-ketoprostaglandin $F_{1\alpha}$ and β-thromboglobulin in antigen-induced asthma before and after indomethacin treatment. *Br. J. Clin. Pharmacol.* 1985; **19**:459–470.
7 Lanes S, Stevenson JS, Codias E, Hernandez A, Sielczak MW, Wanner A, Abraham WM. Indomethacin and FPL-57231 inhibit antigen-induced airway hyperresponsiveness in sheep. *J. Appl. Physiol.* 1986; **61**:864–872.
8 Chung KF, Aizawa, H, Becker AB, Frick O, Gold WM, Nadel JA. Inhibition of antigen-induced airway hyperresponsiveness by a thromboxane synthetase inhibitor (OKY-046) in allergic dogs. *Am. Rev. Respir. Dis.* 1986; **134**:258–261.
9 Hall RA, Gillard J, Guindon Y, Letts G, Champion E, Ethier D, Evans J, Ford-Hutchinson AW, Fortin R, Jones TR, Lord A, Morton HE, Rokach J, Yoakim C. Pharmacology of L-655,240 (3-[1-(4-chlorobenzyl)-5-fluoro-3-methyl-indol-2-yl] 2,2-dimethylpropanoic acid); a potent, selective thromboxane/prostaglandin endoperoxide antagonist. *Eur. J. Pharmacol.* 1987; **35**:193–201.
10 Piechuta H, Ford-Hutchinson AW, Letts GL. Inhibition of allergen-induced bronchoconstriction in hyperreactive rats as a model for testing 5-lipoxygenase inhibitors and leukotriene D_4 receptor antagonists. *Agents Actions* 1987; **2**:69–74.
11 McFarlane CS, Hamel R, Ford-Hutchinson AW. Effects of a 5-lipoxygenase inhibitor (L-651,392) on primary and late pulmonary responses to *Ascaris* antigen in the squirrel monkey. *Agents Actions* 1987; **22**:63–68.
12 Letts GL, McFarlane CS. Thromboxane A_2 and airway responsiveness to acetylcholine aerosol in the conscious primate. In *Mechanisms in Airways: Pharmacology, Physiology, Management*, Progress in Clinical and Biological Research, vol. 263, AR Liss, New York 1988, pp. 91–98.
13 Ford-Hutchinson AW. Leukotrienes: their formation and role as inflammatory mediators. *Fed. Proc.* 1985; **44**:25–29.
14 Ford-Hutchinson AW, Bray MA, Doig MV, Shipley M, Smith MJH. Leukotriene B, a potent chemokinetic and aggregating substance released from polymorphonuclear leukocytes. *Nature* 1980; **286**:264–265.
15 Piper PJ. Formation and action of leukotrienes. *Physiol. Rev.* 1984; **64**:744–761.
16 Jones TR, Young R, Champion E, Charette L, Denis D, Ford-Hutchinson AW, Frenette R, Gauthier J-Y, Guindon Y, Kakushima M, Masson P, McFarlane C, Piechuta H, Rokach J, Zamboni R, Dehaven RN, Maycock A, Pong SS. L-649,923, sodium (βS*,γR*)-4-(3-4-acetyl-3-hydroxy-2-propylphenoxy)propylthio)-γ-hydroxy-β-methylbenzene-butanoate, a selective orally active leukotriene receptor antagonist. *Can. J. Physiol. Pharmacol.* 1986; **64**:1068–1075.
17 Jones TR, Guindon Y, Young R, Champion E, Charette L, Denis D, Ethier D, Hamel R, Ford-Hutchinson AW, Fortin R, Letts G, Masson P, McFarlane C, Piechuta H, Rokach J, Yoakim C, DeHaven RN, Maycock A, Pong SS. L-648,051, sodium 4-[3-(4-acetyl-3-hydroxy-2-propylphenoxy)propylsulfonyl]-γ-oxo-benzenebutanoate: a leukotriene D_4 receptor antagonist. *Can. J. Physiol. Pharmacol.* 1986; **64**:1535–1542.
18 Guindon Y, Girard Y, Maycock A, Ford-Hutchinson AW, Atkinson JG, Bélanger PC, Dallob A, DeSousa D, Dougherty H, Egan R, Goldenberg MM, Ham E, Fortin R, Hamel P, Lau CK, Leblanc Y, McFarlane CS, Piechuta H, Thérien M, Yoakim C, Rokach J. L-651,392 a novel, potent and selective 5-lipoxygenase inhibitor. *Adv. Prostaglandin Thromboxane Leukotriene Res.* 1987; **17**:554–557.
19 Bélanger P, Maycock A, Guindon Y, Bach T, Dallob AL, Dufresne C, Ford-Hutchinson AW, Gale PH, Hopple S, Lau CK, Letts LG, Luell S, McFarlane CS, MacIntyre E, Meurer R, Miller DK, Piechuta H, Riendeau D, Rokach J, Rouzer C, Scheigetz J. L-656,224 (7-chloro-2-

[(4-methoxyphenyl)methyl]-3-methyl-5-propyl-4-benzofuranol; a novel, selective, orally active 5-lipoxygenase inhibitor. *Can. J. Physiol. Pharmacol.* 1987; **65**:2441–2448.

20 Letts LG, McFarlane C, Piechuta H, Ford-Hutchinson AW. Effects of a 5-lipoxygenase inhibitor (L-651-392) and leukotriene D_4 antagonist (L-649,923) in two animal models of immediate hypersensitivity reactions. *Adv. Prostaglandin Thromboxane Leukotriene Res.* 1987; **17**:1007–1011.

21 Abraham WM, Wanner A, Stevenson JS, Chapman GA. The effect of an orally active leukotriene D_4/E_4 antagonist, LY171883, on antigen-induced airway responses in allergic sheep. *Prostaglandins* 1986; **31**:457–467.

22 Garceau D, Ford-Hutchinson AW. The role of leukotriene D_4 as a mediator of allergic conjunctivitis in the guinea-pig. *Eur. J. Pharmacol.* 1987; **134**:285–292.

23 Garceau D, Ford-Hutchinson AW, Charleson S. 5-Lipoxygenase inhibitors and allergic conjunctivitis reactions in the guinea pig. *Eur. J. Pharmacol.* 1987; **143**:1–7.

24 Barnes N, Piper PJ, Costello J. The effect of an oral leukotriene antagonist L-649,923 on histamine and leukotriene D_4-induced bronchoconstriction in normal man. *J. Allergy Clin. Immunol.* 1987; **79**:816–821.

25 Britton JR, Hanley SP, Tattersfield AE. The effect of an oral leukotriene D_4 antagonist L-649,923 on the response to inhaled antigen in asthma. *J. Allergy Clin. Immunol.* 1987; **79**:811–816.

Discussion session

HOLGATE: as you are probably aware, we have been working with another thromboxane receptor antagonist, GR-32 191.

This drug seems to have a similar *in vitro* profile to the compound that you are using, but maybe slightly less potent. In humans, we have shown that *in vivo* it blocks PGD_2-induced bronchoconstriction and produces about a 30- to 35-fold displacement of the PGD_2 dose–response curve. When given prior to allergen challenge in asthma as a single dose it removes about 30% of the early response, has little effect on the late reaction but as yet we do not know what it does to hyperresponsiveness. Do you think that PGD_2 could be involved in the mechanism of hyperresponsiveness to allergen challenge in a similar way that you were postulating for thromboxane?

FORD-HUTCHINSON: I really have no idea on that. Obviously you could look at that in humans, something you could not do with thromboxane A_2.

QUESTIONER (unidentified): thank you for the data you have shown on the 5-lipoxygenase inhibitor. May I raise one question concerning that compound. I think that selectivity in the field of 5-lipoxygenase inhibitor is a very critical issue. Could you give us some idea of what the selectivity of that compound is with respect to other redox enzymes, for example, NADPH oxidase or myeloperoxidase? Does it also interfere with those enzymes?

FORD-HUTCHINSON: no. L-656 224 does not inhibit myeloperoxidase or catalase. There is a complete paper on this compound in the *Canadian Journal of Pharmacology and Physiology* (1987; **65**:2441–2448) which summarizes a lot of these selectivity experiments [19].

QUESTIONER (unidentified): in your squirrel monkey where you have got hyperreactivity can you tell us whether the U-44069 is altering baseline mechanics. Are you measuring hyperreactivity from a different resting level?

FORD-HUTCHINSON: no, we are waiting until the baseline parameters are completely back to normal before doing a second dose–response curve, so yes, it is from a normal baseline. It is also a conscious animal so it has all its reflexes intact.

SAME QUESTIONER (unidentified): that really is my second question. Is the hyperreactivity sensitive to atropine?

FORD-HUTCHINSON: this is something we have not examined yet.

QUESTIONER (unidentified): what is the persistence of this hyperreactivity?

FORD-HUTCHINSON: we also have not looked at that aspect yet.

QUESTIONER (unidentified): you could come back the next day.

FORD-HUTCHINSON: I know it is back to normal within 2 weeks. We have not done it earlier because we cannot carry out any manipulations on these animals for 2 weeks after use according to the guidelines of our animal-care committee.

WILLHELMS (Mannheim): I would like to learn more about the activity pattern of other antiasthma compounds in your rat model. Have you any experiments with well-known compounds?

FORD-HUTCHINSON: the model is sensitive to cromoglycate, theophylline and β-agonists.

3

Leukotriene inhibitors and antagonists in asthma

S.T. HOLGATE AND G.D. PHILLIPS

Introduction

In 1938, while conducting studies on the effects of cobra venom on guinea pig lung tissue, Feldberg and Kellaway discovered the release of a lysolecithin-like substance in the perfusate which had a potent contractile action on gastrointestinal smooth muscle. When compared with the effects of histamine, the contraction was slow in onset and sustained, and accordingly the activity was named 'slow-reacting substance' (SRS) [1]. The same group of investigators later demonstrated that SRS could be anaphylactically generated [2] and after identifying it as a lipid mediator, Brocklehurst named the activity 'slow-reacting substance of anaphylaxis' (SRS-A) [3].

Over the following years there occurred a growing interest in SRS-A as one of the increasing numbers of pro-inflammatory mediators implicated in the pathogenesis of IgE-dependent and other forms of acute allergic reaction. Particular attention was focused on the potent contractile activities of SRS-A on isolated animal and human airway preparations and its cardiovascular effects in reducing cardiac output, causing peripheral vasoconstriction and increasing vascular permeability. Biologically generated SRS-A, whether isolated from exudate anaphylactically generated within the rat peritoneal cavity or from challenged guinea pig lung or human lung tissue, was shown to have similar physicochemical properties in being an acid lipid, soluble in organic solvents and containing sulphur [4]. By the late 1970s, work in several laboratories suggested that SRS-A had characteristics of an arachidonic acid metabolite and indeed its generation and release from rat basophil leukaemia cells and lung tissue could be enhanced in the presence of exogenous arachidonic acid [5].

Physicochemically, SRS-A could be differentiated from the known eicosanoids in having an unusual ultraviolet spectrum of a peak absorption at 283 nm with shoulders on either side [6].

In 1979, the preliminary structure of SRS-A was published as a unique lipid containing three conjugate double bonds and a peptide attached to the 6 position of the arachidonic acid backbone [7] (Fig. 3.1). The complete chemical synthesis of leukotriene (LT) C by Corey and coworkers greatly

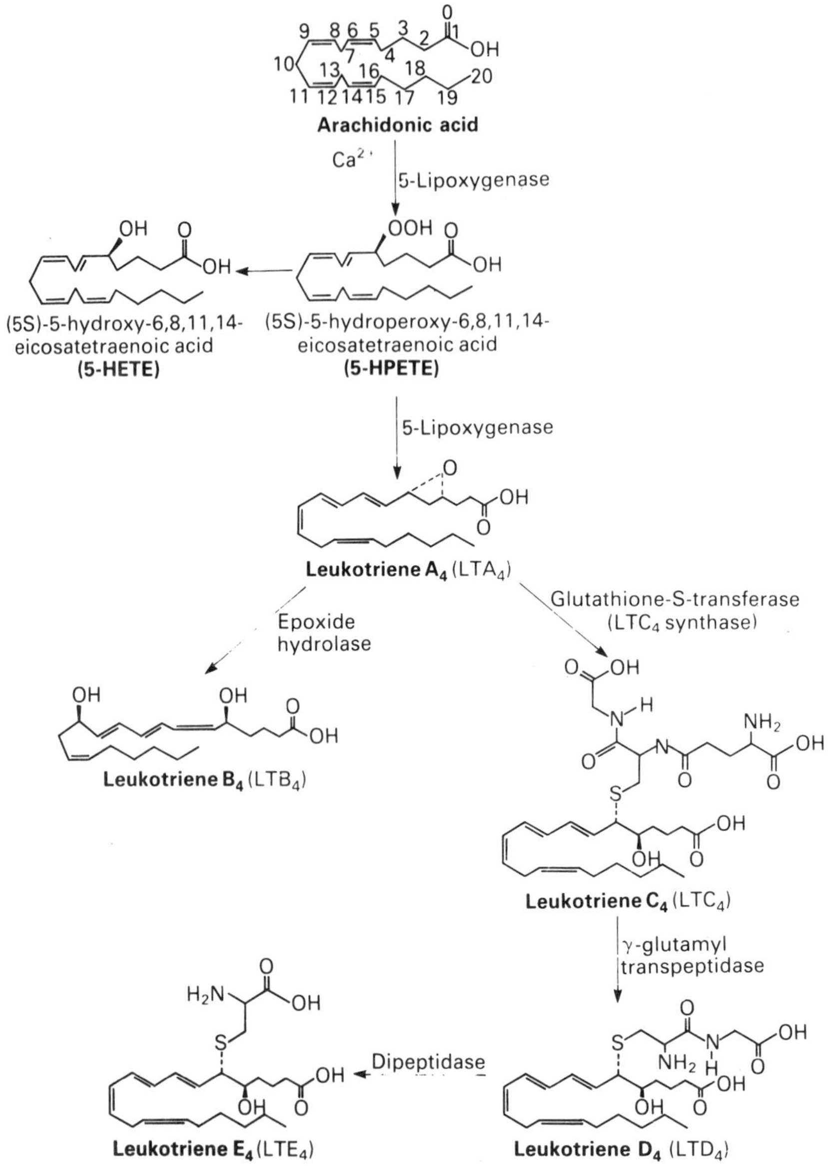

Fig. 3.1 The 5-lipoxygenase (5-LO) pathway of arachidonic acid metabolism.

facilitated further experimental work with this molecule, and the characterization of two further components of SRS-A [8]. Slow-reacting substance of anaphylaxis, generated by rat basophil leukaemia cells, during rat peritoneal anaphylaxis and following allergen challenge of human lung tissue, was shown to comprise the glutathione adduct and named LTC_4. Removal of a glutamic acid residue produces the dipeptide, glycine–cysteinyl derivative LTD_4—the major product generated by guinea pig lung tissue—whilst LTE_4 represents the terminal product in the degradation sequence formed by removal of glycine, and contains the single amino acid, cysteine (Fig. 3.1). All three sulphidopeptide leukotrienes had contractile activities against animal and human airway smooth muscle with an order of potency of $LTD_4 \geqslant LTC_4 > LTE_4$. A similar spectrum of activity was seen for all three leukotrienes in increasing vascular permeability, causing coronary artery vasoconstriction and reducing cardiac output [9–11].

The pharmacological actions of the sulphidopeptide leukotrienes on human airways *in vitro* and *in vivo*

The clear association of leukotriene pharmacology with that of SRS-A stimulated considerable interest in relation to airway obstruction in asthma [12]. The generation of these mediators with anaphylactic challenge of human lung *in vitro* [13] and the finding that LTC_4 is released by human lung mast cells [14], macrophages [15] and eosinophils [16], focused attention on this molecule as a possible key mediator in the pathogenesis of asthma. Studies on isolated human bronchus showed that LTC_4 and LTD_4 were up to 1000 times more potent on a molar basis than histamine as contractile agonists [17]. Increased potency was also demonstrated on parenchymal lung tissue when compared with the more central airways [17, 18]. Preferential activity on the peripheral airways was confirmed by *in vivo* animal studies, showing a greater potency of LTC_4 and LTD_4 on pulmonary dynamic compliance than on airway resistance [19].

In 1982, Holroyde and coworkers administered synthetic LTC_4 by aerosol to a group of normal volunteers and confirmed its potent bronchoconstrictor activity in humans [20]. This was followed by a series of studies confirming the bronchoconstrictor activity of LTC_4 in normal subjects and describing a similar response to LTD_4 [21–23]. On the other hand, LTE_4 was shown to be considerably less potent than its two precursor molecules [24], although later work showed that this derivative had the rather unusual effect of enhancing the airway response to other bronchoconstrictor mediators such as histamine [25, 26].

Leukotrienes C_4 and D_4 were also shown to cause bronchoconstriction when inhaled by patients with asthma, and were of similar potency to that previously described for non-asthmatic airways [24, 27–29]. However, emerging from these studies was the consistent finding that subjects with asthma failed to demonstrate the levels of augmented response to LTC_4 and LTD_4 that they displayed to histamine, so that leukotriene responsiveness, relative to the degree of non-specific airway hyperreactivity, in this way differed markedly from non-asthmatic subjects. The relative lack of hyperresponsiveness to the sulphidopeptide leukotrienes in asthma led Drazen and Austen and coworkers to suggest that these molecules might be implicated in the pathogenesis of bronchial hyperresponsiveness and provided some support for their proposed central importance as effector mediators of asthma [30].

The case for developing drugs active against sulphidopeptide leukotrienes

The chemical identification of leukotrienes as a potent class of pro-inflammatory mediators and the subsequent demonstration of their release following allergen challenge of asthmatic lung tissue *in vitro* [31] provided the background for interest in these compounds as targets for developing drugs of potential therapeutic benefit in the treatment of such allergic diseases as asthma and rhinitis. With improved assay techniques, increased concentrations of LTC_4 have been detected in lavage fluid from the nasal cavity of atopic rhinitic subjects following local allergen challenge [32]. However, there has been difficulty in demonstrating the *in vivo* generation of sulphidopeptide leukotrienes in human asthma. Although several studies have demonstrated quite high levels of histamine and prostaglandin (PG) D_2 in bronchoalveolar lavage fluid [33, 34] in association with allergen-induced immediate bronchoconstriction, there has been some difficulty demonstrating increased release of the leukotrienes [35]. This difficulty extends to the measurement of LTC_4 and its metabolites in peripheral blood following allergen challenge of the airways, or during natural exacerbations of asthma [36]. In the few studies in which increased circulating concentrations of LTC_4 or LTD_4 have been reported [37], validation of the assay procedures used has been questioned. The recovery of sulphidopeptide leukotrienes from the airways of rodents by bronchoalveolar lavage is poor [38] and has led other investigators to look at the urine as a source of more stable leukotriene metabolites [39]. In rodents, a major route for the metabolism and excretion of sulphidopeptide leukotrienes is the hepatobiliary system with the generation of *N*-acetyl LTE_4 [40]. In humans, the major route of excretion of sulphidopeptide leukotriene is the kidney, with LTE_4 being the major product [39]. There is evidence for increased concentrations of LTE_4 in bronchoalveolar lavage (BAL) fluid following allergen provocation of the airways of patients with asthma [41] but at the time of writing there have been no published reports of increased urinary excretion of LTE_4.

The biological effects of the sulphidopeptide leukotrienes are mediated through specific receptors on target cells. In guinea pig lung membrane preparations, two receptor subclasses have been described, one exhibiting some specificity for LTC_4 and the other jointly for LTD_4 and LTE_4 [40,

42–45]. There is doubt about the existence of a specific LTC_4 receptor on cells of human origin [46] since the cell binding site previously attributed to a functional receptor has been identified as the enzyme glutathione-S-transferase which is known to have a high affinity for binding LTC_4 [47, 48]. There is, however, good evidence for the existence of the LTD_4/LTE_4 receptor on smooth muscle cells, and the development of selective antagonists directed against this has provided one approach for novel drug development in asthma [49].

An alternative approach has been to develop enzyme inhibitors of the leukotriene generation pathway. The last 5 years has seen the detailed dissection of the calcium-dependent 5-lipoxygenase (5-LO) enzyme system which is responsible for converting arachidonic acid to the first intermediate in the leukotriene pathway, 5-hydroperoxy-6,8,11,14-eicosatetraenoic acid (5-HPETE) [50]. This unstable intermediate may be subsequently metabolized to 5-hydroxy-6,8,11,14-eicosatetraenoic acid (5-HETE) and is also a substrate for the 5-LO enzyme forming LTA_4 [51], which in turn may be stereospecifically hydrated to LTB_4 or undergo an adduction reaction in the presence of a particulate glutathione-S-transferase to form the 6-glutathionyl derivative, LTC_4 [52]. Sequential cleavage of glutamic acid and glycine by γ-glutamyl transpeptidase and dipeptidases respectively leads to the formation of the other components of SRS-A—LTD_4 and LTE_4 [53].

Since LTB_4 has been shown to be a highly potent chemotactic mediator for neutrophils and, to a lesser degree, eosinophils, inhibitors of its formation could be seen to confer an advantage in controlling an inflammatory reaction. Thus, a number of pharmaceutical companies have attempted to synthesize selective inhibitors of the 5-LO pathway with the prime objective of inhibiting the release of more than one class of lipid mediator, a principle that has proved successful in the development of cyclooxygenase inhibitors as non-steroidal anti-inflammatory drugs. Although the development of inhibitors of glutathione-S-transferase represents another potential target for drug development, LTC_4 synthetase appears to be only one component of a highly complex series of glutathione-S-transferase enzymes with differing substrate specificities [54, 55].

Receptor antagonists of sulphidopeptide leukotrienes

Long before the structural identification of the sulphidopeptide leukotrienes, research workers at Fisons had discovered a compound, FPL-55 712, which proved to be a potent inhibitor of the contractile activities of SRS-A on isolated animal and human airway smooth muscle and when administered intravenously to animals (Fig. 3.2) [49, 56]. Throughout the 1970s, FPL-55 712 was used as a pharmacological tool for confirming a biological activity

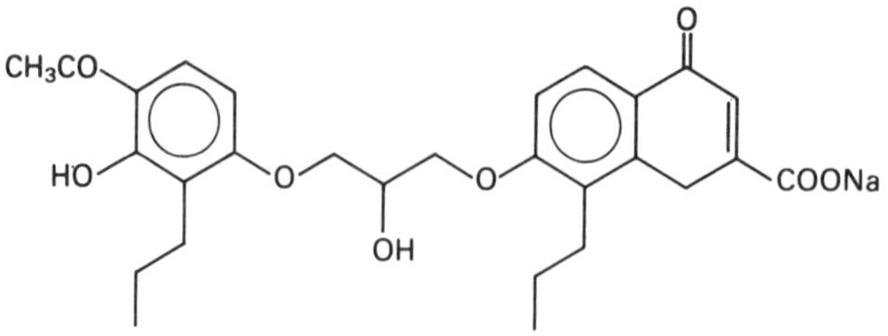

FPL–55 712

Fig. 3.2 Structure of the Fisons SRS-A antagonist FPL-55 712.

of SRS-A. FPL-55 712 is poorly absorbed when administered orally and has a short plasma half-life. By the intravenous route, Feniak *et al.* [57] demonstrated substantial inhibition of LTD_4-induced change in pulmonary mechanics in anaesthetized guinea pigs, and by aerosol a similar but shorter-duration protection was shown [58]. Thus, although promising in its initial pharmacology, FPL-55 712 was not a suitable candidate for development as a therapeutic agent in man.

Since the first description of FPL-55 712 as an SRS-A antagonist, attempts to synthesize other inhibitors and antagonists have continued. In many respects the discovery of FPL-55 712 occurred 'out of its time' because only with the knowledge of the chemical nature of the components of SRS-A and their receptor interactions has it been possible to further develop this theme with structure–activity studies. Thus, propionic acid analogues (FPL-57 231 and 59 257) were developed that were approximately equipotent with FPL-55 712 *in vitro*, but, although FPL-59 257 had a longer duration of action than FPL-55 712 when injected intravenously into guinea pigs [59], problems with toxicity precluded the further development of these drugs. Other derivatives of FPL-55 712 including the tetrahydrocarbazole derivative (oxarbazole) [60] and imidodisulphamides [61], proved to be potent leukotriene antagonists but lacked selectivity. More promise was held for the phenylaminosulphonyl substitute, SKF-88 046. *In vitro* and *in vivo* studies in guinea pigs, showed this drug to be an LTD_4, but not an LTC_4, antagonist and it is approximately equipotent with FPL-55 712, but biologically more stable [62].

Two further compounds that have stimulated considerable interest as leukotriene antagonists are L-649 923 and LY-171 883, both of which have been modelled on the original FPL-55 712 structure (Fig. 3.3). *In vitro* and *in vivo* studies in animal models showed that both of these compounds exhibit antagonistic activity at the LTD_4 receptor, and initial evaluation indicated a low level of toxicity [63, 64]. Of particular interest was the discovery that both drugs were active by oral administration. At high concentrations, both LY-171 883 and L-649 923 shared with FPL-55 712 the property of inhibiting cyclic nucleotide phosphodiesterase, but in all other respects LY-171 883 and

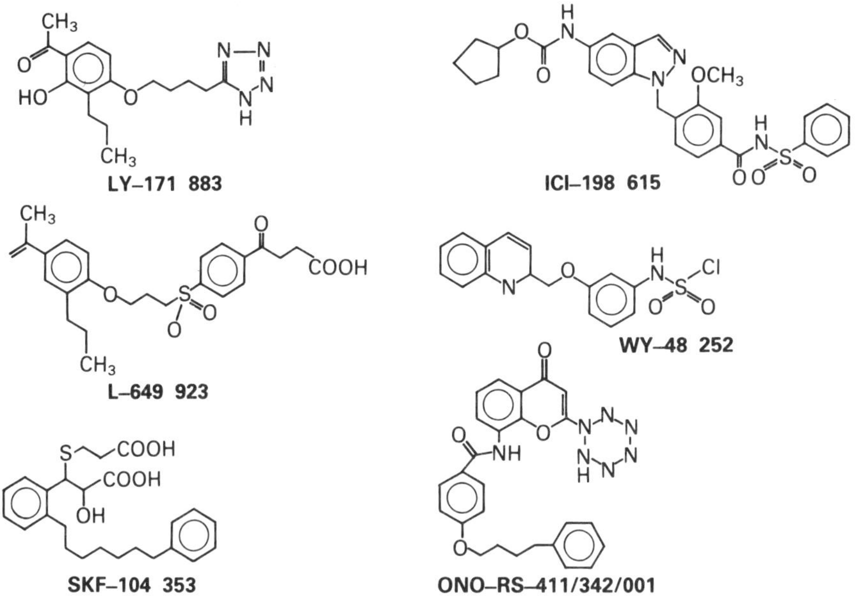

Fig. 3.3 Structure of the orally active LTD_4 antagonists.

L-649 923 held considerable promise as orally active leukotriene antagonists suitable for therapeutic development in man.

The chemical identity of SRS-A having been defined, the way was clear to develop structural analogues of this class of mediators to further define the receptor characteristics of the leukotrienes and develop potent and selective antagonists. Included among these are the diacetylene analogue of LTE_4 and the 2-nor-leukotriene and desamino-2-nor-leukotriene analogues [58, 65–67]. Leukotriene antagonistic activity has also been reported for a wide variety of other molecules including oxpyrrolidinocarboxylic acids, phenylamino-carbonyl propenoic acids, dibenzothiepin acids and hydroxypyrroline-2,5-diones. The specificity and potency of this diverse group of compounds are low and the receptor basis for their actions as leukotriene receptors is poorly defined.

Evaluation of leukotriene antagonists in humans

In an open study, inhalation of aerosolized FPL-55 712 and FPL-59 257 in a concentration of 5×10^{-3} M inhibited the cough and bronchoconstrictor responses produced in non-atopic normal volunteers by inhaled LTC_4 [20]. In six patients with ragweed sensitivity, aerosolized FPL-55 712 failed to protect against the acute bronchoconstrictor effect of allergen challenge when airway calibre was followed as specific conductance (sGaw) [68]. An uncontrolled study in which aerosolized FPL-55 712 was administered four times daily for 7 days to four patients with chronic asthma who had entered an unstable phase, produced improvement in baseline measurements of forced expiratory volume in 1 s (FEV_1) in two of the subjects [69]. FPL-59 257 administered as an aerosol of 5×10^{-3} M afforded partial protection against the immediate allergen-provoked bronchoconstriction in two out of five asthmatic subjects, suggesting that sulphidopeptide leukotrienes may play a limited role in the pathogenesis of this response [70].

The oral activity of L-649 923 presented a unique opportunity for investigating the *in vivo* activity of a leukotriene antagonist in humans. In a double-blind, placebo-controlled study, Barnes *et al.* [71] treated 11 non-asthmatic normal volunteers with a single oral dose of 1000 mg of L-649 923 approximately 60 min prior to inhalation challenge of the airways, on one occasion with histamine and on the other with LTD_4. Airway calibre was monitored both as sGaw and maximum expiratory flow rate 70% below total lung capacity ($\dot{V}max_{30}$) derived from a partial expiratory flow–volume manoeuvre. By administering progressively increasing concentrations of inhaled agonist, dose–response curves of the concentration of inhaled agonist against the percentage fall in index of pulmonary function were constructed. Since L-649 923 did not significantly influence baseline measurements of pulmonary function nor alter the slope of the agonist dose–response curves, it was possible to describe the position of these curves in terms of a provocation concentration (*PC*) of agonist reducing sGaw and $\dot{V}max_{30}$ by 35 and 30% of baseline respectively. Figure 3.4 illustrates the mean data for the measurement

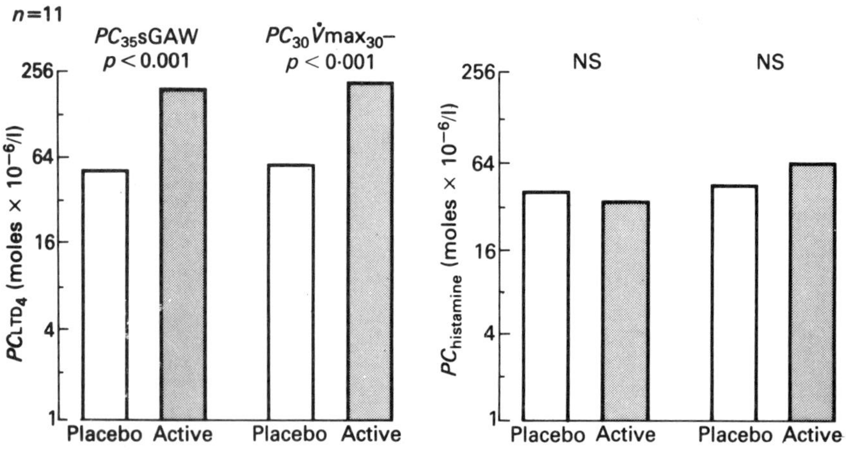

Fig. 3.4 Effect of a single 1000 mg dose of L-649 963 on the sGaw and $\dot{V}max_{30}$ responses to LTD_4 challenge in normal volunteers. The height of the bars represents the geometric mean *PC* value after placebo and active drug treatment for each of the two indices of airway calibre. NS = not significant. (From reference 71 by permission of the publishers.)

of sGaw and $\dot{V}max_{30}$. When compared with oral placebo, L-649 923 produced a geometric mean 3.8-fold displacement to the right of the LTD_4 dose–response curve for both pulmonary function measurements, without having any significant effect on the airway response to inhaled histamine. During this study the authors commented on the high degree of gastrointestinal side effects noted with this drug, particularly abdominal discomfort and diarrhoea [71].

In a study of similar design, the activity of oral LY-171 883 was observed against the bronchoconstrictor effect of inhaled LTD_4 [72]. In 11 subjects, LY-171 883 administered on separate occasions in three oral doses of 50, 200 and 400 mg 2 h before LTD_4 inhalation challenge in 12 non-asthmatic volunteers produced progressive displacements to the right of the LTD_4 dose–response curves. However, when compared with oral placebo this protection only reached statistical significance for the highest dose of LY-171 883 producing a geometric mean (range) 4.6- (0.7–16.2) and 6.3- (0.4–20.7) fold displacement of the LTD_4 dose–response curve when airway calibre was measured as FEV_1 and $\dot{V}max_{30}$ respectively (Fig. 3.5). In contrast to the study with L-649 923, gastrointestinal side effects with single oral doses of LY-171 883 were not noted.

The next step in the investigation of these drugs was to observe their activity against allergen-provoked bronchoconstriction in patients with asthma. In the same dose used for the LTD_4 challenge, L-649 923 (1000 mg), administered 2 h prior to challenge, afforded a small but significant protection

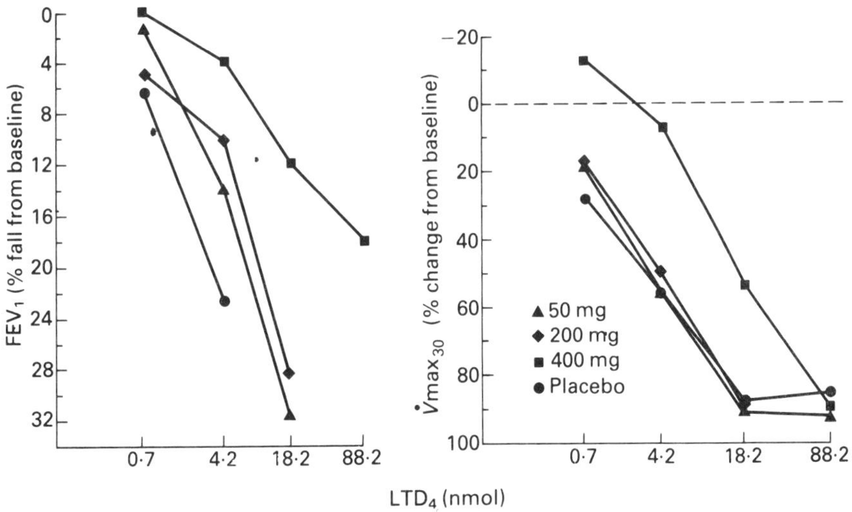

Fig. 3.5 Displacement of LTD_4 dose–response curves by three oral doses of 50, 200 and 400 mg LY-171 883 in a normal volunteer. (Data from reference 72 by permission of the publishers.)

against the immediate bronchoconstrictor response to inhaled allergen when airway calibre was measured as FEV_1, peak expiratory flow (PEF) or $\dot{V}max_{25}$ but not specific airway conductance (Fig. 3.6) [73]. The time of peak drug activity during the early response was after maximum bronchoconstriction had developed, which is in accordance with the kinetics of sulphidopeptide leukotriene generation from allergen-stimulated human lung cells [14] and contrasts with the activity of potent and selective H_1-histamine receptor antagonists such as terfenadine and astemizole. These drugs produce their maximum inhibition during the first 10 min of allergen challenge when mast-cell histamine release is maximum [74]. Prostaglandin D_2, another bronchoconstrictor mediator derived from activated mast cells, is approximately 30 times more potent than histamine as a contractile agonist of human airways. Prior treatment of patients with a cyclooxygenase inhibitor, flurbiprofen [75], or the contractile prostaglandin receptor antagonist, GR-32 191 [76], inhibits allergen-induced bronchoconstriction maximally between 10 and 30 min after challenge, corresponding to the kinetics of PGD_2 generation and release. Separate studies in atopic asthma have also shown that LY-171 883 administered as a single oral dose of 80 mg 2 h prior to challenge has some inhibitory effect on the immediate airway response to allergen [77], the maximum drug activity again being observed during the later time points of the immediate response, compatible with the kinetics of sulphidopeptide leukotriene release within the airways. Some inhibitory activity of LY-171 883 has also been shown on immediate bronchoconstriction provoked by exercise [78] and cold-air inhalation [79]. In the study by Britton *et al.* [73] L-649 923 had no effect on the allergen-induced, late-phase bronchoconstriction which more closely reflects the inflammatory components of clinical asthma.

Although the *in vitro* and animal pharmacology of L-649 923 and LY-171 883 looked promising, their efficacy in protecting against LTD_4 and

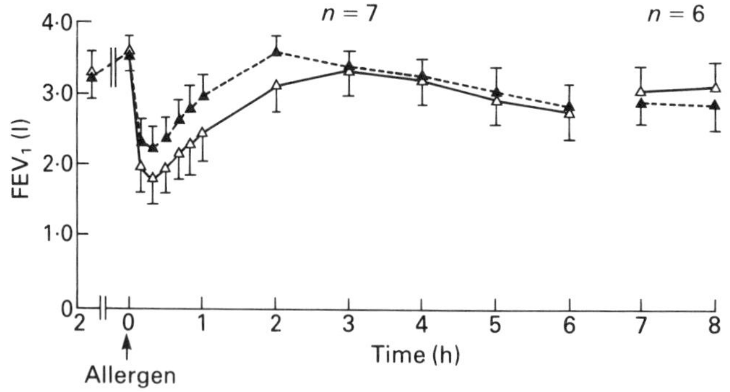

Fig. 3.6 Effect of a single oral dose of 1000 mg L-649 963 on the allergen-provoked immediate fall in FEV_1 in seven atopic asthmatic subjects. (From reference 73 by permission of the publishers.)

allergen *in vivo* challenge was disappointing. Part of the reason for this may relate to the bioavailability of compounds of this class at the relevant leukotriene receptors in the airway. A more promising approach has been to develop antagonists for administration by inhalation, a route of delivery that has proved so effective with other drugs for the treatment of asthma. This route of administration not only allows for more drug to be delivered to the relevant site of mediator release, but also widens the therapeutic index. The three- to four-fold displacement of the LTD_4 dose–response curve by L-649 923 and L-171 883 may be compared with the 30–50-fold protection afforded by the more recently introduced selective H_1-histamine receptor antagonists against the airway effects of histamine. With the availability of such potent compounds as terfenadine and astemizole, it has proved possible to dissect the contribution of histamine to bronchoconstriction provoked by a number of different stimuli, and more recently some of these drugs have been shown to have clinical activity in the treatment of some forms of asthma. Thus, if a future serious attempt is to be made in determining the contribution of sulphidopeptide leukotrienes to the pathogenesis of asthma, far more potent drugs will be required which, in the authors' view, should preferably be administered via the inhaled route.

Clinical activity of leukotriene antagonists in asthma

Apart from the pilot study with FPL-55 712 [69], there has been only one large clinical study investigating the efficacy of a leukotriene antagonist in the treatment of clinical asthma. In the multicentre placebo controlled study reported by Cloud *et al.*, LY-171 883 was administered to 137 patients with mild to moderate asthma whose current treatment consisted of inhaled or oral bronchodilators alone [80]. The active drug was administered in a dose of 600 mg orally twice daily for a period of up to 6 weeks, and patients were followed with respect to symptoms recorded by diary card, morning and night time PEF and FEV_1 measurements at clinic visit, and patient and investigator assessment of treatment preference. After the 6 week treatment period, LY-171 883 was shown to be significantly better than placebo in increasing mean PEF, reducing bronchodilator usage and improving overall symptom scores. However, although statistically significant, the changes observed were small and of only marginal clinical importance. In a double-blind, placebo-controlled study, LY-171 883 was administered to 78 theophylline-controlled asthmatics at doses of 100, 400 or 800 mg daily for 8 weeks [81]. Small but again significant improvements in day and night chest symptoms were noted only in the group receiving the 800 mg dose. While it can be concluded from these trials that a leukotriene antagonist such as LY-171 883 has potential for the treatment of asthma, until drugs of greater potency and efficacy are developed the usefulness of this therapeutic approach cannot be evaluated further. More potent and selective compounds of the acetophenone series such as WY-44 329, RO-233 544 and SC-39 070 have been developed as leukotriene antagonists and are currently undergoing evaluation as potential therapeutic agents [81, 82]. A number of pharmaceutical companies are in the process of investigating leukotriene antagonists which are modelled on analogues of sulphidopeptide leukotrienes such as SKF-104 353 (Fig. 3.3) or which differ in other ways from the original lead molecule of FPL-55 712 examples of which are ICI-198 615, WY-48 252 and ONO-RS-411 347 001 [83–85]. Only if these compounds can be shown to possess greater potency than existing agents in inhibiting the constrictor effects of inhaled leukotrienes in normal and asthmatic subjects, can further clinical trials be seriously considered.

5-Lipoxygenase inhibitors

From a consideration of the biochemical sequence depicted in Fig. 3.1, a prime target for developing pharmacological agents which are leukotriene inhibitors would be the synthesis of compounds which inhibit 5-LO activity. Although this concept is attractive, from a practical standpoint it has proven difficult to target drugs specifically to this enzyme. Indeed, the majority of the early compounds that were referred to as 5-LO inhibitors (Fig. 3.7) had a variety of activities against other biological functions. For example, 5,8,11,14-eicosatetraynoic acid (ETYA), in addition to inhibiting the release of 5-LO products, also suppresses the generation of 12-hydroxyeicosatetraenoic acid (12-HETE) from platelets and both BW-755c and benoxaprofen are dual inhibitors of 5-LO and cyclooxygenase. What has emerged from directed research into inhibitors of the 5-LO enzyme system are a range of compounds with widely differing chemical structure many of which are antioxidants.

Several different factors have confounded the progression of research into the development of selective and potent 5-LO inhibitors. First, many drugs with a putative anti-5-LO activity can be shown only to display this activity against this enzyme function in intact cells, and are inactive or weakly active against the isolated enzyme. One example of this is ETYA which is approximately 17 times more active against the rat basophil leukaemia 5-LO when incubated with intact cells than against the partially purified enzyme [86]. Second, the activity of putative 5-LO inhibitors in attenuating the generation of products such as the leukotrienes may vary enormously from one cell type to another. Third, inhibition of 5-LO may have the effect of diverting arachidonic acid along alternative oxidative pathways, e.g. cyclooxygenase, 12-LO, thereby distorting the observed pharmacological action of the drug.

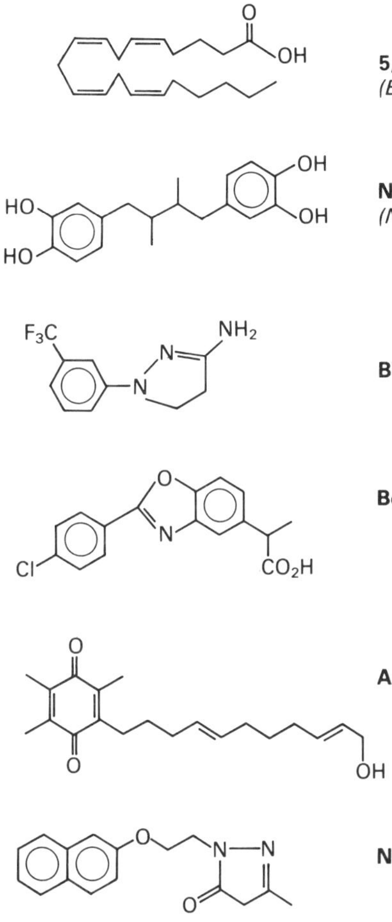

	$\approx IC_{50}$ (μM)
5,8,11,14-eicosatetraynoic acid *(ETYA)*	3–50
Nordihydroguaioretic acid *(NDGA)*	2
BW–755c	4–60
Benoxaprofen	100
AA–861	0·8
Nafazatrom	3

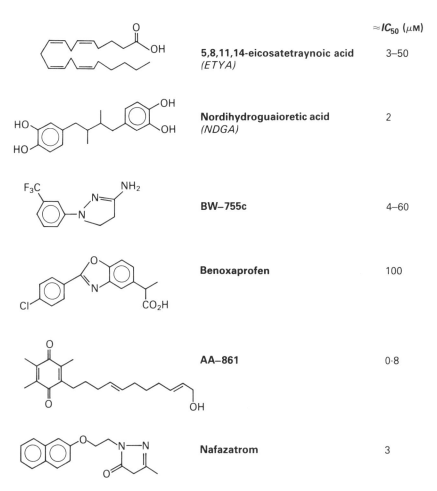

Fig. 3.7 Chemical structure and approximate IC_{50} values for inhibitors of 5-LO.

Fourth, the discovery that 5-LO activity itself may be regulated by products of the arachidonic acid cascade complicates interpretation of LO inhibitors. In this respect, both 5-HETE and 15-HETE are feedback inhibitors of 5-LO in some cells and, therefore, any drug-induced changes in these endogenous negative-feedback loops could have a profound influence on the effect of a putative 5-LO inhibitor [87, 88]. Finally, compartmentalization of arachidonate within functionally different pools within a cell, each of which may be mobilized by different stimuli, has greatly complicated the development of 5-LO drug screens which are predictive of *in vivo* activity. For example, benoxaprofen has been shown to inhibit 5-LO function in rabbit and guinea pig polymorphs stimulated with the calcium ionophore A-23 187, but failed to suppress LTC_4 release [89]. In the same study. Walker and Harvey observed that aspirin and diclofenac inhibited ionophore-stimulated 5-LO in rabbit polymorphs but augmented 5,12-di-HETE production by guinea pig polymorphs. Since many drug screens utilize the incorporation of radiolabelled arachidonic acid into oxidative products, great caution should be exercised in interpreting the results of such experiments since this procedure may not reflect the relevant endogenous pool of substrate utilized when the enzyme is activated under physiological conditions [90].

Variables such as those considered above, together with *in vivo* metabolization and limited drug bioavailability, have frustrated the development of effective 5-LO inhibitors. Some of the earlier compounds such as ETYA, nordihydroguaioretic acid (NDGA) and BW-755c (Fig. 3.7) failed early toxicological studies. Benoxaprofen was the first of a class of drugs with dual cyclooxygenase and 5-LO activity to enter clinical trial in asthma. Considerable enthusiasm was generated when the drug was shown to attenuate late-phase, allergen-provoked bronchoconstriction in patients with asthma [91] but further clinical evaluation of this drug in asthma or any other disease was stopped on account of hepatotoxicity and other adverse effects.

Based on the potential usefulness of a 5-LO inhibitor, the Upjohn company developed a compound, U-60 257 (piriprost) (Fig. 3.8). This PGI_2 analogue was active in inhibiting the release of 5-LO products of arachidonic acid from both guinea pig and human lung when challenged by the calcium ionophore A-23 187 or by reversed anaphylaxis [92, 93]. On challenging human dispersed lung cells with A-23 187, piriprost produced a dose-related inhibition of LTC_4, LTB_4 and 5-HETE generation [94]. In addition, the compound produced inhibition of thromboxane release yet potentiated the release of mast-cell derived PGD_2 by mechanisms probably involving interference with endogenous feedback mechanisms and the diversion of arachidonate along alternative metabolic pathways [95]. Recent

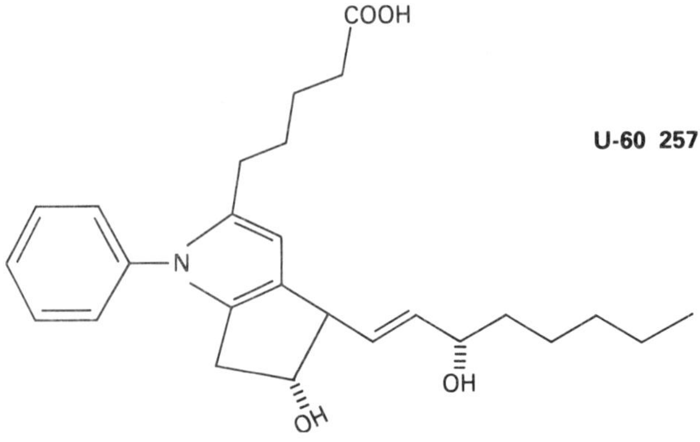

U-60 257 (piriprost)

Fig. 3.8 Structure of the Upjohn compound, piriprost (U-60 257).

studies have questioned the site of action of piriprost as being the 5-LO enzyme system since it is only active against the generation of this enzyme's products in intact cells and not when the enzyme is studied in isolation. Piriprost has been shown to be a weak inhibitor of glutathione-S-transferase and is better regarded as an inhibitor of leukotriene release rather than as a specific enzyme inhibitor. In anaesthetized and ventilated guinea pigs, both intravenous and inhaled piriprost inhibited antigen-induced bronchoconstriction that became more apparent if the animals were pretreated with propranolol, atropine and indomethacin to isolate and accentuate the leukotriene-mediated component [96]. Piriprost administered by inhalation inhibited acute antigen-induced bronchoconstriction in *Ascaris*-sensitized monkeys [97] and therefore held promise as an active drug for the treatment of allergen-provoked asthma. However, when similar studies were undertaken in atopic asthma, piriprost administered as a single inhaled dose of 1 mg had no significant effect on the immediate bronchoconstriction provoked by either inhaled allergen or by exercise [98]. One possible reason for this negative result was the rapid *in vivo* metabolism of this prostaglandin analogue, thereby limiting its bioavailability. An additional factor could be the adverse effect of diverting arachidonic acid to other bronchoconstrictor mediators of the cyclooxygenase pathway such as PGD_2, thereby reducing any beneficial effect of the drug against the generation of 5-LO products.

Another compound thought to have at least some inhibitory activity against 5-LO was nafazatrom. However, in a study where 3 g of nafazatrom were administered as a single oral dose 3 h prior to allergen challenge, no overall effect on bronchoconstriction was observed [99]. When blood from the actively treated subjects was challenged *ex vivo* with calcium ionophore A-23 187 at the time when peak drug concentrations occurred, no inhibitory effect on leucocyte LTB_4 generation could be shown suggesting that lack of *in vivo* activity was most likely the result of rapid metabolism and lack of bioavailability.

Other drugs currently being investigated as 5-LO inhibitors in asthma include AA-861 (antioxidant) and REV-5901 (substrate analogue) which are putatively orally active, and U-66 856—a 5-LO inhibitor developed for inhaled use. Since efficacy has not so far been demonstrated for any 5-LO inhibitor in experimental provocation systems in asthma, none have progressed to the point of investigation in the clinic. At this time, our limited knowledge regarding the mediator components of late-phase bronchoconstriction and acquired bronchial hyperreactivity do not preclude the possibility that a 5-LO inhibitor could be therapeutically effective in clinical asthma but be without effect in an acute-challenge setting. Moreover, chronic drug dosing may have effects beyond that achieved by any single drug dose as exemplified by the beneficial effects of corticosteroids in asthma.

Concluding remarks

Over the last 10 years enormous progress has been made in understanding the role of leukotrienes in the pathogenesis of acute and subacute allergic reactions in humans, but the extent to which their activity contributes to the airway inflammation of asthma still remains to be determined. Until leukotriene antagonists and inhibitors which are bioavailable and potent become available, further prediction on the contribution of these mediators to the pathogenesis of asthma is speculative and, therefore, the potential of this class of mediators as a target for effective drug development will remain controversial.

The recognition that selective recruitment of leucocytes from the circulation into the airways is an important component of the inflammatory response in asthma has focused some attention on those factors which promote adherence of these cells to the vascular endothelium and on their subsequent chemotactic migration. The possibility remains that LTB_4, in being a potent chemotactic factor, may present an additional target for drug development in asthma on the assumption that this mediator comprises a component of the chemotactic stimulus for polymorphonuclear leucocyte recruitment during allergen-induced, late-phase reactions and natural exacerbations of asthma [100]. Whether or not this prediction holds true will have to await the development of specific LTB_4 antagonists of proven efficacy. Another important mediator showing preferential chemotactic activity towards eosinophils is platelet-activating factor (PAF-acether) [101]. This ether-linked phospholipid, when inhaled by normal human volunteers, produces a prolonged increase in non-specific bronchial responsiveness, an abnormality of considerable relevance to asthma [102]. The observation that LY-171 883 inhibits a variety of *in vivo* activities produced by PAF-acether [103–105] indicates that at least a portion of PAF-acether-related pathophysiology is mediated by the sulphidopeptide leukotrienes.

The initial drive towards developing drugs active on 5-LO products of arachidonic acid partly stems from the assumption that corticosteroids produce their beneficial effect in asthma by releasing lipocortin which in turn inhibits the phospholipase A_2 reduction of arachidonic acid. Since corticosteroids have been shown to have many other pharmacological activities, their efficacy probably relates more to the wide number of cells that they are able to influence rather than to an action on any single cell or chemical pathway. One therefore has to question the wisdom of developing drugs of potential use for the treatment of asthma which are targeted on one mediator or a single metabolic pathway, since it is highly likely that the contribution of chemical mediators to the overall state of airway function will differ both between patients and within the same patient on separate occasions. If, however, leukotrienes play a key role in the mediation of or interaction with other pro-inflammatory products in the airways, then the future holds promise for the development of a clinically useful class of drugs targeted against these mediators for the treatment of asthma. Clearly further time is necessary to test this prediction.

References

1 Feldberg W, Kellaway CH. Liberation of histamine and formation of lysolecithin-like substance by cobra venom. *J. Physiol.* 1938; **94**:187–226.

2 Kellaway CH, Trethewie ER. The liberation of a slow reacting smooth muscle stimulating substance in anaphylaxis. *Quart. J. Exp. Physiol.* 1940; **30**:121–145.

3 Brocklehurst WE. The release of histamine and formation of a slow reacting substance (SRS-A) during anaphylactic shock. *J. Physiol.* 1960; **151**:416.

4 Orange RP, Murphy RC, Austen KF. Inactivation of slow-reacting substance of anaphylaxis by arylsulfatases. *J. Immunol.* 1974; **113**:316–322.

5 Engineer DH, Niedhauser U, Piper PJ, Sirois P. Release of mediators of anaphylaxis: inhibition of prostaglandin synthesis and the modification of release of slow reacting substance of anaphylaxis and histamine. *Br. J. Pharmacol.* 1978; **62**:61–66.

6 Morris HR, Taylor GW, Piper PJ, Sirois P, Tippins JR. Slow reacting substance of anaphylaxis: purification and characterization. *FEBS Lett.* 1978; **87**:203–206.

7 Murphy RC, Hammarstrom S, Samuelsson B. Leukotriene C: a slow reacting substance (SRS) from murine mastocytoma cells. *Proc. Natl. Acad. Sci. USA* 1979; **76**:4275–4279.

8 Corey EJ, Clark DA, Gota G, Marfat A, Mioskowski C, Samuelsson B, Hammarstrom S. Stereospecific total synthesis of a 'slow reacting substance of anaphylaxis' (SRS-A), leukotriene C_1. *J. Am. Chem. Soc.* 1980; **220**:568–575.

9 Piper PJ, Samhoun MN, Tippins JR, Williams TJ, Palmer MA, Peck MJ. Pharmacological studies on pure SRS-A, SRS and synthetic leukotriene C_4 and D_4. In *SRS-A and Leukotrienes* (Ed. Piper PJ), John Wiley, Chichester 1981, pp. 81–101.

10 Smedgard G, Hedqvist P, Dahlen S, Revanas B, Hammarstrom S, Samuelsson B. Leukotriene C_4 affects pulmonary and vascular dynamics in monkeys. *Nature* 1982; **295**:327–329.

11 Kito G, Okuda H, Ohkawa S, Terao S, Kikuchi L. Contractile activities of leukotrienes C_4 and D_4 on vascular strips from rabbits. *Life Sci.* 1981; **29**:1325–1332.

12 Hammarstrom S. Leukotrienes. *Annu. Rev. Biochem.* 1983; **52**:355–377.

13 Lewis RA, Austen KF, Drazen JM, Clark DA, Marfat A, Corey EJ. Slow reacting substance of anaphylaxis: identification of leukotrienes C-1 and D from human and rat sources. *Proc. Natl. Acad. Sci. USA* 1980: **77**:3710–3714.

14 MacGlashan DW Jr, Schleimer RP, Peters SP *et al.* Generation of leukotrienes by purified human lung mast cells. *J. Clin. Invest.* 1982; **70**:747–751.

15 Damon M, Chavis C, Godard P, Michel EB, Crastes de Paulet A. Purification and mass spectrometry identification of LTD_4 synthesized by human alveolar macrophages. *Biochem. Biophys. Res. Commun.* 1983; **111**:518–524.

16 Shaw RJ, Walsh GM, Cromwell O, Moqbel R, Spry CJF, Kay AB. Activated human eosinophils generate SRS-A leukotrienes following physiological (IgG-dependent) stimulation. *Nature* 1985; **316**:150–152.

17 Dahlen SE, Hedqvist P, Hammarstrom S, Samuelsson B. Leukotrienes are potent constrictors of human bronchi. *Nature* 1980; **288**:484–486.

18 Creese BR, Temple DM. The mediators of allergic contraction of human airway smooth muscle: a comparison of bronchial and lung parenchymal strip preparations. *Clin. Exp. Pharmacol. Physiol.* 1986; **13**:103–111.

19 Drazen JM, Austen KF, Lewis RA. Comparative airway and vascular activities of leukotrienes C and D *in vivo* and *in vitro*. *Proc. Natl. Acad. Sci. USA* 1980; **77**:4354–4358.

20 Holroyde MC, Altounyan REC, Cole M, Dixon M, Elliott EV, Bronchoconstriction produced in man by leukotrienes C and D. *Lancet* 1981; **2**:17–18.
21 Weiss JW, Drazen JM, Coles N. *et al.* Bronchoconstrictor effects of leukotriene C in humans. *Science* 1982; **216**:196–198.
22 Barnes NC, Piper PJ, Costello JF. Comparative effects of inhaled leukotriene C_4, leukotriene D_4 and histamine in normal human subjects. *Thorax* 1984; **39**:500–504.
23 Smith LJ, Greenberger PA, Patterson R, Krell RD, Bernstein PR. The effect of inhaled leukotriene D_4 in humans. *Am. Rev. Respir. Dis.* 1985; **131**:368–372.
24 Davidson AE, Lee TH, Scanlon PD *et al.* Bronchoconstrictor effects of leukotriene E_4 in normal and asthmatic subjects. *Am. Rev. Respir. Dis.* 1987; **135**:333–337.
25 Lee TH, Austen KF, Corey EJ, Drazen JM. Leukotriene E_4-induced airway hyperresponsiveness of guinea pig tracheal smooth muscle to histamine and evidence for three separate leukotriene receptors. *Proc. Natl. Acad. Sci. USA* 1984; **81**:4922–4925.
26 Arm JP, Spar BW, Lee TH. Leukotriene E_4 (LTE_4) enhances airway histamine responsiveness in asthmatic subjects. *Thorax* 1987; **42**:220 (Abstract).
27 Griffin M, Weiss JW, Leitch AG *et al.* Effects of leukotriene D on human airways in asthma. *N. Engl. J. Med.* 1983; **308**:436–439.
28 Bisgaard H, Groth S, Marsden F. Bronchial hyperreactivity to leukotriene D_4 in exogenous asthma. *Br. Med. J.* 1985; **290**:1468–1471.
29 Adelroth E, Morris MM, Hargreave FE, O'Byrne PM. Airway responsiveness to leukotrienes C_4 and D_4 and to methacholine in patients with asthma and normal controls. *N. Engl. J. Med.* 1986; **315**:480–484.
30 Drazen JM, Austen KF. Leukotrienes and airway responses. *Am. Rev. Respir. Dis.* 1987; **136**:985–998.
31 Dahlen SE, Hansson G, Hedqvist P, Bjorck J. Granston E, Dahlen B. Allergen challenge of lung tissue from asthmatics elicits bronchial contractions that correlates with release of leukotrienes C_4, D_4 and E_4. *Proc. Natl. Acad. Sci USA* 1983; **80**:1712–1716.
32 Naclerio RM, Proud D, Togius AG, Adkinson NF Jr, Meyers DA, Kagey-Sobotka A, Plant M, Norman PS, Lichtenstein LM. Inflammatory mediators in late antigen-induced rhinitis. *N. Engl. J. Med.* 1985; **313**:65–70.
33 Metzger WJ, Zavala D, Richerson HB *et al.* Local allergen challenge and bronchoalveolar lavage of allergic asthmatic lungs. *Am. Rev. Respir. Dis.* 1987; **135**:433–440.
34 Murray JJ, Tonnel AB, Brash AR, Roberts CJ, Cosset P, Worknan R, Capron A, Oates JA. Release of prostaglandin D_2 into human airways during acute antigen challenge. *N. Engl. J. Med.* 1986; **315**:800–804.
35 Cromwell O. Mediators in allergen-induced asthma. In *The Allergic Basis of Asthma* (Ed. Kay AB), Ballière Tindall, London 1988, pp. 197–216.
36 Okudo T, Takahashi H, Sumitomo M, Shindoh K, Suzuki S. Plasma levels of leukotrienes C_4 and D_4 during wheezing attacks in asthmatic patients. *Int. Arch. Allergy Appl. Immunol.* 1987; **84**:149–155.
37 Isono T, Koshihara Y, Murota S-I, Fukuda Y, Furukawa S. Measurement of immunoreactive leukotriene C_4 in blood of asthmatic children. *Biochem. Biophys. Res. Commun.* 1985; **130**:486–490.
38 Norris AA, Entwhistle N, Wilkinson D, Jackson DM. Recovery of leukotrienes from isolated rat lungs. *Br. J. Pharmacol.* 1988: In press.
39 Orning L, Kaijser L, Hammarstrom S. *In vivo* metabolism of leukotriene C_4 in man. Urinary excretion of leukotriene E_4. *Biochem. Biophys. Res. Commun.* 1985; **130**:214–220.
40 Orning L, Noren E, Gustafsson B, Hammarstrom S. *In vivo* metabolism of leukotriene C_4 in germfree and conventional rats. *J. Biol. Chem.* 1986; **261**:766–771.
41 Lam S, Chan H, LeRiche JC, Chan-Yeung M, Salari H. Release of leukotrienes in patients with bronchial asthma. *J. Allergy Clin. Immunol.* 1988; **81**:711–717.
42 Cheng JB, Lang D, Bewta A, Townley RG. Tissue distribution and functional correlation of [^{3}H] leukotriene C_4 and [^{3}H] D_4 binding sites in guinea-pig uterus and lung preparations. *J. Pharmacol. Exp. Ther.* 1985; **232**:80–87.

43 Mong S, Wa HL, Scott MO *et al.* Molecular heterogeneity of leukotriene receptors: correlation of smooth muscle contraction and radioligand binding in guinea-pig lung. *J. Pharmacol. Exp. Ther.* 1985; **234**:316–325.
44 Mong S, Wu HL, Hogaboom GK, Clark MA, Stadel JM, Crooke ST. Regulation of ligand binding to leukotriene D_4 receptors: effects of medications and guanine nucleotides. *Eur. J. Pharmacol.* 1984; **106**:241–253.
45 Mong S, Wu HL, Clark MA, Stadel JM, Gleason JG, Crooke ST. Identification of leukotriene D_4 specific binding sites in the membrane preparations isolated from guinea pig lung. *Prostaglandins* 1984; **28**:805–822.
46 Muccitelli RM, Tucker SS, Hay DWP, Torphy TJ, Wasserman MA. Is the guinea pig trachea a good *in vitro* model of human large and central airways? Comparison on leukotriene-, methacholine-, histamine- and antigen-induced contractions. *J. Pharmacol. Exp. Ther.* 1987; **243**:467–473.
47 Sun FF, Chau LY, Spur B, Corey EJ, Lewis RA, Austen KF. Identification of a high affinity leukotriene C_4-binding protein in rat liver cytosol as glutathione S-transferase. *J. Biol. Chem.* 1986; **261**:8540–8546.
48 Chau LY, Hoover RL, Austen KF, Lewis RA. Subcellular distribution of leukotriene C_4 binding units in cultured bovine endothelial cells. *J. Immunol.* 1986; **137**:1985–1992.
49 Augstein J, Farmer JB, Lee TB, Sheard P, Tattersall ML. Selective inhibitor of slow-reacting substance of anaphylaxis. *Nature* 1973; **245**:215–217.
50 Rouzer CA, Samuelsson B. On the nature of the 5-lipoxygenase reaction in human leukocytes: enzyme purification and requirement for multiple stimulatory factors. *Proc. Natl. Acad. Sci. USA* 1985; **82**:6040–6044.
51 Rouzer CA, Shimizu T, Samuelsson B. Single protein from human leukocytes possesses 5-lipoxygenase and leukotriene A_4 synthetase activities. *Proc. Natl. Acad. Sci. USA* 1986; **83**:857–861.
52 Jakschik BA, Harper T, Murphy RC. Leukotriene C_4 and D_4 formation by particulate enzymes. *J. Biol. Chem.* 1982; **257**:5346–5349.
53 Yoshimoto T, Soberman RJ, Lewis RA, Austen KF. Isolation and characterization of leukotriene C_4 synthetase of rat basophilic leukemia cells. *Proc. Natl. Acad. Sci. USA* 1985; **82**:8399–8403.
54 Lewis RA, Austen KF. Regulatory modulators of the 5-lipoxygenase pathway. In *Allergy and Inflammation* (Ed. Kay AB), Academic Press, London 1987, pp. 71–81.
55 Alin P, Jensson H, Guthenberg C, Danielsson DH, Tahir MK, Mannervik. Purification of major basic glutathione transferase isoenzymes from rat liver by use of affinity chromatography and fast protein liquid chromatofocussing. *Anal. Biochem.* 1985; **146**:313–320.
56 Sheard P, Holroyde MC, Ghelani AM, Bantick JR, Lee TB. Antagonists of SRS-A and leukotrienes. *Adv. Prostaglandin Thromboxane Leukotriene Res.* 1982; **9**:229–235.
57 Feniak L, Kennedy I, Whelan CJ. The contractile action of slow reacting substance of anaphylaxis (SRS-A) on rat isolated fundic strip and guinea-pig isolated ileum and its antagonism by FPL 55712. *J. Pharm. Pharmacol.* 1982; **34**:586–588.
58 Sheard P, Holroyde MC, Bantick JR, Lee TB. SRS-A antagonists. In *Development of Anti-asthma Drugs* (Eds Buckle DR, Smith H.), Butterworth, London 1984, pp. 133–158.
59 Sheard P, Holroyde MC, Ghelani AM, Bantick JR, Lee TB. Antagonists of SRS-A leukotrienes. *Adv. Prostaglandin Thromboxane Leukotriene Res.* 1982; **9**:229–235.
60 Mielens ZE. Evaluation of the anti-asthmatic potential of oxarbazole in guinea pigs. *Pharmacology* 1978; **17**:323–329.
61 Ali FE, Dandridge PA, Gleason JG, Krell RD, Kruse CH, Lavanchy PG, Snader KM. Imidodisulfamides, 1. Novel class of antagonists of slow-reacting substance of anaphylaxis. *J. Med. Chem.* 1982; **25**:947–952.
62 Gleason JG, Krell RD, Weichman BM, Ali FE, Berkowitz B. Comparative pharmacology and antagonisms of synthetic leukotrienes on airway and vascular smooth muscle. *Adv. Prostaglandin Thromboxane Leukotriene Res.* 1982; **9**:243–250.

63 Jones TR, Young R, Champion E *et al.* L-649,923, sodium (beta-S*-, gamma-R*)-4-(3-(4-acetyl-3-hydroxy-2-propylphenoxy)-propylthio)-gamma-hydroxy-beta-methylbenzene-butanoate, a selective, orally active leukotriene receptor antagonist. *Can. J. Physiol. Pharmacol.* 1986; **64**:1068–1075.
64 Fleisch JH, Rinkema LE, Haisch KD *et al.* LY171883, 1-<2-hydroxy-3-propyl-4-<4-(1H-tetrazol-5-yl)butoxy>phenyl>ethanone, an orally active leukotriene D_4 antagonist. *J. Pharmacol. Exp. Ther.* 1985; **233**:148–157.
65 Marshall WS, Goodson T, Swanson-Bean D, Rinkema LE, Haisch DK, Fleisch JH. Structure–activity relationships among a series of acetophenones: potent LTD_4 antagonists. *Prostaglandins and Leukotrienes, their Biochemistry, Mechanism of Action and Clinical Applications.* Abstract 262, 1984.
66 Gleason JG, Ku TW, McCarthy ML *et al.* 2-nor-leukotriene analogs; antagonists of the airway and vascular smooth muscle effects of leukotriene C_4, D_4 and E_4. *Biochem. Biophys. Res. Commun.* 1983; **117**:732–739.
67 Perchonock CD, Vzinskas I, Ku TW *et al.* Synthesis and LTD_4-antagonist activity of desamino-2-nor-leukotriene analogs. *Prostaglandins* 1985; **29**:75–81.
68 Ahmed T, Greenblatt W, Birch S, Marchette B, Wanner A. Abnormal mucociliary transport in allergic patients with antigen-induced bronchospasm. Role of slow reacting substance of anaphylaxis. *Am. Rev. Respir. Dis.* 1981; **124**:110–114.
69 Lee TH, Walport MJ, Wilkinson AH, Turner-Warwick M, Kay AB. Slow-reacting substance of anaphylaxis antagonist FPL 55712 in chronic asthma (letter). *Lancet* 1981; **ii**:304–305.
70 Altounyan REC, Cole M. Leukotrienes in the lung and cardiovascular system: therapeutic implications. In *Proceedings of the XI International Congress of Allergology and Clinical Immunology* (Ed. Kerr JW), Macmillan, London 1983, pp. 271–277.
71 Barnes NC, Piper PJ, Costello JF. The effect of an oral leukotriene antagonist L-649,923 on histamine and leukotriene D_4-induced bronchoconstriction in normal man. *J. Allergy Clin. Immunol.* 1987; **79**:816–821.
72 Phillips GD, Rafferty P, Holgate ST. LY-171883 as an oral leukotriene D_4 antagonist in non-asthmatic subjects. *Thorax* 1987; **42**:723 (Abstract).
73 Britton JR, Hanley SP, Tattersfield AE. The effect of an oral leukotriene D_4 antagonist L-649.923 on the response to inhaled antigen in asthma. *J. Allergy Clin. Immunol.* 1987; **79**:811–816.
74 Rafferty P, Beasley R, Holgate ST. The contribution of histamine to immediate bronchoconstriction provoked by inhaled allergen and adenosine 5′-monophosphate in atopic asthma. *Am. Rev. Respir. Dis.* 1987; **136**:369–373.
75 Curzen N, Rafferty P, Holgate ST. Cyclooxygenase inhibition and H_1-histamine receptor antagonism alone and in combination on allergen-induced bronchoconstriction in man. *Thorax* 1987; **42**:946–952.
76 Beasley RCW, Featherstone RL, Church MK, Rafferty P, Varley JG, Harris A, Robinson C, Holgate ST. The effect of a prostaglandin antagonist GR32191 on PGD_2 and allergen-induced bronchoconstriction. *J. Appl. Physiol.* Submitted.
77 Fuller RW, Black PN, Dollery CT. Effect of oral LY 171883 on inhaled and intradermal antigen and LTD_4 in atopic subjects. *Br. J. Clin. Pharmacol.* 1988; **25**:626P (Abstract).
78 Glovsky MM, Shaker G, Kebo D *et al.* The effect of LY 171883 on exercise-induced asthma in man. *Allergol. Immunopathol.* 1987; **15**:292 (Abstract).
79 Data on file, Eli Lilly Company, Earl-Wood, Bucks, UK.
80 Cloud M, Enas G, Kemp J, Platts-Mills T, Altman L, Townley R *et al.* Efficacy and safety of LY 171883 in patients with mild chronic asthma. *J. Allergy Clin. Immunol.* 1987; **79**:256 (Abstract).
81 Dillard RD, Carr FP, McCullough D *et al.* Leukotriene receptor antagonists. 2: The ((tetrazol-5-ylaryl)oxyl) methyl-acetophenone derivatives. *J. Med. Chem.* 1987; **30**:911–918.
82 Marshall WS, Goodson T, Callinass GJ. Leukotriene receptor antagonists. 1: Synthesis and structure activity relationships of alkoxyacetophenone derivatives. *J. Med. Chem.* 1987; **30**:682–689.

83 Hay DWP, Muccitelli RM, Tucker SS. Pharmacologic profile of SK and F 104353: a novel, potent and selective peptidoleukotriene receptor antagonist in guinea pig and human airways. *J. Pharmacol. Exp. Ther.* 1987; **243**:474–481.
84 Musser JH *et al.* Peptide leukotriene antagonists. *Agents Actions* 1987; **22**:59–62.
85 Piwinski JT, Kreutner W, Green MJ. Pulmonary and anti-allergy agents. *Annu. Rep. Med. Chem.* 1987; **22**:73–84.
86 Falkenheim SF, MacDonald H, Huber MM, Koch D, Parker CW. Effect of the 5-hydroperoxide of eicosatetraenoic acid and inhibitors of the lipoxygenase pathway on the formation of slow reacting substance by rat basophilic leukaemia cells; direct evidence that slow reacting substance is a product of the 5-lipoxygenase pathway. *J. Immunol.* 1980; **125**:163–168.
87 Arai Y, Shimoji K, Konno M, Konishi Y, Okuyama S, Iguchi S, Jayashi S, Myomoto T, Toda M. Synthesis and 5-lipoxygenase inhibitory activities of eicosanoid compounds. *J. Med. Chem.* 1983; **26**:72–78.
88 Goetzl EJ. Selective feedback inhibition of the 5-lipoxygenation of arachidonic acid in human T-lymphocytes. *Biochem. Biophys. Res. Commun.* 1981; **101**:344–350.
89 Walker JR, Harvey J. Action of anti-inflammatory drugs on leukotriene and prostaglandin metabolism: relationships to asthma and other hypersensitivity reactions. In *Side Effects of Anti-inflammatory/Analgesic Drugs* (Eds Rainsford RD, Velo GP), Raven Press, New York 1984, pp. 227–238.
90 Harvey J, Holgate ST, Peters BJ, Robinson C, Walker JR. Oxidative transformations of arachidonic acid in human dispersed lung cells: disparity between utilization of endogenous and exogenous substrate. *Br. J. Pharmacol.* 1985; **86**:417–426.
91 Fairfax AJ, Hanson JM, Morley J. The late reaction following bronchial provocation with house dust mite allergen. Dependence upon arachidonic acid metabolism. *Clin. Exp. Immunol.* 1983; **52**:393–398.
92 Smith RN, Sun FF, Bowman BJ, Iden SS, Smith HW, McGuire JC. Effect of 6,9-deepoxy-6,9-phenylimino-6,8-prostaglandin I_1 (U-60,257), an inhibitor of leukotriene synthesis, on human neutrophil function. *Biochem. Biol. Res. Commun.* 1982; **109**:943–949.
93 Bach MK. Prospects for the inhibition of leukotriene synthesis. *Biochem. Pharmacol.* 1984; **33**:515–521.
94 Robinson C, Holgate ST. Ionophore-dependent generation of eicosanoids in human dispersed lung cells: modulation by 6,9-deepoxy-6,9-(phenylimino)-6,8-prostaglandin I_1 (U-60,257). *Biochem. Pharmacol.* 1986; **35**:1903–1908.
95 Holgate ST, Robinson C. 6,9-deepoxy-6,9-(phenylimino)-6,8-prostaglandin I_1 (U-60,257) stimulates prostaglandin D_2 and inhibits thromboxane B_2 release from ionophore challenged human lung cells. *Br. J. Pharmacol.* 1984; **83**:603–605.
96 Bach MK, Griffin RL, Richards IM. Inhibition of the presumably leukotriene-dependent component of antigen-induced bronchoconstriction in the guinea-pig by piriprost (U-60,257). *Int. Arch. Allergy Appl. Immunol.* 1985; **77**:264–266.
97 Johnson HG, McNee ML Bach MK, Smith HW. The activity of a new novel inhibitor of leukotriene synthesis in rhesus monkey *Ascaris* reactors. *Int. Arch. Allergy Appl. Immunol.* 1983; **70**:167–173.
98 Mann JS, Holgate ST. Effect of piriprost (U-60,257) a novel leukotriene inhibitor on allergen- and exercise-induced asthma. *Thorax* 1986; **41**:746–752.
99 Fuller RW, Maltby N, Richmond R, Dollery CT, Taylor GW, Ritter W, Philipp E. Oral nafazatrom in man: effect on inhaled antigen challenge. *Br. J. Clin. Pharmacol.* 1987; **23**:677–681.
100 Wardlaw AJ, Kurihara K, Moqbel R, Walsh GM, Kay AB. Eosinophils in allergic and non-allergic asthma. In *The Allergic Basis of Asthma* (Ed. Kay AB), Baillière Tindall, London 1988, pp. 15–36.
101 Wardlaw AJ, Moqbel R, Cromwell O, Kay AB. Platelet-activating factor. A potent chemotactic and chemokinetic factor for human eosinophils. *J. Clin. Invest.* 1986; **78**:1701–1706.
102 Cuss FM, Dixon CMS, Barnes PJ. The effect of inhaled platelet activating factor on

pulmonary function and bronchial responsiveness in man. *Lancet* 1986; **ii**:189–192.
103 Calhoun W, Chang J, Carlson RP. Effect of related anti-inflammatory agents and other drugs on zymosan, arachidonic acid, PAF and carrogeenan-induced paw oedema in the mouse. *Agents Actions* 1987; **21**:306–307.
104 Carlson RP, O'Neill-Davis L, Chang J. Pharmacologic modulation of PAF-induced mortality in mice. *Agents Actions* 1987; **21**:379–381.
105 Stahl GL, Lefer DJ, Lefer AM. PAF-acethèr induced cardiac dysfunction in the isolated perfused guinea pig heart. *Arch. Pharmacol.* 1987; **336**:459–463.

Discussion session

QUESTIONER (unidentified): without speaking about the data and its interpretation, you obviously paid some attention to the biphasic effect in two of the illustrations in the pretreated group. Is there any explanation for this?

HOLGATE: no not at all. I think it is a high noise to signal ratio. This clinical trial is a difficult one to interpret because it was carried out in the patients with the most mild disease and I really feel that it was not a fair test of activity against the drug. If investigators decide to go into a disease setting like asthma with a drug, they should try to pick patients in whom a clinical response could be revealed more cleanly.

LEE: could I just comment on the LY study because we have had the opportunity to study this drug in bronchial asthma. In fact, in subjects with rather more severe asthma than the ones in the multicentre trial. Our design was very similar to the multicentre trial with the additional parameter of bronchial hyperresponsiveness to histamine being monitored at the beginning and at the end of the study. We were not able to show any clinical benefit at all in any of the variables including β_2-agonist usage. We were also unable to demonstrate any significant change in non-specific responsiveness during a 6-week period in 20 asthmatic subjects.

HOLGATE: that is 20 or 10 who received the treatment?

LEE: 10 treatment and 10 placebo.

HOLGATE: the study group was fairly small.

LEE: yes it was small.

DREBORG (Sweden): I agree with you when you say that this is not a fair test of the drug. I think you have to include more severe asthma in trials to give a drug a fair chance.

HOLGATE: I absolutely agree.

KAY: could you comment on the 1-week run-in period. Was the salbutamol usage comparable between the two groups in the 1-week period and was the 1-week period representative of those asthmatics over an extended period of time?

HOLGATE: the trial really should have had a minimum of a 2-week run-in and on stopping the drug they should have followed up after the end of the trial.

In terms of the matching of the β_2-agonist usage between the two groups, as far as I could glean from the information that was given to me (by the Eli Lilly Company), they were well matched.

FORD-HUTCHINSON: as I understand the trial, the patients were divided into two groups—those who were using more than 25 puffs per week and those who were using less. It was only in the ones who were using the higher amounts of β-agonists that there was a significant reduction in β-agonist usage. I do not think it represents a very large number of patients.

HOLGATE: I did not know about that.

4

Platelet-activating factor and biologically active acetylated phospholipids in asthma

J.-M. MENCIA-HUERTA, D. HOSFORD AND P. BRAQUET

Introduction

Since the initial description of platelet-activating factor (PAF-acether) in the beginning of the 1970s, several lines of evidence indicate that the molecules identified as either 1-O-hexadecyl- or 1-O-octadecyl-2-acetyl-*sn*-glycero-3-phosphocholine are in fact members of a large family of compounds, namely the biologically active acetylated phospholipids (BAAL). When the structure of PAF* was described [1–3], synthetic molecules soon became available and were investigated for biological activity in various experimental models. The use of C_{16} or C_{18} PAF-acether molecules allowed the description of various platelet-independent effects including hypotension, neutropenia, increase in vascular permeability, gastric and intestinal ulcerations and more recently a regulatory role in immune processes [reviewed in 4–7]. This large spectrum of biological activities implies that the synthetic molecules stimulate a variety of cell types and organs and soon the name of PAF-acether appeared to be too restrictive to define the factor fully. More recently, various molecular species of BAAL have been shown to be generated at the time of cell stimulation, mostly varying in the length and degree of unsaturation of the alkyl or acyl chain, as well as in the size and charge of the polar head group [8–16]. Although some of these BAAL are much less active on platelets than PAF-acether, they could represent molecules directed towards specific tissues, and thus may be interacting with specific binding sites. Determination of receptor subtypes, specific for BAAL other than PAF-acether, may be a difficult goal to reach since most of the binding studies have been carried out with the

* In this review the term 'PAF-acether' will be used to describe the factor(s) synthesized by various cell types and whose structure has been determined as a 1-O-hexadecyl- or 1-O-octadecyl-2-acetyl-*sn*-glyceryl-3-phosphocholine. In addition, natural molecules possessing platelet-stimulating activity but whose structure is not defined will be named PAF-acether. Finally, the term BAAL is given to the whole family of molecules bearing an alkyl or acyl chain (regardless of the length and degree of unsaturation) at the *sn* 1 position of the glycerol, an acetate at the *sn* 2 position and at least a phosphate group at the *sn* 3 position of the glycerol and presenting biological activity, regardless of their platelet-activating properties.

only available C_{16} ^{3}H-PAF-acether. This compound may not represent the actual ligand for all the considered binding sites, especially those at the bronchopulmonary level. In the present review we will consider the possibility that a bias was introduced in the development of this field of research leading to the concept that C_{16} or C_{18} alkyl-acetyl-*sn*-glyceryl-3-phosphocholine molecules are the archetypes of the large family of BAAL.

Molecular heterogeneity of BAAL

During the 1970s and the beginning of the 1980s, the crude material generated by various tissue or cell types was defined by its capacity to aggregate platelets (or to induce the release reaction), independently from the other known activating agents such as ADP and arachidonic acid metabolites [17–25]. For many years, further characterization of the generated material was done almost exclusively by the determination of the rate of flow (R_F) or retention time in thin-layer or (straight-phase) high-performance liquid chromatography (HPLC) respectively, and the sensitivity of semipurified material to various lipases and phospholipases [26]. During these very productive years, however, a bias may have been introduced, leading to the selection of platelet-aggregating or -activating substances since this was the initial assay of the compounds that were further characterized by physicochemical methods. During the past few years, however, several groups have directed their efforts towards a more detailed characterization of the various lipid species generated by various cell types and tissues regardless of their platelet-activating activity. These studies involved sophisticated physicochemical methods such as improved HPLC analyses, gas chromatography (GC) [27] coupled or not to mass spectrometry, and, more specifically, fast-atom bombardment/mass spectrometry [16]—the only method giving the structure of the considered molecule(s) directly without requiring derivatization. Some of these various molecules have been chemically synthesized or purified to sufficient extent to assess their biological activity, once again on platelets [reviewed in 28, 29] or platelet-dependent processes such as bronchoconstriction in the guinea pig [30]. Among the various molecules generated at the time of cell stimulation, those exhibiting activity on platelets have been shown to be limited in number and possess alkyl chains with a length of between 16 and 18 carbons (Fig. 4.1). Increasing or decreasing this alkyl chain length will produce platelet-activating molecules with markedly decreased specific activity. Although the presence of the ether oxide group at the *sn* 1 position of the glycerol gives maximal activity on platelets, molecules with an ester bound are biologically active and are generated at the time of polymorphonuclear leucocyte stimulation [14]. For this latter cell type, molecules bearing an ester

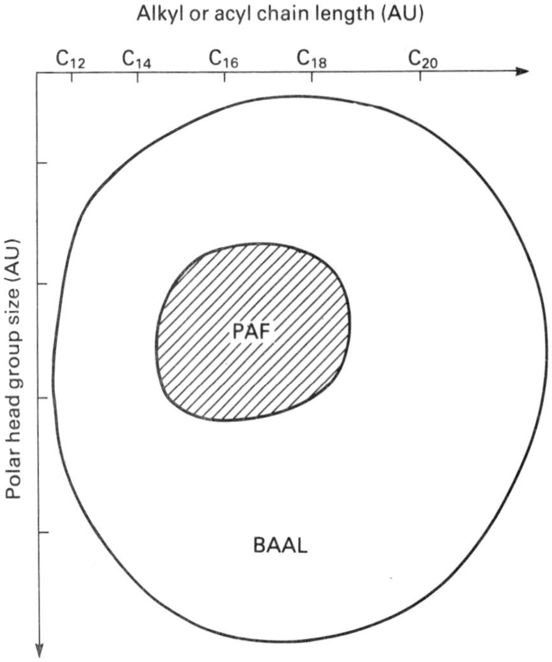

Fig. 4.1 Diagrammatic representation of the molecular heterogeneity of BAAL generated at the time of cell stimulation. This figure indicates that the molecules exhibiting high specific activity on platelets are represented in the centre circle labelled PAF. The specific activity on platelets decreases from the centre to the periphery. In this and the following figures, the upper scale indicates the length of the alkyl or acyl chain at the *sn* 1 position of the glycerol, irrespective of the degree and position of possible double bond(s). The left-hand scale arbitrarily indicates the size and charge of the polar head group at the *sn* 3 position of the glycerol. All scales are in arbitrary units (AU).

bound at the *sn* 1 position of the glycerol may represent up to 50% of the generated material. The presence of an alkenyl link at this *sn* 1 position gives a molecule almost as active as PAF-acether. However, the relationship between position of possible double bond(s) within the alkyl (or acyl) chain and biological activity on platelets and on platelet-independent phenomena is not fully established.

In addition, altering the acetate moiety at the *sn* 2 position of the glycerol dramatically changes the activity on platelets. Indeed, only the propionyl analogue of PAF-acether is as active as the original molecule, whereas, in contrast, the butyryl analogue does not exhibit significant biological activity. The presence of the ester linkage is required for full expression of the activity on platelets, although ethoxy and methoxy analogues of PAF-acether still express a potent aggregating property (Fig. 4.2 and Table 4.1).

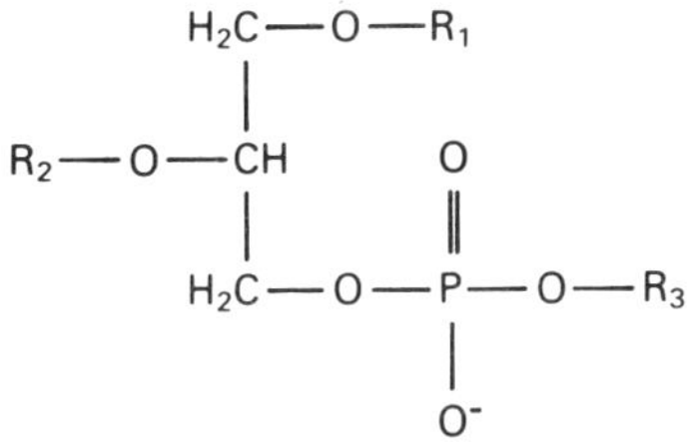

Fig. 4.2 Structure of BAAL.

Table 4.1 Platelet-activating activity of BAAL. (Data from reference 31 by permission of the publishers.)

R_1 Chain length	R_1 Linkage	R_3	ED_{50} (nM) for platelet stimulation
12:0	Ether	Phosphocholine	1.80
14:0	Ether	Phosphocholine	1.30
15:0	Ether	Phosphocholine	0.14
16:0	Ether	Phosphocholine	0.14
18:0	Ether	Phosphocholine	0.52
18:1	Ether	Phosphocholine	0.28
12:0	Ester	Phosphocholine	700.00
16:0	Ester	Phosphocholine	46.00
18:0	Ester	Phosphocholine	700.00
16:0/18:0	Ether	Phosphodimethylethanolamine	0.40
16:0/18:0	Ether	Phosphomonomethylethanolamine	3.50
16:0	Ether	Phosphoethanolamine	1020.00
18:0	Ether	Phosphoethanolamine	9940.00

Due to technical difficulties, the relationship between the volume and charge of the polar head group and the activity on platelets has not been extensively studied. In one report, Coeffier *et al.* [30] examined various analogues substituted at the *sn* 3 position for both their platelet-activating and secretory activity and their bronchoconstrictor effect. A good correlation between the actions of the PAF-acether analogues on platelets and bronchoconstriction was observed, a fact in keeping with the well-known platelet dependency of the latter phenomenon [5, and see below]. In this study, 1-O-octadecyl-2-O-acetyl-*sn*-glyceryl-3-phosphoryl-*N*-methylmorpholinoethanol was shown to be about two times more active than C_{18} PAF-acether in inducing platelet aggregation and secretion. However, the presence of the phosphate group esterified at the *sn* 3 position of the glycerol is a prerequisite for the activity on platelets. 1-O-alkyl-2-acetyl-*sn*-glycero-3-phosphatidic acid is also a potent platelet-aggregating agent [8, 31], suggesting that the choline head group is not mandatory for biological activity. Interestingly, the presence of a piperydinium instead of a phosphocholine polar head-group has been shown to generate a molecule more potent than PAF-acether on platelets.

Generally, most of these structure–activity relationship studies have been performed with synthetic molecules without considering the possible forms of the BAAL generated at the time of cell stimulation. Indeed, for many of the compounds produced at the time of cell stimulation the specific activity on platelets and platelet-independent processes is at present unknown. The knowledge of these specific activities, as well as the possible presence of receptor subtypes for each of these molecules, are prerequisites for the development of tissue-specific antagonists. In addition, some of these molecules exhibiting low biological activity on platelets such as 1-O-alkyl-2-acetyl-*sn*-glyceryl-3-phosphatidyl ethanolamine have been shown to influence the rate of catabolism of PAF-acether by the acetylhydrolase [32]. Thus, besides the intrinsic specific biological activities of the various BAAL, one may also have to determine that expressed when the molecule is in a complex mixture.

Although they vary considerably, it is reasonable to speculate that these various substances belong to the same family since they probably share the same biosynthetic pathways. Indeed, the so-called alkyl glyceryl phosphocholine—acetyl coenzyme A (CoA) acetyltransferase—exhibits no specificity at the *sn* 1 position for the lipid substrate and uses propionyl CoA as well as acetyl CoA as the cosubstrate [33, 34]. Less is known about the specificity of the enzymes involved in the cytosine diphosphate (CDP) choline phosphotransferase pathway. However, the large spectrum of molecules generated by kidney cells where this biosynthetic pathway is operational suggests a lack of specificity similar to that observed for the remodelling pathway [reviewed in 34].

Role of PAF-acether in asthma

The implication of PAF-acether in allergic diseases stems from the fact that the molecule was initially described as originating from IgE-sensitized rabbit basophils [35], i.e. a cell type involved in this pathological condition. Despite the fact that the generation of PAF-acether from purified, IgE-activated human basophils deserves confirmation [36], several lines of evidence suggest the participation of the autacoid in asthma.

The generation of PAF-acether during anaphylactic shock is supported by the specific desensitization of platelets to this autacoid following antigen administration to sensitized rabbits [37–40]. In addition, intravenous injection of natural or synthetic C_{18} PAF-acether to guinea pig induces a dose-dependent bronchoconstriction related to the effects of the mediator on platelets [41]. Indeed, platelet depletion or pretreatment of guinea pigs with prostaglandin (PG) I_2 or sulphinpyrasone (two procedures that prevent platelet activation) suppresses the bronchomotor effect of PAF-acether [42, 43]. In addition, treatment of guinea pigs with aspirin (which blocks thromboxane synthesis) in combination with mepyramine and methysergide (which inhibit the action of histamine and serotonin released from the activated platelets) abrogates the PAF-acether-induced bronchoconstriction [44]. However, the decrease in the number of circulating leucocytes and hypotension are still observed, both phenomena being unrelated to platelet activation [reviewed in 5]. In contrast, when administered by aerosol, PAF-acether evokes a platelet-independent bronchoconstriction that is blocked by aspirin alone [45] suggesting that in this case the phospholipid recruits primarily cyclooxygenase arachidonic metabolites.

Platelet-activating factor also exerts direct effects on lung tissues since injection of the mediator via the pulmonary artery of isolated perfused guinea pig lungs induces a potent bronchoconstriction that is blocked by aspirin [46]. However, recent work indicates that the *in vitro* sensitivity of guinea pig lung tissue to various agonists, including PAF-acether, is influenced by the state of immunization of the animals [47–49]. Indeed, lungs from actively sensitized guinea pigs exhibit enhanced bronchoconstrictor response, secrete more cyclooxygenase metabolites and release histamine on stimulation with PAF-acether, leukotriene (LT) D_4 or arachidonic acid [47–49]. Whereas the PAF-acether-induced bronchoconstriction in lung from control guinea pigs is inhibited by cyclooxygenase blockers [50], these drugs have no action when lungs from actively sensitized animals are used. Recent results from Pretolani *et al.* [48, 49] indicate that in this latter case leukotrienes play a major role in the PAF-acether-induced bronchoconstriction since the dual cyclooxygenase and lipoxygenase blocker, BW-755c, is inhibitory. In addition, the data by Pretolani *et al.* [48, 49] also demonstrate that the immune system is involved in the bronchopulmonary responsiveness, a result in keeping with the hypothesis that lymphocytes are involved in the pathogenesis of asthma. This hypothesis has been recently clinically supported by Corrigan *et al.* [51] showing that lymphocytes from patients with acute severe asthma exhibit increased expression of interleukin-2 receptor (CD-25), class II (HLA-DR) and very late-activation (VLA-1) antigens.

Surprisingly, although various PAF-acether antagonists abrogate the response to the phospholipid mediator when lungs originate from non-immunized animals, no effects of these drugs are observed when lungs from actively sensitized guinea pigs are used [52]. However, in lungs from both control and actively sensitized guinea pigs, these PAF-acether antagonists prevent oedema formation induced by PAF-acether, once again possibly indicating the presence of receptor subtypes for the autacoid.

Effect of PAF-acether on bronchial reactivity and lung morphology

Non-specific increase in airway responsiveness to various agonists such as histamine, methacholine and atmospheric pollutants is a characteristic feature of asthma [53–55]. Recently, the involvement of PAF-acether in bronchial hyperreactivity has been suggested by experiments showing that the administration of the mediator to guinea pigs leads to a long-lasting increase in the responsiveness to histamine [56]. Such hyperresponsiveness is prevented by various drugs including corticosteroids, disodium cromoglycate and theophylline, suggesting a complex set of events [56]. In humans also, PAF-acether induces long-lasting bronchial hyperreactivity to methacholine [57].

Since asthmatic patients are repeatedly in contact with the allergen, which in turn may lead to the generation of PAF-acether, the *in vivo* effects of long-term infusion of the phospholipid mediator in the guinea pig have been investigated [58, 59]. This has been achieved using osmotic Alzet minipumps containing C_{16} PAF-acether (or the solvent alone) placed under the back skin of guinea pigs and connected to the jugular vein. Guinea pigs receiving PAF-acether (7.2 mg/kg) for 2 weeks exhibit an increased response to intravenous injections of histamine as compared with animals treated with the solvent alone. In these experiments, the amplitude of the response rather than the sensitivity of the pulmonary tissue to histamine is affected by PAF-acether infusion. The effect of PAF-acether infusion is not observed in animals also treated orally with the PAF-acether antagonist, BN-52 021 (15 mg/kg, twice a day). In contrast, the *in vitro* reactivity to histamine of lung parenchymal strips from PAF-acether-infused guinea pigs, treated or not with BN-52 021, is not increased as compared with those from control animals. Similar hyperresponsiveness induced by PAF-acether was observed by Metzger *et al.* [60] in the rabbit. These authors, however, administered the autacoid by aerosol instead of using the intravenous route. At variance to our results, Metzger's group demonstrated that the *in vivo* hyperresponsiveness is also observed *in vitro* using lung parenchymal strips. This difference may be related to the route of administration of the mediator.

The *in vivo* alterations of lung responsiveness observed following long-term infusion of PAF-acether in the guinea pig and the rabbit are associated with changes in lung morphology [58–60]. Indeed, lungs from guinea pigs receiving the mediator presented marked hypertrophy of the Reissessem muscles associated with muciparous hyperplasia and bronchial obstruction, similar to those observed in asthmatic patients. In the rabbit also, administration of PAF-acether by aerosol evokes smooth-muscle cell hypertrophy [60]. This may also contribute to the fact that the hyperresponsiveness is observed *in vitro* in the rabbit but not in the guinea pig. Indeed, in this latter species, smooth-muscle cells are already well developed in control animals as compared with the rabbit lung which is relatively poor in this cell type. Therefore, investigating lung parenchymal strips' responses after PAF-acether treatment is identical to comparing the response of one tissue almost devoid of smooth-muscle cells (and thus exhibiting no response) to that of an organ in which this cell has greatly proliferated. Thus, this difference in the smooth-muscle cell content may explain the hyperresponsiveness observed *in vitro* in the rabbit [60].

Platelet-activating factor has been shown to increase mucus secretion in the ferret [61, 62], suggesting that the mediator could have exerted a direct effect on mucus cells in the guinea pig and the rabbit. In contrast, the hypertrophy of bronchial smooth muscles, which may explain the *in vivo* hyperresponsiveness observed during histamine provocation of the PAF-acether-treated animals, is probably indirect. Secretion of platelet-derived growth factor from platelets or other cytokines from various cell types may be involved in this process. In addition, in agreement with Vargaftig's group [63, 64] eosinophil infiltration and mast-cell hyperplasia are observed in lungs from PAF-acether-treated guinea pigs [58, 59]. The latter effect of the mediator is probably related to its action on lymphocytes [reviewed in 65] whereas that on eosinophils could reflect the chemotactic activity of PAF-acether for this cell type [66].

Involvement of PAF-acether in acute anaphylactic reactions

Up to now, most of the evidence implicating PAF-acether in acute anaphylactic reactions is indirect and has been obtained in experimental models of allergic asthma. However, these animal models sometimes poorly reflect the pathology of allergic reactions in the human. Indeed, in the guinea pig, anaphylactic bronchoconstriction involves the release of histamine from various stores and is therefore markedly inhibited by antihistamine drugs [50], a result at variance with the poor effectiveness of such drugs in bronchial asthma in humans. However, on the basis of the data obtained in animal models, a role for PAF-acether in allergic asthma has been suggested [reviewed in 5]. This hypothesis has been recently strengthened by the use of specific PAF-acether antagonists, either chemically related or not to the structure of the autacoid and acting at the receptor level [reviewed in 5, 6, and see below]. Up to now, only a limited number of studies investigating the effect of PAF-acether antagonists on antigen-induced bronchopulmonary alterations in animal models are available. However, one has to take into account the fact that the pharmacology of these alterations depends upon the species, the protocol of immunization and the route of administration at the time of antigen challenge. Thus, until experimentations involving the comparison of PAF-acether antagonists in defined animal models are available, i.e. where the problems related to bioavailability of the drugs have been solved, no definitive conclusion can be drawn. Therefore, only the major features observed using PAF-acether antagonists will be reported here.

Antigen challenge of guinea pigs passively sensitized with autologous IgE-rich antiserum for a 10-day period evokes a bronchoconstriction associated with only leucopenia which is blocked by the PAF-acether antagonists, BN-52021 [67] or WEB-2086 [68]. In addition, the bronchoconstriction induced by an aerosol of antigen is reduced by BN-52021 or Ro 19-3704 [69].

Antigen challenge of guinea pigs passively sensitized with heterologous serum from rabbit containing mostly antigen-specific IgG induces a bronchoconstriction associated with thrombocytopenia, neutropenia and hypotension [70]. BN-52021 [71, 72], as well as WEB-2086 [73], markedly reduce the antigen-induced bronchoconstriction. Also, BN-52021 prevents the antigen-induced thrombocytopenia but is poorly effective on the neutropenia. The ginkgolide mixture, BN-52063, containing the PAF-acether antagonists BN-52020, BN-52021 and BN-52022 (weight ratio 2:2:1), is also effective in preventing the effect of antigen administration in guinea pigs passively sensitized with heterologous serum [74].

When animals are actively sensitized with heterologous proteins such as ovalbumin, the modulation of the bronchopulmonary alterations observed on antigen challenge depends upon the protocol of immunization [47–49, 75]. When the animals are sensitized using a protocol favouring the production of IgE, poor protection by either BN-52021 or WEB-2086 is afforded [reviewed in 5]. In contrast, when the animals are sensitized so as to produce primarily IgG, a 50–60% reduction of the antigen-induced decrease in dynamic compliance, and an increase in pulmonary resistance by BN-52021 are observed [76, 77]. This is also supported by the recent clinical studies by Guinot *et al.* [78] demonstrating an improvement in pulmonary functions during treatment with BN-52063.

Effect of PAF-acether antagonists in experimental models of allergic asthma

Platelet-activating factor is released, along with lyso-PAF, from lung tissue stimulated with the specific antigen, although the mediator has not been fully characterized in all cases [79–81]. This antigen-induced release of lyso-PAF is inhibited by glucocorticosteroids [82], an observation in keeping with the phospholipase A_2 dependency of the generation of the PAF-acether precursor, lyso-PAF. In addition, in humans, the presence of PAF-acether in lung fluid following anaphylactic shock has been demonstrated recently [83].

Marked differences are noted when studying the effects of various unrelated PAF-acether antagonists in experimental models of asthma. For some compounds, no *in vivo* activity has been reported. Since these substances are potent PAF-acether antagonists *in vitro*, this lack of effect could be related to a poor bioavailability, a rapid rate of degradation and/or elimination of these drugs. Among the products that are active *in vivo* in preventing PAF-acether-induced alterations, some of them, such as SRI-63 441 [84] as well as 48 740-RP, have minimal effect on antigen-evoked bronchopulmonary changes. Besides directly inhibiting antigen-induced alterations, some PAF-acether antagonists such as BN-52 021 and WEB-2086 also impair the development of bronchial hyperresponsiveness. In contrast, kadsurenone which inhibits antigen-induced bronchoconstriction in the guinea pig fails to prevent the subsequent bronchial hyperresponsiveness. Since all these PAF-acether antagonists are very potent in inhibiting PAF-acether-induced alterations and PAF-acether binding to its specific binding sites, several hypotheses can be raised to try to explain these discrepancies. The simplest one involves the presence of receptor subtypes and this will be considered below. Interestingly, most of the PAF-acether antagonists possessing a structure related to that of PAF-acether are poorly effective in these models. This is the case for BN-52 111 and BN-52 115 and this difference may be due to an unfavourable ratio of agonist/antagonist activity. Unfortunately, no data are yet available for CV-6209 (Takeda Pharmaceutical Company) which also belongs to this series of antagonists structurally related to PAF-acether.

Another explanation may involve the action of some PAF-acether antagonists on protease activity. Indeed, PAF-acether injection into animals markedly increases blood protease activity [85] within 30 s. After 10 min, this effect of PAF-acether vanishes and normal values of protease activity are measured in blood. This effect of PAF-acether is probably related to its degranulating activity on polymorphonuclear leucocytes [86], although in an *in vivo* model this action of the autacoid could be indirect. Interestingly, BN-52 021 and WEB-2086 efficiently block the increase in blood protease activity whereas BN-52 111 and BN-52 115 (PAF-acether-related structures) fail to do so. Thus, the inhibitory effect of some PAF-acether antagonists on protease activity could explain their efficiency in preventing the development of bronchial hyperresponsiveness. Further studies are required, however, to confirm this hypothesis.

Recently, eosinophil infiltration in lung tissue has been suggested to participate in the development of bronchial hyperresponsiveness. Indeed, this cell type is present in the lung after PAF-acether or antigen stimulation [63, 64]. Since eosinophils secrete various phlogistic constituents which may play a role in bronchopulmonary alterations, the question of whether PAF-acether is chemotactic for this cell type has been addressed. Consistent to what is observed *in vivo* [63, 64, 87], works by Wardlaw and Kay [66] have demonstrated that PAF-acether is a potent chemotactic factor for both eosinophils and neutrophils. In contrast, LTB_4 appears to be poorly active, if at all, on the former cell type. In addition, Rabier *et al.* [88] have demonstrated that the sensitivity of eosinophils from allergic patients to PAF-acether is increased. For both eosinophils from normal and allergic subjects, PAF-acether-induced chemotaxis is inhibited by BN-52 021 [88], as is the weal and flare reaction in human skin [89]. Thus, the possible effects of PAF-acether antagonists on eosinophil chemotaxis represent a very attractive explanation for the efficiency of some of them in preventing bronchial hyperresponsiveness. However, this has been challenged recently by the fact that SRI-63 441, which is weakly active on bronchial hyperresponsiveness, is efficient in inhibiting eosinophil chemotaxis. Thus, if the lack of effect of this drug *in vivo* is not related to a poor bioavailability, other hypotheses will have to be raised to explain the differences in the potency of the various PAF-acether antagonists.

Possible heterogeneity of PAF-acether receptors

Among the various possibilities that may explain the differences in the potency of the PAF-acether antagonists, heterogeneity at the receptor level is now accepted. This latter hypothesis implies that various subtypes of PAF-acether receptors may be present on the same cell type or that different receptors specific for BAAL of defined structure are present on tissue (Fig. 4.3). Indeed, this latter hypothesis has been recently supported by data from Hwang [90] showing that the binding sites for the autacoid on human platelets and leucocytes are different, as assessed using various unrelated PAF-acether antagonists. In this study, ONO-6240 has been shown to inhibit PAF-acether binding to platelets with a median effective dose (ED_{50}) eight times lower than that required for human polymorphonuclear leucocytes. Data from our institute also indicate that BN-52 111 (and BN-52 115) is more active on this latter cell type whereas BN-52 021 mostly inhibits the effect of PAF-acether on lung and stomach (Fig. 4.4). Heterogeneity at the receptor level is also indicated by the work of Valone [91] demonstrating that the inhibitory effects of BN-52 021 and kadsurenone depend on the cell type studied. Work by Ukena *et al.* [92] using labelled 52 770-RP and WEB-2086 has demonstrated that the number and the affinity of the putative receptor(s) on human platelets for these antagonists depend on the drug. Finally, Hayashi *et al.* [93] have shown that some BAAL are more active than PAF-acether in activating macrophages. However, in this study it is unclear whether the increased activity of these compounds as compared with PAF-acether is related to a lower rate of catabolism. Thus, until the PAF-acether

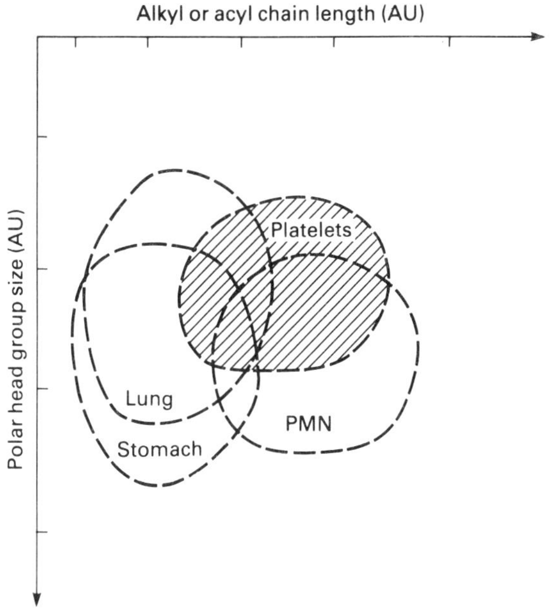

Fig. 4.3 Diagrammatic distribution of the possible activity of BAAL on various cell types and tissues. In this and the following figure, the representation has been randomly chosen and does not intend to attribute a defined group of molecules to one target. It simply intends to indicate that the activity of some of these molecules may not overlap totally.

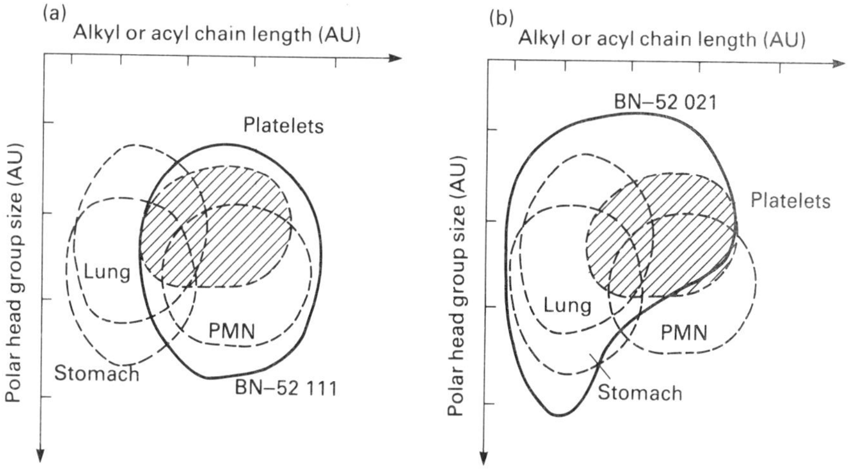

Fig. 4.4 Diagrammatic distribution of the possible activity of BAAL on various cell types and tissues and interference with (a) BN-52111 (or BN-52115), (b) BN-52021.

(or BAAL) receptor(s) is fully characterized at the molecular level no definite conclusion can be reached, although the evidence already available strongly suggests that the presence of receptor subtypes is likely.

Concluding remarks

From the present data it is obvious that the alkyl or acyl glycerophospholipids potentially exhibiting high specific activity on platelets are only a limited number of the molecules generated at the time of cell stimulation. Indeed, the material generated by various cell types has been shown to vary extensively at the *sn* 1 position of the glycerol. A short chain esterified at the *sn* 2 position of the glycerol appears to be mandatory since only molecules containing an acetyl moiety have been shown to be generated. In addition, the phosphocholine head group appears to be the major molecular species generated at the time of cell or tissue stimulation. However, on the basis of the prevailing procedures for characterization and identification, molecules other than those bearing an acetate and a phosphocholine head group may have been rejected.

It is noteworthy that both BN-52021 and WEB-2086 were discovered from empiricism and were not designed on the basis of the structure of the PAF-acether molecule. In contrast, most of the compounds synthesized on the basis of the structure of PAF-acether (1-O-hexadecyl- or octadecyl-2-O-acetyl-*sn*-glycero-3-phosphocholine), although very potent PAF-acether antagonists, are poorly effective in asthma-related experimental models. This may suggest that they actually do not interfere with specific sites of recognition for the molecule generated at the time of antigen challenge. In contrast, BN-52021 and WEB-2086 effectively block antigen-induced allergic manifestations in some experimental models, irrespective of their relative potency on platelets. In addition, recent experimental evidence from our institute indicates that BN-52021 is more active than many other PAF-acether antagonists in preventing the binding of tritiated PAF-acether to lung membrane preparations. An alternative explanation for these discrepancies is that the molecules generated at the time of antigen challenge in the lung are BAAL and not true PAF-acether. These molecules may interact with binding sites other than those for PAF-acether that also recognize BN-52021 and WEB-2086 with a higher affinity than the antagonists designed on the basis of the PAF-acether structure.

Thus, the precise knowledge of the molecules involved in physiopathological situations such as asthma is now a prerequisite for the development of new antiallergic drugs. Indeed, differences in the specific activity of the BAAL on various cell and tissue targets, as well as a possible heterogeneity at the receptor level, may explain the large differences observed in the potency of the PAF-acether antagonists, which may, in fact, be tissue specific.

References

1 Demopoulos CA, Pinckard RN, Hanahan DJ. Platelet-activating factor. Evidence for 1-O-alkyl-2-acetyl-*sn*-glyceryl-3-phosphorylcholine as the active component (a new class of lipid chemical mediators). *J. Biol. Chem.* 1979; **254**:9355–9358.

2 Benveniste J, Tencé M, Varenne P, Bidault J, Boullet C, Polonsky J. Semi-synthesis and proposed structure of platelet-activating factor (PAF): PAF-acether an alkyl ether analog of lysophosphatidylcholine. *C. R. Acad. Sci.* 1979; **289D**:1037–1040.

3 Hanahan DJ, Demopoulos CA, Liehr J, Pinckard RN. Identification of platelet-activating factor isolated from rabbit basophils as acetyl glyceryl ether phosphorylcholine. *J. Biol. Chem.* 1980; **255**:5514–5516.

4 Sturk A, Ten Cate JW, Hosford D, Mencia-Huerta JM, Braquet P. Synthesis, catabolism and pathophysiological role of platelet-activating factor. *Adv. Inflamm. Res.* 1988; In press.

5 Braquet P, Touqui L, Shen TY, Vargaftig BB. Perspectives in platelet-activating factor research. *Pharmacol. Rev.* 1987; **39**:97–145.

6 Braquet P, Chabrier PE, Mencia-Huerta JM. The promise of platelet-activating factor antagonists. *ISI Atlas of Sciences* 1988; In press.

7 Hosford D, Mencia-Huerta JM, Braquet P. The role of platelet-activating factor and structurally related alkyl phospholipids in immune and cytotoxic processes. *Adv. Inflamm. Res.* 1988; In press.

8 Ludwig JC, Pickard RN. Diversity in the chemical structures of neutrophil-derived platelet-activating factors. In *New Horizons in Platelet Activating Factor Research* (Eds Winslow CM, Lee JL), John Wiley & Sons, New York 1987, pp. 59–71.

9 Mueller HW, O'Flaherty JT, Wykle RL. Biosynthesis of platelet-activating factor in rabbit polymorphonuclear neutrophils. *J. Biol. Chem.* 1983; **258**:6213–6218.

10 Mueller HW, O'Flaherty JT, Wykle RL. The molecular species distribution of platelet-activating factor synthesized by rabbit and human neutrophils. *J. Biol. Chem.* 1984; **259**:14554–14559.

11 Oda M, Satouchi K, Yasunaga K, Saito K. Molecular species of platelet-activating factor generated by human neutrophils challenged with ionophore A23187. *J. Immunol.* 1985; **134**:1090–1093.

12 Pinckard RN, Jackson EM, Hoppens C, Weintraub ST, Ludwig JC, McManus LM, Mott GE. Molecular heterogeneity of platelet-activating factor produced by stimulated human polymorphonuclear leukocytes. *Biochem. Biophys. Res. Commun.* 1984; **122**:325–332.

13 Ramesha CS, Pickett WC. Species-specific variations in the molecular heterogeneity of the platelet-activating factor. *J. Immunol.* 1987; **138**:1559–1563.

14 Satouchi K, Oda M, Yasunaga K, Saito K. Evidence for the production of 1-acyl-2-acetyl-*sn*-glyceryl-3-phosphorylcholine concomitantly with platelet-activating factor. *Biochem. Biophys. Res. Commun.* 1985; **128**:1409–1417.

15 Tessner T, Wykle R. Stimulated human neutrophils synthesize an ethanolamine plasmologen analog of platelet-activating factor. *Fed. Proc.* 1987; **46**:2033.

16 Weintraub ST, Ludwig JC, Mott GE, McManus LM, Lear C, Pinckard RN. Fast atom bombardment mass spectrometric identification of molecular species of platelet-activating factor produced by stimulated human polymorphonuclear leukocytes. *Biochem. Biophys. Res. Commun.* 1985; **129**:868–876.

17 Mencia-Huerta JM, Benveniste J. Platelet-activating factor and macrophages. I. Evidence for the release from rat and mouse peritoneal macrophages and not from mastocytes. *Eur. J. Immunol.* 1979; **9**:409–415.

18 Mencia-Huerta JM, Ninio E. Biosynthesis and release of PAF-acether by mouse bone marrow-derived mast cells. In *Platelet-Activating Factor and Related Lipid Mediators* (Ed. Snyder F), Plenum Press, New York 1987, pp. 425–445.
19 Mencia-Huerta JM, Lewis RA, Razin E, Austen KF. Antigen-initiated release of platelet-activating factor (PAF-acether) from mouse bone marrow-derived mast cells sensitized with monoclonal IgE. *J. Immunol.* 1983; **131**:2958–2964.
20 Lynch JM, Lotner GZ, Betz SJ, Henson PM. The release of a platelet-activating factor by stimulated rabbit neutrophils. *J. Immunol.* 1979; **123**:1219–1226.
21 Jouvin-Marche E, Grzych JM, Boulet C, Capron M, Benveniste J. Formation of PAF-acether by human eosinophils. *Fed. Proc.* 1984; **43**:1924.
22 Chignard M, Le Couédic JP, Tencé M, Vargaftig BB, Benveniste J. The role of platelet-activating factor in platelet aggregation. *Nature* 1979; **275**:799–800.
23 Chignard M, Le Couédic JP, Delautier D, Benveniste J. Formation of PAF-acether and of another aggregating phospholipids by human platelets. *Fed. Proc.* 1983; **42**:659.
24 Mencia-Huerta JM, Benveniste J. Platelet-activating factor (PAF-acether) and macrophages. II: Phagocytosis-associated release of PAF-acether from rat peritoneal macrophages. *Cell. Immunol.* 1981; **57**:281–292.
25 Roubin R, Tencé M, Mencia-Huerta JM, Arnoux B, Benveniste J. A chemically defined monokine: macrophage-derived platelet-activating factor (PAF-acether). In *Lymphokines* (Ed. Pick E), Academic Press, New York 1983, pp. 249–276.
26 Tencé M, Polonsky J, Le Couédic JP, Benveniste J. Release, purification and characterization of platelet-activating factor (PAF). *Biochimie* 1980; **62**:251–259.
27 Bossant MJ, Farinotti M, Mencia-Huerta JM, Benveniste J, Mahusier G. Characterization and quantification of PAF-acether (platelet-activating factor) as heptafluorobutyrate derivatives of 1-O-alkyl-2-acetyl-*sn*-glyceryl by capillary column gas chromatography with electron capture detection. *J. Chromatogr.* 1987; **423**:In press.
28 Godfroid JJ, Braquet P. PAF-acether specific binding sites: 1. Quantitative SAR study of PAF-acether isosteres. *Trends Pharmacol. Sci.* 1986; **7**:368–373.
29 Braquet P, Godfroid JJ. PAF-acether specific binding sites: 2. Design of specific antagonists. *Trends Pharmacol. Sci.* 1986; **7**:397–403.
30 Coeffier E, Borrel MC, Lefort J, Chignard M, Broquet C, Heymans F, Godfroid JJ, Vargaftig BB. Effects of PAF-acether and structural analogues on platelet activation and bronchoconstriction in guinea-pigs. *Eur. J. Pharmacol.* 1986; **131**:179–188.
31 Pinckard RN, Ludwig JC, McManus LM. *Platelet-activating Factors. Inflammation: Basic Principles and Clinical Correlates* (Eds Gallin JI, Goldstein IM, Snyderman R), Raven Press, New York 1988, pp. 139–167.
32 Stafforini DM, Prescott S, McIntyre T. Human plasma platelet-activating factor acetyl-hydrolase, purification and properties. *J. Biol. Chem.* 1987; **262**:4223–4230.
33 Snyder F. Enzymatic pathways for platelet-activating factor, related alkyl glycerolipids and their precursors. In *PAF and Related Lipid Mediators* (Ed. Snyder F), Plenum Press, New York 1987, pp. 89–113.
34 Ninio E, Mencia-Huerta JM, Heymans F, Benveniste J. Biosynthesis of platelet-activating factor. I. Evidence for an acetyl-transferase activity in murine macrophages. *Biochim. Biophys. Acta* 1982; **710**:23–31.
35 Benveniste J, Henson PM, Cochrane CG. Leukocyte-dependent histamine release from rabbit platelets: the role for IgE, basophils and a platelet-activating factor. *J. Exp. Med.* 1972; **136**:1356–1377.
36 Mencia-Huerta JM, Hosford D, Braquet P. Acute and long term effects of platelet-activating factor. *Clin. Exp. Allergy* 1988; In press.
37 Pinckard RN, Halonen M, Palmer JD, McManus LM, Shaw JO, Henson PM. Intravascular aggregation and pulmonary sequestration of platelets during IgE-induced systemic anaphylaxis in the rabbit: abrogation of lethal anaphylactic shock by platelet depletion. *J. Immunol.* 1987; **119**:2185–2193.

38 Halonen M, Lohman IC, Dunn AM, McManus LM, Palmer JD. Participation of platelets in the physiologic alterations of the AGEPC response and of IgE anaphylaxis in the rabbit. *Am. Rev. Respir. Dis.* 1985; **131**:11–17.
39 Halonen M, Palmer JD, Lohman IC, McManus LM, Pinckard RN. Respiratory and circulatory alterations induced by acetyl glyceryl ether phosphorylcholine, a mediator of IgE anaphylaxis in the rabbit. *Am. Rev. Respir. Dis.* 1980; **122**:915–924.
40 Halonen M, Palmer JD, Lohman IC, McManus LM, Pinckard RN. Differential effects of platelet depletion on the physiologic alterations of IgE anaphylaxis and acetyl glyceryl ether phosphorylcholine infusion in the rabbit. *Am. Rev. Respir. Dis.* 1981; **124**:416–421.
41 Vargaftig BB, Lefort J, Chignard M, Benveniste J. Platelet-activating factor induces a platelet-dependent bronchoconstriction unrelated to the formation of prostaglandin derivatives. *Eur. J. Pharmacol.* 1980; **65**:185–192.
42 Vargaftig BB, Lefort J, Wal F, Chignard M. Role of the metabolites of arachidonate in platelet-dependent and independent experimental bronchoconstriction. *Bull. Eur. Physiopathol. Respir.* 1981; **17**:723–736.
43 Chignard M, Wal F, Lefort I, Vargaftig BB. Inhibition by sulphinpyrazone of the platelet-dependent broncho-constriction due to platelet-activating factor (PAF-acether) in the guinea pig. *Eur. J. Pharmacol.* 1982; **78**:71–79.
44 Vargaftig BB, Lefort J, Wal F, Chignard M, Medeiros MC. Non-steroidal anti-inflammatory drugs if combined with anti-histamine and anti-serotonin agents interfere with the bronchial and platelet effects of platelet-activating factor (PAF-acether). *Eur. J. Pharmacol.* 1982; **82**:121–130.
45 Garcez Do Carmo L, Cordeiro R, Lagente V, Lefort J, Randon J, Vargaftig BB. Failure of a combined anti-histamine and anti-leukotriene treatment to suppress passive anaphylaxis in the guinea-pig. *Int. J. Immunopharmacol.* 1986; **8**: 985–995.
46 Lefort J, Rotilio D, Vargaftig BB. The platelet-independent release of thromboxane A_2 by PAF-acether from guinea-pigs involves mechanisms distinct from those for leukotriene. *Br. J. Pharmacol.* 1984; **82**:565–575.
47 Lefort J, Malanchère E, Pretolani M, Vargaftig BB. Immunization induces bronchial hyperreactivity and increased mediator release from guinea-pig lungs. *Br. J. Pharmacol.* 1986; **89**:768P.
48 Pretolani M, Lefort J, Dumarey C, Vargaftig BB. Lipoxygenase metabolites and guinea-pig lung hyperresponsiveness. *FASEB J.* 1988; **2**:5562.
49 Pretolani M, Lefort J, Dumarey C, Vargaftig BB. Role of the immunization in the development of lung hyperresponsiveness in the guinea-pig. *Am. Rev. Respir. Dis.* 1988; In press.
50 Detsouli A, Lefort J, Vargaftig BB. Histamine and leukotriene-independent guinea-pig anaphylactic shock unaccounted for by PAF-acether. *Br. J. Pharmacol.* 1985; **84**:801–810.
51 Corrigan CJ, Hartnell A, Kay AB. T lymphocyte activation in acute severe asthma. *Lancet* 1988; **i**:1129–1132.
52 Pretolani M, Lefort J, Vargaftig BB. Interference of PAF-acether antagonists with the hyperresponsiveness to PAF-acether of lung from actively sensitized guinea-pigs. *Prostaglandins* 1988; **38**:800.
53 Cockcroft DW, Killian DN, Mellon JJA. Bronchial reactivity to inhaled histamine: a method and clinical survey. *Clin. Allergy* 1977; **7**:235–243.
54 Fish JE, Rosenthal RR, Batra G. Airway responses to methacholine in allergic and non-allergic subjects. *Am. Rev. Respir. Dis.* 1976; **13**:579–586.
55 Spector S, Farr R. Bronchial inhalation challenge with antigens. *J. Allergy Clin. Immunol.* 1979; **64**:580–586.
56 Mazzoni L, Morley J, Page CP, Sanjar S. Induction of hyperreactivity by platelet-activating factor in the guinea-pig. *J. Physiol.* 1985; **365**:107P.
57 Cuss FM, Dixon CMS, Barnes PJ. Effects of inhaled platelet-activating factor on pulmonary function and bronchial responsiveness in man. *Lancet* 1986; **ii**:189.

58 Touvay C, Vilain B, Pfister A, Pignol B, Mencia-Huerta JM, Braquet P. Functional and morphologic alterations of guinea-pig lung induced by chronic infusion of platelet-activating factor support a role for the mediator in bronchial hyperresponsiveness. *New Trends in Lipid Mediator Research* 1988; In press.

59 Touvay C, Vilain B, Pfister AP, Coyle AJ, Page CP, Mencia-Huerta JM, Braquet P. Effect of chronic administration of PAF on airway responsiveness in the guinea-pig. *Am. Rev. Respir. Dis.* 1988; **137**:283.

60 Metzger WJ, Ogden-Ogle C, Atkinson LB. Chronic platelet-activating factor (PAF) aerosol challenge induces *in vivo* and *in vitro* bronchial airway hyperresponsiveness. *FASEB J.* 1988; **2**:5560.

61 Hahn HL, Hartlage J, Lang M. Effect of platelet activating factor on submucosal gland morphology—comparison to acetylcholine. *Fed. Proc.* 1986; **45**:3085 (Abstract).

62 Wirtz H, Lang M, Sannwald U, Hahn H-L. Mechanism of platelet-activating factor (PAF)-induced secretion of mucus from tracheal submucosal glands in the ferret. *Fed. Proc.* 1986; **45**:418.

63 Lellouch-Tubiana A, Lefort J, Pirotzky E, Vargaftig BB, Pfister A. Ultrastructural evidence for extra-vascular platelet recruitment in the lung upon intravenous injection of PAF-acether to guinea-pigs. *Br. J. Exp. Pathol.* 1985; **66**:345–355.

64 Lellouch-Tubiana A, Lefort J, Pfister A, Vargaftig BB. Interaction between granulocytes and platelets with the guinea-pig lungs in passive anaphylactic shock. Correlation with PAF-acether-induced lesions. *Int. Arch. Allergy Appl. Immunol.* 1987; **83**:198–205.

65 Braquet P, Rola-Plesczcynski M. The role of PAF in immunological responses. *Immunol. Today* 1987; **8**:345–352.

66 Wardlaw AJ, Moqbel R, Cromwell O, Kay AB. Platelet activating factor. A potent chemotactic and chemokinetic factor for human eosinophils. *J. Clin. Invest.* 1986; **78**:1701–1706.

67 Lagente V, Touvay C, Randon J, Desquant S, Cirino M, Vilain B, Lefort J, Braquet P, Vargaftig BB. Interference of the PAF-acether antagonist BN 52021 with passive anaphylaxis in the guinea-pig. *Prostaglandins* 1987; **33**:265–274.

68 Pretolani M, Lefort J, Malenchère E, Vargaftig BB. Antiallergic activity of WEB 2086, a new potent platelet-activating factor (PAF) antagonist. *International Congress of Pharmacology*, Sydney, 1987 (Abstract).

69 Lagente V, Desquant S, Hadvary F, Cirino M, Lellouch-Tubiana A, Vargaftig BB. Interference of the PAF antagonist Ro 19-3704 with PAF- and antigen-induced bronchoconstriction in the guinea-pig. *Br. J. Pharmacol.* 1988; In press.

70 Braquet P. Involvement of PAF-acether in various immune disorders using BN 52021 (Ginkgolide B): a powerful PAF-acether antagonist isolated from Ginkgo biloba L. *Adv. Prostaglandin Thromboxane Leukotriene Res.* 1986; **16**:179–198.

71 Braquet P, Etienne A, Touvay C, Bourgain RH, Lefort J, Vargaftig BB. Involvement of platelet-activating factor in respiratory anaphylaxis demonstrated by PAF-acether inhibitor, BN 52021. *Lancet* 1986; **i**:1501–1502.

72 Touvay C, Coyle A, Vilain B, Page CP, Braquet P. Effect of BN 52021 on antigen-induced bronchial hyperreactivity in guinea-pigs. *Fed. Proc.* 1987; **46**:5155.

73 Casals-Stenzel J, Muacevic G, Heuer H. Modulation of the endotoxin shock and anaphylactic lung reaction by WEB 2086, a new potent antagonist of PAF. *Second International Conference on Platelet-activating Factor and Structurally Related Alkyl Ether Lipids*, Gatlinburg TN, USA, October 1986, p. 108 (Abstract).

74 Braquet P. The ginkgolides: potent platelet-activating factor antagonists isolated from *Ginkgo biloba L.*: chemistry, pharmacology and clinical applications. *Drugs of the Future* 1987; **12**:643–699.

75 Pretolani M, Page CP, Lefort J, Lagente V, Vargaftig BB. Pharmacological modulation of the respiratory and haematological changes accompanying active anaphylaxis in the guinea-pig. *Eur. J. Pharmacol.* 1986; **125**:403–409.

76 Berti F, Omini C, Rossoni G, Braquet P. Protection by two ginkgolides, BN 52020 and

BN 52021, against guinea-pig lung anaphylaxis. *Pharmacol. Res. Commun.* 1986; **18**:775–793.
77 Berti F, Rossoni G. PAF Ginkgolides and active anaphylactic shock in lung and heart. In *The Ginkgolides: Chemistry, Biology, Pharmacology and Clinical Sciences* (Ed. Braquet P) J.R. Prous Scientific Publications, 1988, In press.
78 Guinot P, Brambilla C, Duchier J, Braquet P, Bonvoisin B, Cournot A. Effect of BN 52063, a specific PAF-acether antagonist, on bronchial provocation test to allergen in asthmatic patients: a preliminary study. *Prostaglandins* 1987; **34**:723–731.
79 Fitzgerald MF, Moncada S, Parente L. The anaphylactic release of platelet-activating factor from perfused guinea-pig lungs. *Br. J. Pharmacol.* 1986; **88**:149–153.
80 Parente L, Fitzgerald MF, Moncada S. Anaphylactic release of platelet-activating factor and eicosanoids from guinea-pigs sensitized to ovalbumine aerosol. *Adv. Prostaglandin Thromboxane Leukotriene Res.* 1987; **17**:171–176.
81 Rotilio D, Lefort J, Destouli A, Vargaftig BB. Absence de contribution du PAF-acether à la reponse anaphylactique pulmonaire *in vitro* chez le cobaye. *J. Pharmacol.* 1983; **14**:1.
82 Parente L, Fitzgerald MF, Flower RJ, De Nucci G. The effect of glucocorticoids on lyso-PAF formation *in vitro* and *in vivo*. *Agents Actions* 1985; **17**:312–313.
83 Gateau O. Acute pulmonary and cardiovascular effects of platelet-activating factor in man: relevance in anaphylactic shock. *Prostaglandins* 1987; **34**:182.
84 Stenzel H, Hummer B, Hahn HL. Effect of the PAF antagonist SRI 63 441 on the allergic response in awake dogs with natural asthma. *Agents Actions* **21**:253–260.
85 Etienne A, Guilmard C, Thonier F, Hecquet F. Interference of Ginkgolides on normal and pathological plasma protein activities in the rat. In *Ginkgolides: Chemistry, Biology, Pharmacology and Clinical Perspectives* (Ed. Braquet P), J.R. Prous Scientific Publications, 1988, pp. 465–476.
86 Goetzl EJ, Derian CK, Tauber AL, Valone FH. Novel effects of 1-O-hexadecyl-2-acyl-*sn*-glycero-3-phosphocholine mediators on human leukocyte functions: delineation of the specific roles of the acyl substituents. *Biochem. Biophys. Res. Commun.* 1980; **94**:881–888.
87 Lellouch-Tubiana A, Lefort J, Simon MT, Pfister A, Vargaftig BB. Eosinophil recruitment into guinea-pig lungs after PAF-acether and allergen administration; modulation by prostacyclin, platelet depletion and selective antagonists. *Am. Rev. Respir. Dis.* 1988; In press.
88 Rabier M, Damon M, Mencia-Huerta JM, Braquet P, Michel FB, Godard P. PAF-acether chemotactic activity on neutrophils in bronchial asthma. Effect of specific PAF-acether antagonists. *Prostaglandins* 1988; **35**:795.
89 Guinot P, Braquet P, Duchier J, Cournot A. Inhibition of PAF-acether induced weal and flare reaction in man by a specific PAF-antagonist. *Prostaglandins* 1986; **32**:160–163.
90 Hwang SB. Identification of a second putative receptor of platelet-activating factor from polymorphonuclear leukocytes. *J. Biol. Chem.* 1988; **263**:3225–3233.
91 Valone FH. Identification of platelet-activating factor receptors in P388 D_1 murine macrophages. *J. Immunol.* 1988; **140**:2389–2394.
92 Ukena D, Dent G, Birke FW, Robaut C, Sybrecht GW, Barnes PJ. Radioligand binding of antagonists of platelet-activating factor to intact human platelets. *FEBS Lett.* 1988; **228**:285–289.
93 Hayashi H, Kudo I, Inoue K, Onozaki K, Tsushima S, Nomura H, Nojima S. Activation of guinea-pig peritoneal macrophages by platelet-activating factor (PAF) and its antagonists. *J. Biochem.* 1985; **97**:1737–1745.

Discussion session

KAY: can I ask you about the passive sensitization experiments in the guinea pig? This is presumably an IgE- or IgG1-mediated reaction and so it would be my understanding that this would be largely a mast cell-mediated event. Yet you find that your PAF-acether antagonist is inhibiting eosinophil infiltration and so how could you easily account for that amount of PAF-acether coming from mast cells? It would seem to me unlikely.

BRAQUET: yes it is unlikely.

KAY: is there an alternative explanation then for where the PAF-acether is coming from in that type of manipulation?

BRAQUET: I think that nobody knows. At the beginning we thought it was a specific effect of BN-52 021 and therefore there was a lot of controversy between people who claimed that it was specific or non-specific. WEB-2086 now gives exactly the same result. It is difficult to know where the PAF-acether is coming from. We have no idea.

5

Pharmacological modulation of mediator generation and the allergic response by fish-oil diets

J.P. ARM AND T.H. LEE

Introduction

The possibility that leukotriene compounds might have a significant role in a variety of inflammatory disorders is suggested by the extensive pro-inflammatory effects demonstrated by these substances, the demonstration that many inflammatory cells generate leukotrienes and the finding that the leukotrienes can be detected in complex biological fluids *in vivo*. A novel approach to limiting leukotriene synthesis and biological activities is through the provision of alternative substrate fatty acids.

In order to appreciate the effects of alternative fatty acids on the functions of the metabolites of the 5-lipoxygenase (5-LO) pathway, it is appropriate to review initially the steps in the biosynthesis and metabolism of these compounds.

Biosynthesis and metabolic inactivation of arachidonic acid-derived mediators

Arachidonic acid released from membrane phospholipids by the action of phospholipase A_2 [1–4] or by the combined action of a phosphoinositide-specific phospholipase C_1 and a 1,2-diacylglycerol lipase [5–7] during cell activation may be re-esterified or oxidatively metabolized by the enzymes of the cyclooxygenase or lipoxygenase pathway. The cyclooxygenase pathway leads through intermediates to the formation of thromboxane (Tx) A_2 and prostaglandins. The 5-LO generates 5*S*-hydroperoxy-6-*trans*-8,11,14-*cis*-eicosatetraenoic acid (5-HPETE) [8] that is converted by the same 5-LO to 5,6-oxido-7,9-*trans*-11,14-*cis*-eicosatetraenoic acid (leukotriene A_4; LTA_4) [9–13]. Alternatively, 5-HPETE is reduced to its alcohol 5-hydroxy-6,8,11,14-eicosatetraenoic acid (5-HETE). Leukotriene A_4 is processed by an epoxide hydrolase to 5*S*,12*R*-dihydroxy-6,14-*cis*-8,10-*trans*-eicosatetraenoic acid (LTB_4) [8] or by LTC_4 synthetase to 5*S*-hydroxy-6*R*-*S*-glutathionyl-7,9-*trans*-11,14-*cis*-eicosatetraenoic acid (LTC_4) [14–17].

Leukotriene C_4 is cleaved by γ-glutamyl transpeptidase to 5*S*-hydroxy-6*R*-*S*-cysteinylglycyl-7,9-*trans*-11,14-*cis*-eicosatetraenoic acid (LTD_4) and by a dipeptidase to 5*S*-hydroxy-6*R*-*S*-cysteinyl-7,9-*trans*-11,14-*cis*-eicosatetraenoic acid (LTE_4) [18–20, 14, 15]. Leukotriene A_4 also undergoes non-enzymatic hydrolysis to 5*S*,12*R*- and 5*S*,12*S*-dihydroxy-6,8,10-*trans*-14-*cis*-eicosatetraenoic acid diastereoisomers (5*S*,12*R*- and 5*S*,12*S*-dihydroxy-6-*trans*-LTB_4 respectively) and to minor products, 5,6-dihydroxy-eicosatetraenoic acid diastereoisomers [21].

In addition to the bioconversion of LTC_4 to LTD_4 and to LTE_4, there is another mechanism for modifying the structures and functional activities of the sulphidopeptide leukotrienes, which depends upon the triggering of the respiratory burst [22]. Human neutrophils (PMN) and eosinophils metabolize LTC_4, LTD_4 and LTE_4 through an extracellular H_2O_2–peroxidase–chloride-dependent reaction to six products that elute as three doublets in reverse-phase high-performance liquid chromatography (RP-HPLC). More than 90% of the metabolites are composed of the 6-*trans*-LTB_4 diastereoisomers and the subclass-specific diastereoisomeric leukotriene sulphoxides. Leukotriene B_4 is metabolized in human PMN to the substantially less active omega metabolites, 20-hydroxy- and 20-carboxy-LTB_4 respectively [23, 24].

Fatty acids derived from fish lipid: *in vitro* actions and 5-lipoxygenase pathway products

The two major types of polyunsaturated fatty acids prominent in marine fish oils are eicosapentaenoic acid (EPA) (20:5, n−3) and docosahexaenoic acid (DCHA) (22:6, n−3) (Fig. 5.1). These fatty acids are incorporated into tissue

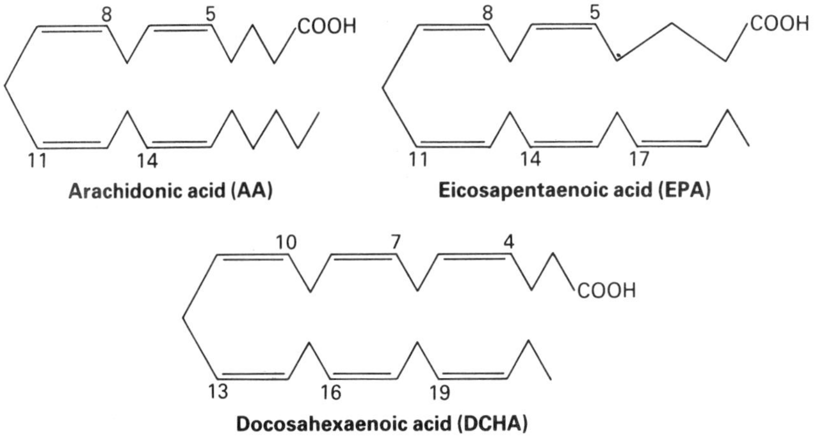

Fig. 5.1 The structures of arachidonic acid, eicosapentaenoic acid and docosahexaenoic acid.

phospholipids of mammals who consume a predominantly fish-oil enriched diet. Eicosapentaenoic acid and DCHA competitively inhibit the conversion of arachidonic acid by the cyclooxygenase pathway to prostanoid metabolites [25, 26]. To the extent they are formed, the endoperoxide and TxA_3 derived from EPA are substantially less active than the arachidonic acid-derived counterparts in eliciting aggregation of human platelets [27, 28]. Docosahexaenoic acid is not metabolized to any cyclooxygenase product. With respect to the oxidative metabolism of EPA and DCHA by the 5-LO cascade, EPA is converted sequentially by specific enzymes to LTB_5, LTC_5, LTD_5 and LTE_5 (Fig. 5.2). Docosahexaenoic acid is metabolized only to the 7- and 4-hydroperoxy DCHA and their reduction products, 7- and 4-hydroxy DCHA respectively. Leukotriene B_5 is substantially less active than LTB_4 in a number of pro-inflammatory functions [29–32] (Table 5.1), but LTC_5 and LTC_4 are equiactive in constricting non-vascular smooth muscle [33, 34]. Thus, EPA is capable of inhibiting the elaboration of inflammatory mediators

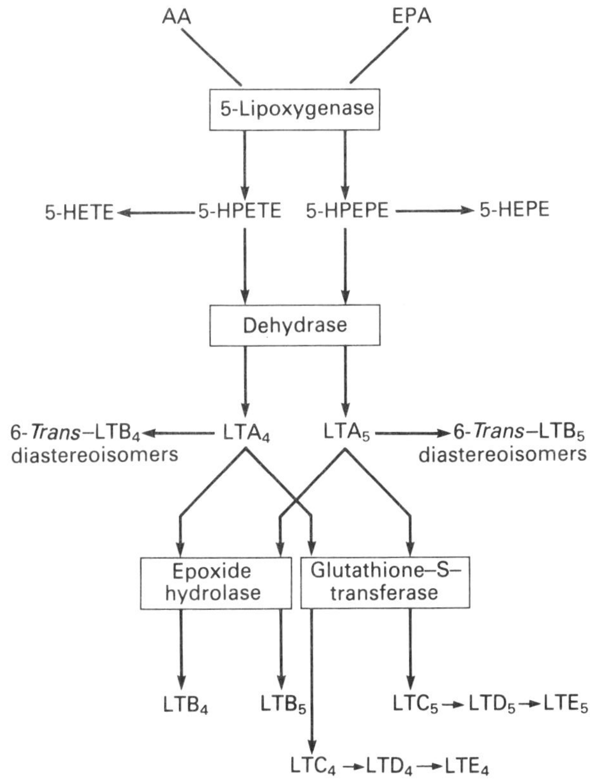

Fig. 5.2 The metabolism of arachidonic acid (AA), eicosapentaenoic acid (EPA) and docosahexaenoic acid (DCHA) by the 5-lipoxygenase pathway.

Table 5.1 Comparative biological activities of LTB_4 and LTB_5. The activities of LTB_4 have been designated a value of 100

	LTB_4	LTB_5
Neutrophils		
Chemotaxis	100	10
Chemokinesis	100	3
Complement receptor enhancement (CR_1 and CR_3)	100	1
Aggregation	100	5
Lysosomal enzyme release	100	1
Vascular		
Enhanced bradykinin induced vascular permeability	100	1

by the cyclooxygenase pathway and is metabolized to LTB_5 with attenuated biological activity.

Fatty acids derived from fish lipid; *in vitro* incorporation and effects on the generation of 5-lipoxygenase pathway products and PAF-acether

The capacity of non-esterified EPA and DCHA to modulate the oxidative metabolism of phospholipid-derived arachidonic acid by the 5-LO pathway in human PMN has recently been compared [35]. In these *in vitro* experiments, exogenous fatty acids (5–40 μg/ml) did not alter the release of arachidonic acid from ionophore-activated PMN. The metabolism of low concentrations of exogenous EPA (5 and 10 μg/ml) not only generated EPA-derived products, but also stimulated the formation of the major 5-LO pathway-derived metabolites, except LTB_4, from membrane-derived arachidonic acid. Leukotriene B_4 production was diminished throughout the EPA dose response, beginning at 5 μg/ml EPA and reaching 50% suppression at 10 μg/ml and 84% suppression at 40 μg/ml. Docosahexaenoic acid did not stimulate the metabolism of membrane-derived arachidonic acid, did not inhibit LTB_4 generation and was not a substrate for leukotriene formation. Thus, in contrast to DCHA, EPA attenuated the generation of LTB_4 and was converted to LTB_5.

Human leucocytes generate platelet-activating factor (PAF-acether), a lipid mediator of inflammation, from membrane alkyl phospholipids through the release of arachidonic acid or other fatty acids from the 2 position and subsequent acetylation [36]. In separate experiments, the effects of *in vitro* incubation with fish-oil fatty acids on calcium ionophore-stimulated PAF-acether generation in human monocytes were investigated [37]. An optimally effective concentration of arachidonic acid of 1 μg/ml resulted in a 64% increase of calcium ionophore-induced PAF-acether generation from

7.75 ± 0.78 ng/10^6 cells for untreated monolayers to 12.70 ± 1.21 ng/10^6 cells (mean $\pm$ SEM). Treatment of monolayers with EPA at the optimal concentration of 1 μg/ml decreased PAF-acether generation by 28%. However, treatment of monocyte monolayers with DCHA did not appreciably affect PAF-acether generation. The changes in PAF-acether release with each fatty acid added *in vitro* paralleled those in total PAF-acether generation; the percentage PAF-acether release remained unaffected.

Leukotriene B_4 and PAF-acether are pro-inflammatory mediators. The capacity of EPA to inhibit their generation could be considered as an anti-inflammatory effect.

Fatty acids derived from fish lipid: *in vivo* incorporation and effects on neutrophil function and on the *ex vivo* generation of 5-lipoxygenase pathway products and PAF-acether

Murine mastocytoma cells grown in the peritoneal cavities of mice fed an EPA enriched diet have approximately 50% of their phospholipid arachidonic acid replaced by EPA [38]. When these cells were removed and challenged *in vitro* with the calcium ionophore A-23 187, they generated LTB_4, LTB_5, LTC_4 and LTC_5. There was a 12-fold reduction of LTC (LTC_4 and LTC_5) and a 17-fold reduction of LTB (LTB_4 and LTB_5) generation when compared with mastocytoma cells grown in mice on a control diet. Thus an EPA enriched diet not only promoted the formation of LTB_5 but it also inhibited the elaboration of LTB_4 and LTC_4 from membrane-derived arachidonic acid.

The effects of dietary fish-oil fatty acids on the 5-LO pathway activity of PMN and monocytes have recently been studied in seven normal human subjects, who supplemented their usual diets for 6 weeks with daily doses of Max EPA containing 3.2 g of EPA and 2.2 g of DCHA [39]. The diet increased the EPA content in PMN and monocytes more than seven-fold without changing the quantities of arachidonic acid and DCHA. When the PMN were activated *in vitro* with the ionophore A-23 187, the release of arachidonic acid and its metabolites was reduced by a mean of 37% and the maximum generation of the major 5-LO metabolites, including LTB_4, was reduced by a mean of 48%. When monocyte monolayers were activated with the ionophore A-23 187, the release of arachidonic acid and its metabolites was reduced by a mean of 39% and the generation of LTB_4 was suppressed by 58%. In addition, the generation of PAF-acether was inhibited by approximately 50% [37].

The adherence of PMN to endothelial cell monolayers which had been pretreated with LTB_4 was inhibited completely and their average chemotactic response to LTB_4 was inhibited by 70% as compared with values determined before the diet was started [39]. The margination of leucocytes to endothelial surfaces is the initial step in the recruitment of cells by a chemotactic stimulus to an inflammatory focus. Thus, the impairment of leucocyte function caused by the dietary incorporation of fish-oil fatty acids into membrane phospholipids would be expected to be anti-inflammatory. This effect would be amplified by the substantial suppression of the biosynthesis of arachidonic acid-derived metabolites and PAF-acether. The leucocyte biochemical and functional suppression had recovered by 6 weeks after the diet was discontinued.

Comparison of the effects of non-esterified and esterified eicosapentaenoic acid on arachidonic acid metabolism in human neutrophils

A comparison of the effects of non-esterified [35] and esterified EPA [39] on arachidonic acid metabolism in human PMN demonstrates interesting differences. Non-esterified EPA does not alter arachidonic acid release from membrane phospholipids, whereas esterified EPA causes a substantial inhibition of phospholipase A_2. When PMN are activated in the presence of low concentrations of non-esterified EPA, the 5-LO activity is enhanced. This is in sharp contrast to the effect of esterified EPA which suppresses the 5-LO activity. The presence of non-esterified EPA inhibits the epoxide hydrolase, thus selectively suppressing the elaboration of LTB_4. Although there is also inhibition of LTB_4 generation in the presence of esterified EPA, this is not due to a selective inhibition of the epoxide hydrolase, rather it represents decreased substrate availability due to the combined suppression of endogenous fatty acid hydrolysis and 5-LO activity. Since both esterified and non-esterified EPA are likely to exert a regulating influence on the biochemistry and function of leucocytes at the site of an inflammatory response, the results of an anti-inflammatory effect will be complex. This will depend on the relative concentrations of the free and esterified fatty acid and may include an inhibitory effect on the phospholipase A_2 in addition to a modulating influence on the specific enzymes of the cyclooxygenase and lipoxygenase pathways.

Alternative fatty acids derived from fish lipid: *in vivo* incorporation, *in vivo* generation of 5-lipoxygenase pathway products and pharmacological effects

Animal studies

The generation of sulphidopeptide leukotrienes and LTB in response to an IgG-mediated immune complex reaction in the peritoneal cavities of fish-oil fed (FFD) or beef-tallow fed (BFD) rats was used to identify and quantitate products formed from the 5-LO pathway *in vivo* [34]. Fish-oil fed rats incorporated EPA in a ratio to arachidonic acid of approximately 2:1 whereas tissues from the BFD rats contained no detectable EPA. In this animal model, fish-oil supplement decreased the inflammatory potential of the chemotactic leukotrienes generated by effecting the elaboration of two to seven times more LTB_5 compared with LTB_4. The total biological activity of the spasmogenic sulphidopeptide leukotrienes was not altered, since the sum of the quantities of EPA-derived and arachidonic acid-derived products, which were equiactive in constricting non-vascular smooth muscle, was the same as the quantities of arachidonic acid-derived products in the animals in the control diet.

In order to investigate the effects of a fish-oil enriched diet in acute anaphylaxis, pulmonary mechanical responses to intravenous antigen challenge have been evaluated in two groups of mechanically ventilated and anaesthetized guinea pigs [40]. The animals were pretreated with mepyramine to uncover the leukotriene and prostaglandin contributions. One group of animals was fed a fish-oil diet whereas the control group was fed a beef-tallow diet. The FFD animals demonstrated incorporation of EPA and DCHA into the phospholipids of pulmonary tissue with a ratio of EPA and DCHA to arachidonic acid of 2.5 as compared with that of BFD animals of 0.04. Intravenous antigen challenge in both groups of animals elicited a decrease in dynamic compliance (C_{dyn}) and pulmonary conductance (G_L). The decrease in C_{dyn} in the FFD animals was significantly greater than that of the control animals at 1.5–5 min after antigen challenge. Since EPA and DCHA inhibit the cyclooxygenase pathway, these results are consistent with the views that the tissue levels of these fatty acids either partially inhibited the generation of bronchodilator prostaglandins released after antigen challenge and/or diverted EPA and arachidonic acid metabolism from the cyclooxygenase to the lipoxygenase pathway, resulting in the generation of greater quantities of bronchoconstrictor leukotrienes.

Support for the enhancement in 5-LO pathway activity in the presence of EPA comes from studies in guinea pigs. The changes in arterial plasma concentrations of immunoreactive LTB were compared after antigen challenge of two groups of sensitized mepyramine-treated, mechanically ventilated guinea pigs, one fed a diet enriched with fish oil and the other a control diet enriched with beef tallow [41]. The lung tissue of animals fed an FFD for 9–10 weeks incorporated EPA and DCHA to constitute 8–9% of total fatty acid content, whereas these alternative fatty acids constituted <1% of the total fatty acid content of the lung tissue of animals on a BFD. There was a significant ($p<0.02$) maximum increase after antigen challenge in immunoreactive LTB_4 from 0.16 ± 0.04 ng/ml to 0.84 ± 0.25 ng/ml in BFD animals and from 0.47 ± 0.11 to 5.1 ± 1.4 ng/ml immunoreactive LTB (LTB_4 and LTB_5) in FFD animals. Furthermore, the increase in total immunoreactive LTB in mepyramine-treated FFD animals was significantly greater than the increase in LTB_4 in mepyramine-treated BFD guinea pigs at 2–8 min after antigen challenge ($p < 0.05$).

Resolution of arterial plasma immunoreactive LTB from pooled samples by RP-HPLC demonstrated that the sum of LTB_4 and LTB_5 in FFD animals exceeded that of LTB_4 in BFD animals and that the quantity of LTB_4 in the FFD animals was at least as great as that in the BFD animals during anaphylaxis. The products eluting at the retention times of LTB_4 and LTB_5 exhibited the chemotactic activity of their respective synthetic standards.

The combination of indomethacin and mepyramine markedly augmented the antigen-induced increase in arterial plasma immunoreactive LTB_4 concentrations in BFD animals, but had no effect on immunoreactive LTB levels in FFD animals. Limited *in vivo* measurements showing a lesser increase of plasma immunoreactive TxB_2 in the FFD relative to the BFD animals during anaphylaxis, and *ex vivo* measurements showing a decreased LTB_4-stimulated (cyclooxygenase product-dependent) contractile response of pulmonary parenchymal strips from the FFD relative to the BFD animals, provide evidence for blockade in the cyclooxygenase pathway in the FFD animals. The measurements of arterial plasma LTB indicate that indomethacin treatment alone, which inhibits cyclooxygenase activity, and FFD treatment each augment the metabolism of arachidonic acid by the 5-LO pathway in animals pretreated with mepyramine.

These findings in guinea pigs subjected to anaphylaxis provide the first *in vivo* evidence of the augmented presence of 5-LO pathway products in the presence of cyclooxygenase pathway blockade.

Studies in bronchial asthma

Asthma is characterized by airway inflammation, by bronchial hyperresponsiveness to non-specific stimuli, and by episodic and reversible airflow obstruction. Studies in both experimental animals and humans have indicated an association between airway hyperresponsiveness and bronchial inflammation [42–50]. Since airway inflammation may be important in the pathophysiology of bronchial asthma, we have investigated the effects of a fish-oil enriched diet in subjects with bronchial asthma [51].

Subjects received either 18 capsules a day of fish oil (3.2 g EPA and 2.2 g DCHA) or identical placebo capsules containing olive oil in a double-blind fashion. Aspects of PMN function, airway response to both specific and non-specific stimuli and severity of disease were assessed before and after the treatment period. There was incorporation of EPA into PMN phospholipids from barely detectable amounts prior to treatment to comprising 2.6% of total PMN fatty acids following dietary supplementation with fish oil. In addition, PMN from subjects who had received fish oil demonstrated a 50% reduction in the generation of LTB (LTB_4 and LTB_5) in response to calcium ionophore, and a substantial attenuation of their chemotactic responses to formyl-methionyl-leucyl-phenylalanine (FMLP) and LTB_4. In subjects who had received placebo, the phospholipid content of EPA, LTB_4 generation and chemotactic responses of PMN were unchanged. The changes in PMN function in subjects who had ingested EPA were not accompanied by changes in airway non-specific responsiveness, or severity of asthma.

Payan *et al.* have studied the effects of EPA in subjects with severe persistent asthma [52]. Two groups of six subjects received either 0.1 or 4.0 g of purified EPA a day in a double-blind fashion for 8 weeks. Both doses of EPA led to small but significant generation of LTB_5 by PMN and mixed mononuclear leucocytes in response to calcium ionophore, A-23 187. Only high-dose EPA suppressed ionophore-induced LTB_4 generation by PMN and mononuclear leucocytes. High-dose EPA also suppressed PMN, but not mononuclear leucocyte, chemotaxis to C5a, FMLP and LTB_4. Neither dose of EPA attenuated ionophore-induced PAF-acether generation by mixed mononuclear cells. In addition both low-dose and high-dose EPA led to an enhanced T-cell proliferation in response to mitogen. Changes in leucocyte function were not accompanied by changes in severity of asthma as assessed by history, clinical examination or a panel of pulmonary function tests.

We have recently studied the effect of a fish-oil enriched diet on both early and late asthmatic responses to antigen [53]. Whilst there were no changes in the acute airway response to antigen, there was a significant—approximately 35% attenuation—of the late asthmatic response in subjects who had received fish oil. There were no changes in immediate cutaneous responses to antigen, total serum IgE, or airway responses to histamine in the same subjects. In so far as airway inflammation is believed to be central to the pathophysiology of the allergen-induced late asthmatic response, the attenuation of the late-phase reaction by fish oil is consistent with an anti-inflammatory effect.

These studies extend previous observations in normal subjects and demonstrate that a fish-oil enriched diet attenuates PMN and mononuclear cell function in subjects with asthma. The associated attenuation of the late asthmatic response to antigen suggests that these alterations in leucocyte function were sufficient to attenuate the induction of airway inflammation. The lack of any clinical benefit in subjects with either mild disease [51] or in subjects with severe persistent asthma [52] may have been due to insufficient time for regeneration of airway epithelium and resolution of the chronic inflammatory response to effect a change in clinical variables.

Concluding remarks

A fish-oil enriched diet has potential in modulating the humoral and inflammatory components of the allergic response by inhibiting the generation of pro-inflammatory mediators derived from arachidonic acid and by reducing the production of PAF-acether. In addition, EPA suppresses the responses of target cells and tissues. However, the modulating influence of alternative fatty acid substrates on arachidonic acid metabolism may not always be beneficial, as demonstrated in the animal model of antigen-induced acute anaphylaxis. The reason for these discrepant clinical results may be related to the complex interactions between the metabolites of the cyclooxygenase and lipoxygenase pathways. Attenuation of a portion of one or both pathways would alter the relative quantities of end products and may have quite different effects depending upon the type of pathobiological mechanisms involved in the disease process.

Studies in bronchial asthma confirm the anti-inflammatory potential of a fish-oil enriched diet. Dietary supplementation with EPA in subjects with asthma led to changes in leucocyte mediator generation and chemotactic responses. There was also a significant attenuation of the late asthmatic response to inhaled antigen. Thus the potential of a fish-oil enriched diet to modulate the inflammatory component of the allergic response is realized. Further studies are needed to determine the full potential of such diets in effecting changes in the clinical aspects of allergic disease.

Acknowledgements

This work was supported in part by the National Institutes of Health (Fogarty International Fellowship), Medical Research Council (UK) and Asthma Research Council (UK).

References

1 Bills TK, Smith JB, Silver MJ. Selective release of arachidonic acid from the phospholipids of human platelets in response to thrombin. *J. Clin. Invest.* 1977; **60**:1–6.

2 Lapetina EG. Regulation of arachidonic acid production: role of phospholipase C and A_2. *Trends Pharmacol. Sci.* 1982; **3**:115–118.

3 Lapetina EG, Siess W. The role of phospholipase C in platelet responses. *Life Sci.* 1983; **33**:1011–1018.

4 Siess W, Lapetina EG. Properties and distribution of phosphatidylinositol-specific phospholipase C in human and horse platelets. *Biochim. Biophys. Acta* 1983; **752**:329–338.

5 Majerus PW. Arachidonate metabolism in vascular disorders. *J. Clin. Invest.* 1983; **72**:1521–1525.

6 Majerus PW, Prescott SM, Hofmann SL, Neufeld EJ, Wilson DB. Uptake and release of arachidonate by platelets. *Adv. Prostaglandin Thromboxane Leukotriene Res.* 1983; **11**:45–52.

7 Neufeld EJ, Majerus PW. Arachidonate release and phosphatidic acid turnover in stimulated human platelets. *J. Biol. Chem.* 1983; **258**:2461–2467.

8 Borgeat P, Samuelsson B. Arachidonic acid metabolism in polymorphonuclear leukocytes: effects of ionophore A23187. *Proc. Natl. Acad. Sci. USA* 1979; **76**:2148–2152.
9 Borgeat P, Samuelsson B. Arachidonic acid metabolism in polymorphonuclear leukocytes. Unstable intermediate in formation of dihydroxy acids. *Proc. Natl. Acad. Sci. USA* 1979; **76**:3213–3217.
10 Hammarstrom S. Leukotriene C5: a slow reacting substance derived from eicosapentaenoic acid. *J. Biol. Chem.* 1980; **255**:7093–7094.
11 Radmark O, Malmsten C, Samuelsson B, Goto G, Marfat A, Corey EJ. Leukotriene A_I: isolation from human polymorphonuclear leukocytes. *J. Biol. Chem.* 1980; **255**:11828–11831.
12 Shimizu T, Radmark O, Samuelsson B. Enzyme with dual lipoxygenase activities catalyses leukotriene A4 synthesis from arachidonic acid. *Proc. Natl. Acad. Sci. USA* 1984; **81**:689–693.
13 Rouzer CA, Matsumoto T, Samuelsson B. Single protein from human leukocytes possesses 5-lipoxygenase and leukotriene A_4 synthase activities. *Proc. Natl. Acad. Sci. USA* 1986; **83**:857–861.
14 Murphy RC, Hammarstrom S, Samuelsson B. Leukotriene C: a slow reacting substance from murine mastocytoma cells. *Proc. Natl. Acad. Sci. USA* 1979; **76**:4275–4279.
15 Bach MK, Brashler JR, Hammarstrom S, Samuelsson B. Identification of C-1 as a major component of slow reacting substance from rat mononuclear cells. *J. Immunol.* 1980; **125**:115–117.
16 Bach MK, Brashler JR, Morton DR. Solubilisation and characterization of the leukotriene C_4 synthetase of rat basophil leukaemia cells: a novel particulate glutathione-S-transferase. *Arch. Biochem. Biophys.* 1984; **230**:455–465.
17 Yoshimoto T, Soberman RJ, Lewis RA, Austen KF. Isolation and characterisation of leukotriene C_4 synthetase of rat basophilic leukaemia cells. *Proc. Natl. Acad. Sci. USA* 1985; **82**:8399–8403.
18 Orning L, Hammarstrom S, Samuelsson B. Leukotriene D: a slow reacting substance from rat basophilic leukemia cells. *Proc. Natl. Acad. Sci. USA* 1980; **77**:2014–2018.
19 Morris HR, Taylor GW, Piper PJ, Tippins JR. Structure of slow reacting substance of anaphylaxis from guinea pig lung. *Nature* 1980; **285**:104–105.
20 Lewis RA, Drazen JM, Austen KF, Clarke DA, Corey EJ. Identification of the C(6)-S-conjugate of leukotriene A with cysteine as a naturally occurring slow reacting substance of anaphylaxis (SRS-A). Importance of the 11-*cis* geometry for biological activity. *Biochem. Biophys. Res. Commun.* 1980; **96**:271–277.
21 Borgeat P, Samuelsson B. Metabolism of arachidonic acid in polymorphonuclear leukocytes. Structural analysis of novel hydroxylated compounds. *J. Biol. Chem.* 1979; **254**:2643–2646.
22 Lee CW, Lewis RA, Tauber AI, Mehrotra M, Corey EJ, Austen KF. The myeloperoxidase-dependent metabolism of leukotrienes C_4, D_4 and E_4 to 6-*trans*-leukotriene B_4 diastereoisomeric sulfoxides. *J. Biol. Chem.* 1983; **258**:15004–15010.
23 Hansson G, Lindgren JA, Dahlén SE, Hedqvist P, Samuelsson B. Identification and biological activity of novel ω-oxidized metabolites of leukotriene B_4 from human leukocytes. *FEBS Lett.* 1981; **130**:107–112.
24 Camp RD, Woollard PM, Mallet AI, Fincham NJ, Ford-Hutchinson AW, Bray MA. Neutrophil aggregating and chemokinetic properties of a 5,12,20-trihydroxy-6,8,10,14-eicosatetraenoic acid isolated from human leukocytes. *Prostaglandins* 1982; **23**:632–639.
25 Needleman P, Raz A, Minkes NS, Ferendelli A, Sprecher H. Triene prostacyclin and thromboxane biosynthesis and unique biological properties. *Proc. Natl. Acad. Sci. USA* 1979; **76**:944–948.
26 Corey EJ, Shih C, Cashman JR. Docosahexaenoic acid is a strong inhibitor of prostaglandin but not leukotriene biosynthesis. *Proc. Natl. Acad. Sci. USA* 1983; **80**:3581–3584.
27 Dyerberg J, Bang HO, Stofferson E, Moncada S, Vane JR. Eicosapentaenoic acid and prevention of thrombosis and atherosclerosis. *Lancet* 1978; **2**:117–119.
28 Whitaker MO, Wyche A, Fitzpatrick F, Sprecher H, Needleman P. Triene prostaglandins: prostaglandin D_3 and eicosapentaenoic acid as potential antithrombotic substances. *Proc. Natl. Acad. Sci. USA* 1979; **76**:5919–5923.

29 Goldman DW, Pickett WC, Goetzl EJ. Human neutrophil chemotactic and degranulating activities of leukotriene B_5 (LTB_5) derived from eicosapentaenoic acid. *Biochem. Biophys. Res. Commun.* 1983; **117**:282–288.
30 Lee TH, Mencia-Huerta JM, Shih C, Corey EJ, Lewis RA, Austen KF. Characterisation and biological properties of 5,12-dihydroxy derivatives of eicosapentaenoic acid including leukotriene B_5 and the double lipoxygenase product. *J. Biol. Chem.* 1984; **259**:2383–2389.
31 Terano T, Salmon JA, Moncada S. Biosynthesis and biological activity of leukotriene B_5. *Prostaglandins* 1984; **27**:217–232.
32 Lee TH, Sethi T, Crea AEG, Peters W, Arm JP, Horton CE, Walport MJ, Spur BW. Characterisation of leukotriene B_3. Comparison of its biological activities with leukotriene B_4 and leukotriene B_5 in complement receptor enhancement, lysozyme release and chemotaxis of human neutrophils. *Clin. Sci.* 1988; **74**:467–475.
33 Dahlén SE, Hedqvist P, Hammarstrom S. Contractile activities of several cysteine-containing leukotrienes in the guinea pig lung strip. *Eur. J. Pharmacol.* 1982; **86**:207–215.
34 Leitch AG, Lee TH, Ringel EW, Prickett JD, Robinson DR, Pyne SG, Corey EJ, Drazen JM, Austen KF, Lewis RA. Immunologically-induced generation of tetraene and pentaene leukotrienes in the peritoneal cavities of menhaden-fed rats. *J. Immunol.* 1984; **132**:2559–2564.
35 Lee TH, Mencia-Huerta JM, Shih C, Corey EJ, Lewis RA, Austen KF. Effects of exogenous arachidonic, eicosapentaenoic, and docosahexaenoic acids on the generation of 5-lipoxygenase pathway products by ionophore-activated human neutrophils. *J. Clin. Invest.* 1984; **74**:1922–1933.
36 Benveniste J, Tence M, Varenne P, Bidault J, Boullet C, Polansky J. Semi-synthese et structure proposee du facteur activant les plaquettes (PAF): PAF-acether, un alkyl ether analogue de la lysophosphatidylcholine. *C.R. Acad. Sci.* 1979; **289D**:1037–1040.
37 Sperling RI, Robin JL, Kylander KA, Lee TH, Lewis RA, Austen KF. The effects of N-3 polyunsaturated fatty acids on the generation of platelet-activating factor-acether by human monocytes. *J. Immunol.* 1987; **12**:4187–4191.
38 Murphy RC, Pickett WC, Culp BR, Lands WEM. Tetraene and pentaene leukotrienes: selective production from murine mastocytoma cells after dietary manipulation. *Prostaglandins* 1981; **22**:613–622.
39 Lee TH, Hoover RL, Williams JD, Sperling RI, Ravalese J, Spur BW, Robinson DR, Corey EJ, Lewis RA, Austen KF. Dietary enrichment with eicosapentaenoic and docosahexaenoic acids in human subjects impairs *in vitro* neutrophil and monocyte function and leukotriene generation. *N. Engl. J. Med.* 1985; **312**:1217–1224.
40 Lee TH, Austen KF, Leitch AG, Israel E, Robinson DR, Lewis RA, Corey EJ, Drazen JM. The effects of a fish-oil enriched diet on pulmonary mechanics during anaphylaxis. *Am. Rev. Respir. Dis.* 1985; **132**:1204–1209.
41 Lee TH, Drazen JM, Leitch AG, Ravelese J, Corey EJ, Robinson DR, Lewis RA, Austen KF. Enhancement of plasma levels of biologically active leukotriene B compounds during anaphylaxis in guinea pigs pretreated by indomethacin or by a fish-oil enriched diet. *J. Immunol.* 1986; **136**:2575–2582.
42 Marsh WR, Irvin CG, Murphy KR, Behrens BL, Larsen GL. Increases in airway reactivity to histamine and inflammatory cells in bronchoalveolar lavage after the late asthmatic response in an animal model. *Am. Rev. Respir. Dis.* 1985; **131**:857–879.
43 Murphy KR, Wilson MC, Irvin CG *et al.* The requirement for polymorphonuclear leukocytes in the late asthmatic response and heightened airways reactivity in an animal model. *Am. Rev. Respir. Dis.* 1986; **134**:62–68.
44 Chung KF, Becker AB, Lazarus SC, Frick OL, Nadel JA, Gold WM. Antigen-induced airway hyperresponsiveness and pulmonary inflammation in allergic dogs. *J. Appl. Physiol.* 1985; **58**:1347–1353.
45 Cartier A, Thomson NC, Frith PA, Roberts R, Hargreave FE. Allergen-induced increase in bronchial responsiveness to histamine: relationship to the late asthmatic response and change in airway calibre. *J. Allergy Clin. Immunol.* 1982; **70**:170–177.

46 De Monchy JGR, Kauffman HF, Venge P *et al.* Bronchoalveolar eosinophilia during allergen-induced late asthmatic reactions. *Am. Rev. Respir. Dis.* 1985; **131**:373–376.
47 Metzger WJ, Richerson HB, Worden K, Monick M, Hunninghake GW. Bronchoalveolar lavage of allergic asthmatic patients following allergen bronchoprovocation. *Chest* 1986; **89**:477–483.
48 Fabbri LM, Aizawa H, Alpert SE *et al.* Airway hyperresponsiveness and changes in cell counts in bronchoalveolar lavage after ozone exposure in dogs. *Am. Rev. Respir. Dis.* 1984; **129**:288–291.
49 Stelzer J, Geffroy B, Stalborg M, Holtzman MJ, Nadel JA, Boushey HA. Association between airway inflammation and changes in bronchial reactivity induced by ozone exposure in healthy subjects. *Am. Rev. Respir. Dis.* 1984; **129**:A262 (Abstract).
50 O'Byrne PM, Leikauf GD, Aizawa H *et al.* Leukotriene B_4 induces airway hyperresponsiveness in dogs. *J. Appl. Physiol.* 1985; **59**:1941–1946.
51 Arm JP, Horton CE, Mencia-Huerta JM, House F, Eiser NM, Clark TJH, Spur BW, Lee TH. Effect of dietary supplementation with fish oil lipids on mild asthma. *Thorax* 1988; **43**:84–92.
52 Payan DG, Wong YS, Chernov-Rogan T. Alterations in human leukocyte function induced by ingestion of eicosapentaenoic acid. *J. Clin. Immunol.* 1986; **78**:937–942.
53 Arm JP, Horton CE, Eiser NM, Clark TJH, Lee TH. The effects of dietary supplementation with fish oil on asthmatic responses to antigen. *Am. Rev. Respir. Dis.* 1989; In press.

Discussion session

BJORKSTEN (Sweden): have you looked at cell types other than the neutrophil? Is there a change in the lipid composition of other cells?

LEE: in the study of normal subjects we also looked at monocyte function and were able to demonstrate similar changes to those found with neutrophils. Incorporation of EPA was also demonstrated in monocytes. In the asthma studies we did not address monocyte function because there was so much to do already.

BJORKSTEN: did you look at the fatty acid composition of other cell types?

LEE: yes, we looked at the fatty acid composition of monocytes.

BJORKSTEN: how pure were these preparations?

LEE: we separated our neutrophils and monocytes using dextran sedimentation and centrifugation on gradients of Lymphoprep and they were over 90% pure.

MUERS (Leeds): I want to ask you about the severity of the asthmatics you used in your study. It looked to me from the illustration as if the mean peak flow was round about 500 which is pretty high for chronic persistent asthma. I wondered whether you were correct in writing off the data as judged by a study quite small in mild asthma. What about more severely disabled people with peak flows around about 200–300?

LEE: that is a valid criticism of our study. We purposely elected to study mild asthmatics to allow us to do bronchial provocation tests. A study by Ed Goetzl and his colleagues also demonstrated that in mild to severe asthmatics (some of whom were actually on corticosteroid therapy) there was no benefit from a similar diet.

MORLEY (Switzerland): have you biopsied skin sites in asthma or volunteer subjects for direct evidence of anti-inflammatory activity?

LEE: no we have not.

QUESTIONER (unidentified): did you monitor the normal fat intake in the diet of these patients of yours? Presumably this will vary from one individual to the next.

LEE: you are absolutely right. We did not try to do that at all. I will emphasize that our study is internally controlled and each subject is compared with himself. We made no attempt to try to control the diet that the patients were on and we allowed them to eat whatever they liked. We did not anticipate that any person would suddenly decide to become vegetarian in the middle of the study, but your point is a good one.

HOLGATE: did the diet alter the release of PAF-acether from the leucocytes?

LEE: yes.

HOLGATE: secondly, did you measure histamine responsiveness of the airways in this study?

LEE: we measured baseline reactivity but the crucial question is, of course, what happens after antigen challenge and that information is not available yet.

KAY: it could be that properties such as cytotoxicity and adherence are more relevant than chemotaxis to the sort of inflammatory process we envisage in asthma. After the diet, did cells still adhere to glass and endothelial cells and was the skin window response altered?

LEE: we know that in normal volunteers, an identical diet inhibits neutrophil endothelial adherence.

QUESTIONER (unidentified): do you know anything about fatty acid composition in cells of atopic patients and treated patients?

LEE: there is some literature on this subject but we have not studied this.

General discussion (Chapters 1–5)

BJÖRKSTÉN: why did all the patients not improve or have the same tendency after receiving the platelet-activating factor (PAF-acether) antagonist?

BRAQUET: it is difficult to answer your question at this time because we have not yet finished the investigation. We only began the clinical studies 6 months ago. You are right, that certain patients respond very well, certain other ones did not respond so well. It is difficult to know how they differ. I am sure that this kind of treatment will not be used for all kinds of asthma.

LEE: what is the difference between giving a mediator chronically and giving a mediator as an acute dose in the induction of bronchial hyper-responsiveness?

BRAQUET: in the acute situation we did not see increased responsiveness to histamine.

MORLEY: in my opinion acute exposure to PAF-acether does cause hyper-reactivity. Acute exposure to allergen also causes hyperreactivity but whereas the acute response to PAF-acether is abolished by a number of selective PAF-acether antagonists, the acute hyperreactivity to allergen is not. I would expect that if you look at 24 h when you have substantial eosinophil accumulation you have another form of hyperreactivity. I think it is quite possible that you may have inhibition in one situation and not in another.

WASSERMAN: how do you explain eosinophil accumulation in the lung from intravenously administered PAF-acether? Do you see eosinophil accumulation in the lung from aerosolized PAF-acether? My second question relates to PAF-acether inhibitors which block histamine release in the guinea pig. Could you tell me if the guinea pig platelet, like the rabbit platelet, contains histamine?

BRAQUET: we have not yet performed these experiments with aerosolized PAF-acether. Your second question whether guinea pig platelets contain histamine—maybe. Why?

WASSERMAN: because the guinea pig platelet clearly contains histamine. That, in fact, is how PAF-acether was described, as a molecule which induced histamine release from the guinea pig platelet. If you show the inhibition of histamine release in your model, one alternative explanation is that you are blocking platelet degranulation by PAF-acether. Perhaps the histamine release that you detect from that model is from platelets and not from asthma models.

BRAQUET: it is maybe an explanation since when we used prostaglandin (PG) I_2 we did not see the phenomenon.

BACH: PAF-acether is rapidly inactivated when it reaches the surface of the lungs and consequently the aerosol administration may be a non-experiment.

MORLEY: we have published the effect of inhalation of PAF-acether in the guinea pig. It causes selective accumulation of eosinophils which is maximal at 48 h and it is easily the best stimulus of eosinophil accumulation that one can identify. It also causes selective accumulation in the airways if you administer PAF-acether intraperitoneally, subcutaneously or intravenously and so I think chemotaxis may not be particularly relevant to this phenomenon.

WASSERMAN: that was, in fact, my question. How do you explain PAF-acether inducing eosinophil accumulation in the airway when the drug is administered in various sites which, if it is a chemoattractant, would be thought to localize it at the site where it has been administered? If it is given intravenously I could understand first activating eosinophils then stimulating them to random migration. One would then expect that all tissues would show an increased accumulation of eosinophils and that it would not be specific to the airway.

MORLEY: perhaps the activated eosinophil preferentially attaches to certain endothelial cells, notably those in the airways and the gut.

WASSERMAN: it may be quite the opposite. It may be that the PAF-acether is acting on the endothelial cells or the airway.

MORLEY: yes, the point is that the chemotaxis is not an appropriate model for the event and I was really just reporting what happened.

WASSERMAN: it must be chemotaxis. They cannot get there if it is not chemotaxis.

BRAQUET: when we used PAF-acether injection in the guinea pig and checked the mesenteric artery we also found eosinophil infiltration in the mesenteric artery. Therefore I think your question is important because it appears that lung may be more sensitive to eosinophil infiltration. But we also found eosinophils in the mesenteric artery.

LEE: I would like to make just one comment about Dr Ford-Hutchinson's presentation which I found very interesting. His hypothesis was that mediators are released sequentially? Another way in which antagonists of different mediators may work in an animal model is if the mediators act synergistically and the target organ response is a result of the synergism between different mediators. If that is the case then if you remove the effect of one mediator by a specific antagonist you could quite readily down-regulate the target organ response without any sequential action and I think that is something that we need to consider in designing studies such as the ones we are discussing.

JOHNSON (Upjohn): a question for Dr Braquet. Have you had a chance to look at the activation state of the eosinophils which might be in the lung after PAF-acether inhalation or intravenous administration? For example, what kind of lipid mediators might they make?

BRAQUET: *in vitro* PAF-acether is able to elicit the release of leukotrienes from eosinophils. PAF-acether is able to induce its own release but I do not know if this event is involved *in vivo*. We have only based this on eosinophil activation by PAF-acether *in vitro*.

JOHNSON: any difference by staining?

BRAQUET: some degranulation, but it is impossible to tell you what kind of mediators are released *in vivo*.

LEE: the ability of PAF-acether to induce leukotriene release from eosinophils is in the dose range of 10^{-5} M. Do you think that this has any physiological or pathological relevance?

BRAQUET: it is very important to consider the role of PAF-acether in terms of amplification of a response (i.e. priming). For example, if you add PAF-acether at concentrations as little as 10^{-15} to 10^{-13} M to monocytes and then LPS a very significant increase in interleukin (IL) production is observed. Conversely, with PAF-acether alone, very high doses are required. Therefore it is very important to consider the amplifying role of PAF-acether.

VEAL (Newcastle upon Tyne): could I ask Professor Lee if the modulation of plasma lipids is of importance in the effect that he is trying to mediate? In that case two important points arise. One is; is there any change in plasma lipids and, secondly, is olive oil an appropriate placebo? Would the administration of olive oil perhaps invalidate what you are trying to find?

LEE: to answer the first question, we did not measure plasma lipids during this study. The reason we chose olive oil is to give an isocaloric preparation which is identical in taste and appearance. Also, *in vitro*, olive oil has no effect on leucocyte function. I am interested to hear your hypothesis that plasma lipids may have something to do with the severity of asthma, perhaps you would like to expand on that.

VEAL: it is entirely speculative.

LEE: it is not something I had thought about and I do not really see at the moment how plasma lipids could have anything to do with it.

WILHELMS (Mannheim): I would like to put a question to Professor Holgate. You presented an illustration showing the importance of the various mediators in the early asthmatic reaction. You said that about 70% of the change in forced expiratory volume in 1 s (FEV_1) was due to histamine and only 2% to PAF-acether. May I ask how did you arrive at this conclusion? Also, why do you say that antihistamines are helpful in asthma?

HOLGATE: could I take the last point first? To my knowledge there are at least three highly potent H_1-receptor antagonists that have been evaluated in clinical trials, not just in the challenge situation but in ongoing asthma. In each case, efficacy was demonstrated. We have recently completed a study with terfenadine using the drug across the pollen season. Sixty patients entered the trial and the results really look quite good. I think that antihistamines might have a place in the treatment of asthma in certain well-defined situations. My views are changing as we get more information. Going back to the first part of your question, which is a very important one. If I say that histamine contributes 70% and Dr Braquet says PAF-acether accounts for the whole response, how do we reconcile our views? We have different model systems for one thing. For example, in the dog, prostaglandins (particularly thromboxane (Tx) A_2 and PGD_2) are a major component of the airway bronchoconstrictor response and pathway for the expression of bronchial hyperresponsiveness. In humans, we believe that 50% or more of the acute airway response to allergen is due to histamine. Another important factor is the possible influence of one mediator on the target sensitivity produced by a second mediator.

BACH: there is a lot more to be looked at than acute bronchospasm.

WASSERMAN: I would like to respond to the question about change in blood lipids and the effect that might have in inflammatory processes. There are several lines of evidence that would suggest that they may very well be important in inflammatory processes. There is some old work from Huehn's group in Johns Hopkins suggesting that lysophosphatidylcholine is a chemoattractant for lymphocytes and that lysophosphatidylcholine is primarily carried on low-density lipoproteins (LDL) and not high-density lipoproteins (HDL). In addition, oxidized LDL are very potently chemotactic for monocytes and more recent work has shown once again that lysophosphatidylcholine (conceivably through a complicated series of chemical reactions where the phosphatidylcholine polar head group is removed by a unique phospholipase C and then diacylglycerol is regenerated) is itself a chemoattractant for lymphocytes and macrophages. This might explain why LDL and macrophages lead to atherosclerosis. If this were a more general phenomenon for neutrophils and eosinophils and/or if the effects on the endothelium are due to HDL versus LDL it would be very important to look at the effect of various dietary supplements on the HDL/LDL ratio?

KAY: are lipoxins important? What is new regarding their biological properties? Should we be thinking about them in the context of asthma and allergy?

LEE: lipoxins are fairly recently described molecules formed by the 5- and 15-lipoxygenase pathways. One of the major difficulties in studying the biological properties of these molecules is having them in adequate quantities. They are very unstable and isomerize on reverse-phase, high-performance liquid chromatography (RP-HPLC) columns. We have been very fortunate in being able to have a collaboration with a synthetic chemist, Dr Spur, who has produced these molecules for us in purity and in sufficient quantities for biological evaluation. We have been looking at their properties in a number of assay systems. We have shown, using contraction of guinea pig parenchymal strips, that lipoxin A_4 is some 10 000-fold less potent than LTD_4 but it does have a similar time course of action as LTD_4. Antagonists to the LTD_4-receptor (Merck) and FPL-55 712 inhibited the activity of lipoxin A_4 and we have suggested that lipoxin A_4 works through the LTD_4-receptor. Lipoxin B_4 is inactive in contracting airway tissue in guinea pigs. A lot more work needs to be done on them. Based on their biological potency in the guinea pig airway preparation one would not emphasize a pro-inflammatory role for these metabolites in contracting airways. Their *in vivo* relevance and biological role are still to be established.

BACH: I believe very similar data to Professor Lee's have been obtained by Erik Dahlén in Sweden. I think some of this has been published.

MORLEY: if you abolish the capacity of eosinophils to mount defensive actions against parasites by using a PAF-acether antagonist, does this pose a problem for therapeutic use in asthma?

KAY: I suppose that one answer to that question is that if one had asthma *and* a helminthic parasitic infection it would be much more serious if you were taking oral corticosteroids than a PAF-acether antagonist!

II

Modulating the Specific Immune Response

6

Prospects for modulation of the specific immune response in allergic diseases

J.M. DEWDNEY

Summary

Significant progress in the development of allergen-based immunotherapy is dependent upon clearer definition of therapeutic targets and the development of modified allergens in which allergenic epitopes have been rendered unresponsive to cell-bound IgE antibody. Suppression of the established human IgE response remains the most attractive target for specific hyposensitization therapy but modulation of T-cell function by a new generation of T-allergoids holds prospects for additional therapeutic benefits.

The impact of genetic engineering on the treatment of allergic disease has yet to be fully appreciated but it is likely that the ability to isolate and express the genes responsible for the factors which regulate IgE synthesis will, in the longer term, transform both understanding and treatment of allergy. The prospect is not only of restoration of normal control by therapy with recombinant derived proteins or muteins but also of designer pharmaceuticals based on knowledge of receptor structure and the molecular modelling of agonists and antagonists.

Introduction

Significant progress has been made in the past decade in the development of methods for the assessment of immunological changes induced in patients by immunotherapy and in allergen immunochemistry. The major international initiative on allergen standardization has led to the establishment of primary standards for the main allergens [1] and to methods which ensure batch reproducibility. While these advances are clearly supportive of immunotherapy and fully justified in that it is right that allergen-based vaccines should be subjected to the same standards of quality control as other biologicals in clinical use, their contribution to improvement of immunotherapy is questionable. One problem is that although standardization has removed speculation regarding the variability in potency of extracts, the majority of treatments are carried out with allergens that have been modified physically or chemically and such products cannot be calibrated directly against native extract reference preparations. Universal systems of unit labelling are therefore not available. More significantly, effort on standardization cannot address the fundamental problem of immunotherapy which is that the mechanisms by which these products effect clinical improvement in allergic patients remain unknown. It should not surprise us that the administration of immunogenic foreign proteins—allergens—to humans induces immune responses which can be detected at the cellular or immunoglobulin level, but we should not lose sight of the fact that the relationship of these changes to hyposensitization is still speculative. The mechanism underlying adverse reactions to allergy vaccines is better established; immediate reactions involve the interaction of allergen with cell-bound IgE antibody, resulting in the release of mediators of allergic reactions. The direction of much research and development in recent years has therefore been towards the discovery of allergen-based products in which allergenicity has been reduced by physical or chemical means, because the target has been much clearer. Future research must try to define more precisely the therapeutic target in order to enhance the efficacy of immunotherapy without compromising safety.

Two strategies are under investigation in a number of laboratories. The first is an allergen-based approach in which, in continuation and extension of conventional hyposensitization therapy, the biological properties of a variety of modified allergens are being defined. The second strategy involves exploitation of newer information on the regulation of the IgE antibody response by T-cell derived factors, an approach which has come closer to reality by genetic engineering advances. In terms of clinical potential, each approach has special merit and unique problems. Some of these concerns will be outlined to provide a framework for more detailed considerations covered in subsequent papers.

The main therapeutic target for both approaches is suppression of IgE, whether of allergen-specific IgE or IgE-isotype suppression. This target is clearly justified in view of the importance of this antibody class in allergic diseases. In terms of immunomodulation, it is perhaps fortunate that IgE synthesis seems particularly sensitive to modulation, whether by specific or non-specific mechanisms. It is worth noting, however, that suppression of IgE is not the only goal of newer approaches to immunotherapy. Enhancement of allergen-specific IgG subclasses remains an important objective, and modulation of T-cell subsets may have consequences over and above that of isotype suppression.

Allergen-based approaches

Conventional hyposensitization therapy involves the administration of native or modified extracts in a variety of formulations, mostly, but not exclusively, by subcutaneous injection. The aim of modification has been primarily based on the 'allergoid' principle in which immunogenicity is retained but allergenicity as measured by interaction of the allergen moiety with IgE antibody is reduced [reviewed in 2].

The development of new allergen vaccines of this type is now unlikely in the UK. The Committee on Safety of Medicines, in expressing concern about the safety of allergy vaccines in relation to their perceived efficacy, not only now requires additional and very stringent resuscitation facilities to be available before such therapy is carried out but also does not distinguish between vaccines based on native allergen and those modified to reduce immediate reactivity [3].

The issues which arise from the Committee decision have been discussed in recent publications [4–6]. The impact of this decision on future prospects for new approaches to the treatment of patients with allergen-based vaccines is profound. The prime objective in the development of novel allergen-based preparations must now be a virtual lack of allergenicity as measured by ability to interact with IgE antibody on mast cells but with retention of properties, currently poorly understood, likely to lead to clinical benefit. It is uncertain whether this can be achieved or indeed if the pharmaceutical industry will be prepared to invest in the research and development required in the light of unsympathetic regulatory, and to some extent clinical, attitudes.

None the less, it is appropriate to assess progress being achieved with some novel allergen preparations, and to review the extent to which any shows promise of an improved therapeutic ratio in humans, compounded either by better efficacy or lower allergenicity or both. A number of preparations have been described which have as their therapeutic target the suppression of allergen-specific IgE. They include urea-modified allergens, photo-oxidized allergens and allergens coupled to a variety of non-immunogenic polymers [2]. The reports on the properties of these preparations, taken together with our own experience with one such material, allergen covalently conjugated to polysarcosine (poly-*N*-methylglycine), permit analysis of the potential of this approach [7–10].

All studies have shown that the administration of allergen–polymer conjugates to mice by the intravenous route effectively suppresses the induction of the IgE antibody response. Suppression of secondary IgE responses and the established antibody response is less readily achieved, with disparate results being obtained with different polymer preparations. Our allergen–polysarcosine conjugates, for example, were less effective at suppressing established responses than other polymers reported in the literature.

We believe that the most probable explanation of these apparent discrepancies between agents is more quantitative than of fundamental mechanistic significance. Low doses of polymer allergen were used in our studies, in order to mimic the conditions under which it might be possible to carry out clinical tests. Indeed, other conjugates of more impressive activity in mice when tested at realistic doses in man have not fulfilled their promise [2]. The inability of this type of conjugate, therefore, to inhibit established IgE responses when given in doses which must be restricted because of residual allergenicity, might well prove to be a problem which is very difficult to overcome.

It is of course possible that apparently similar polymer allergens have different mechanisms of action and may differentially affect the induction of T-suppressor cells or the suppression of T-helper cells or B-cells. Allergen–polysarcosine conjugates generate relatively transient T-cell suppressor activity, dependent upon a population of anti-ly-2^+ sensitive lymphocytes [9].

It is likely that agents are needed which also generate T-suppressor cell memory but there is no evidence that this can be achieved in humans with currently available agents. Thus, further study is needed to give perspective to apparent discrepancies between suppression of the murine and human IgE response, otherwise inappropriate novel allergens may be progressed for clinical studies.

Care will also have to be exercised in the judgement of reduced allergenicity. A number of reported novel allergens show substantially reduced ability to inhibit the RAST (radioallergosorbent test), in some cases by as much as 1000-fold or more. Two factors should lead to caution in the interpretation and extrapolation of data of these kinds to the allergic patient. Reduction in allergenicity as judged by RAST inhibition is often greater than when based on skin-test reactivity, sometimes by orders of magnitude. Moreover, there may be considerable variation in the sensitivities of individual patients to the same modified allergens, a situation which no doubt reflects not only the marked heterogeneity of the human IgE response to allergens but also differential reactivity of epitopes with the modifying agent. Thus, absolute safety from allergen-triggered allergic responses, including anaphylaxis, cannot yet be guaranteed, but there is some prospect that these different types of novel allergen derivatives may offer the degree of efficacy and safety required.

Three options for further evaluation of allergen–polymer conjugates present themselves. The first is an attempt to minimize the problems potentially associated with retained allergenicity by administration of the conjugate by a route other than parenteral. In mice, our conjugate behaves similarly when given by intranasal administration or intravenously, in that the primary IgE serum antibody response is totally suppressed [11]. A polymer more effective at suppressing the established IgE response and free of significant allergenicity at the nasal mucosa or systemically could offer a clinical opportunity. Thoughtful comments on mechanistic aspects of this approach are to be found in [12]. Recent work has suggested that oral immunotherapy should be more fully evaluated [13] and sublingual administration of allergen has its proponents [14]. Modified allergens with IgE-suppressive properties might be seen as suitable candidates for such studies.

Secondly, it has been suggested, that not only dose levels but also clinical regimens need to be considered if IgE-suppressor allergen–polymer conjugates are to find a place in the clinic [8]. Murine work would lead to a recommendation for continuous therapy throughout periods of environmental exposure to allergens, e.g. during the pollen season, to prevent the seasonal boost of IgE, but again residual allergenicity would have to be taken into account.

Thirdly, perhaps the most important strategy which could be followed based on data and ideas generated from the study of modified allergens of this type can be expressed in terms of a word coined in our laboratories—T-allergoids. The word is meant to define allergen preparations which retain the ability to activate allergen-specific T-cells but which do not interact with cell-bound IgE nor induce or boost IgG antibody. The concept of such differential epitope denaturation with modified allergens is one which offers considerable potential. At the immunochemical level it will be important to understand the structural determinants involved. It may be important also to appreciate that whereas the suppression of IgE synthesis may be the most obvious therapeutic objective, other biological consequences of modulation of T-cell function might well be of clinical significance. For example, T-allergoids might be seen as antigen-driven T-cell probes influencing the influx of inflammatory cells into the nasal mucosa and modulation of T-cell function might be a means of controlling cell traffic and late-phase reactions [15]. T-allergoids might also provide a safe means of inducing T-cell priming prior to very low-dose allergen boosting to enhance the allergen-specific IgG response.

We have carried out initial studies on materials within this general class. In one series of investigations we describe the preparation and properties of allergens covalently bound to the peptide formyl-methionyl-leucyl-phenylalanine (FMLP) [16, 17]. Reactivity of allergens with specific IgE was lost as a function of increased modification with peptide but reactivity with allergen-specific T-cells following accessory cell processing was retained or enhanced. Immunization studies have shown that while allergen-specific T-cells are induced as measured *in vitro* and by the development of delayed hypersensitivity, no allergen-specific antibody is detectable.

In another series of investigations, enzymatically produced allergen fragments have been prepared and evaluated. Chymotryptic fragments of ryegrass with a molecular weight of less than 10 000 were shown to be processed by murine accessory cells and presented to specific T-cells to induce proliferation. Rye-specific T-helper cells were stimulated but the fragments failed to react with human IgE or mouse IgG and did not induce IgG antibody in immunization studies [18, 19].

Thus, preliminary investigations have shown that it is possible to prepare T-allergoids with properties which *a priori* might be of benefit to allergic patients. These conjugates may not themselves be progressed for clinical evaluation but they point the way to novel immunomodulatory procedures which might be developed in the future.

Regulation of IgE synthesis

Rational manipulation of IgE biosynthesis demands better understanding of the regulatory networks underlying suppression and amplification of the production of this important immunoglobulin. Significant advances have been made in recent years and prospects now exist for clinical exploitation as the result not only of fundamental studies but also of the power of genetic engineering technology [20–23].

Already a number of regulatory proteins or lymphokines involved in the control of human IgE synthesis have been cloned and expressed. As a result, it is now clear that the murine suppressing and potentiating factors share a common structural gene and that the different biological properties of these factors are determined by post-translational glycosylation which, in turn, is controlled by two T-cell derived factors, glycosylation-inhibiting factor (GIF) and glycosylation-enhancing factor (GEF). Moreover, the immunoregulatory network in humans appears to parallel that of rodents quite closely, although some differences in both producing and target cells for the factors have been noted. Suppressive factor of allergy (SFA) seems to be analogous to GIF. The gene for this human regulatory factor has been expressed in prokaryotic cells and the production and purification programme now ongoing should supply sufficient recombinant protein for clinical studies. Fundamental studies on immunoregulation by Fc signals and continuing work on B-cell differentiation factors [22–24] are all now approaching clinical exploitation through cloning and expression technology. For example, the low-affinity receptor, FcεRII has recently been cloned and expressed in eukaryotic cells [22, 25]. However, progression of these engineered proteins will be no simple or rapid task. A number of draft regulations cover the safety issues to be addressed for the products of biotechnology. Exploratory work will have to be carried out to determine how these protein molecules can be given to allergic patients, by what route, in what dose and how, in relation to allergen exposure, both allergen-specific and isotype specific factors may be progressed. These problems are similar to those to be overcome in other fields of medicine, in which the exploitation of the products of biotechnology is sought and, as in all areas of therapeutic medicine, the balance of risk and cost to the patient against benefit must be carefully appraised.

The second strategy also depends on a genetic-engineering capability but uses this technology as a tool rather than as a direct means of producing a novel product for clinical use. As discussed, a number of receptors involved in the regulation of IgE synthesis are now recognized and some have been cloned and expressed. Molecular modelling techniques could now assist in the discovery of pharmaceuticals interactive with the receptor and acting, as appropriate, as agonist or antagonist. It can reasonably be anticipated that new generations of drugs will be discovered by these means, and the targets for the control of allergic disease are now sufficiently defined to justify this approach. This strategy has considerable appeal, and the prospect of new, orally active drugs interactive at different sites in the regulatory network of IgE synthesis is within realistic grasp.

Initial advances in this area have been the result of studies using peptides [26] and more will follow now that the constant regions of the human ε chain can be expressed in *Escherichia coli* and will become available for clinical studies [27]. Molecular modelling studies should now permit the discovery of lower molecular weight analogues which should retain a blocking function while opening up the prospect of oral therapy.

In these ways, it seems likely that the currently disparate clinical approaches to the treatment of allergic patients, immunotherapy and pharmacological control, will find a common purpose in the development of new medicines for this patient group.

References

1 Norman PS. Why standardized extracts? *J. Allergy Clin. Immunol.* 1986; **77**:405–406.

2 Moran DM, Wheeler AW. Specific hyposensitization. In *Development of Anti Asthma Drugs* (Eds Buckle DR, Smith H), Butterworths, London 1984, pp. 349–379.

3 Committee on Safety of Medicines. CSM update. Desensitising vaccines. *Br. Med. J.* 1986; **293**:948.

4 Morrow Brown H, Frankland AW. Hyposensitisation. *Br. Med. J.* 1987; **294**:1613–1614.

5 Warner JO, Kerr JW. Hyposensitisation. *Br. Med. J.* 1987; **294**:1179–1180.

6 Norman PS. Fatal misadventures. *J. Allergy Clin. Immunol.* 1987; **79**:572–573.

7 Whittall N, Moran DM, Wheeler AW, Cottam GP. Suppression of murine IgE responses with amino acid polymer/allergen conjugates. I. Preparation of poly-(*N*-methylglycine)/grass pollen extract conjugates using 4-(methylmercapto)phenyl-succinimidyl succinate as coupling reagent. *Int. Arch. Allergy Appl. Immunol.* 1985; **76**:354–360.

8 Wheeler AW, Henderson DC, Garman AJ, Moran DM. Suppression of murine IgE responses with amino acid polymer/allergen conjugates. II. Suppressive activities in adjuvant-induced IgE responses. *Int. Arch. Allergy Appl. Immunol.* 1985; **76**:361–368.

9 Cook RM, Henderson DC, Wheeler AW, Moran DM. Suppression of murine IgE responses with amino acid polymer/allergen conjugates. III. Activity *in vitro*. *Int. Arch. Allergy Appl. Immunol.* 1986; **80**:355–360.

10 Moran DM, Wheeler AW, Henderson DC, Whittall N. Suppression of murine IgE responses with amino acid polymer/allergen conjugates. IV. Suppressive activities in established IgE model systems. *Int. Arch. Allergy Appl. Immunol.* 1986; **81**:357–362.
11 Henderson DC, Wheeler AW, Moran DM. Suppression of murine IgE responses with amino acid polymer/allergen conjugates. V. By intranasal administration. *Int. Arch. Allergy Appl. Immunol.* 1987; **82**:208–211.
12 Holt PG, Sedgwick JD. Suppression of IgE responses following inhalation of antigen. *Immunol. Today* 1987; **8**:14–15.
13 Björkstén B, Dewdney JM. Oral immunotherapy in allergy—is it effective? *Clin. Allergy* 1987; **17**:91–94.
14 Scadding GK, Brostoff J. Low dose sublingual therapy in patients with allergic rhinitis due to house dust mite. *Clin. Allergy* 1986; **16**:483–491.
15 Platts-Mills TAE. Desensitization treatment for hay fever. *Immunol. Today* 1981; **2**:35–37.
16 Cook RM, Wheeler AW, Spackman VM, Musgrove NRJ, Dave YK, Moran DM. Induction of allergen-specific T-cells by conjugates of *N*-formyl-methionyl-leucyl-phenylalanine and rye grass pollen extract. *Int. Arch. Allergy Appl. Immunol.* 1988; **85**:104–108.
17 Wheeler AW, Whittall N, Cook RM, Spackman VM, Moran DM. T cell reactivity of conjugates of *N*-formyl-methionyl-leucyl-phenylalanine and rye-grass pollen allergens. *Int. Arch. Allergy Appl. Immunol.* 1987; **84**:69–73.
18 Standring R, Wheeler AW, Lavender EA, Moran DM. Induction of helper 'T'-cell activity by chymotryptically-cleaved fragments of rye grass pollen extract. *Int. Arch. Allergy Appl. Immunol.* 1988; **86**:1–8.
19 Wheeler AW, Spackman VM, Cottam GP, Moran DM. Retained T cell reactivity of rye grass pollen extract following cleavage with cyanogen bromide and thiocyanonitrobenzoic acid. In preparation.
20 Ishizaka K, Jardieu P, Akasaki M, Iwata, M. T cell factors involved in the regulation of the IgE synthesis. *Int. Arch. Allergy Appl. Immunol.* 1987; **82**:383–388.
21 Katz DH. Immunologic alterations related to expression of the allergic phenotype. *J. Allergy Clin. Immunol.* 1986; **78**:980–987.
22 Gershwin LJ, Gershwin ME. The regulation of the IgE response. *Immunol. Today* 1986; 7(11):328–329.
23 Kishimoto T. B-cell stimulatory factors (BSFs): molecular structure, biological function, and regulation of expression. *J. Clin. Immunol.* 1987; **7**:343–355.
24 Sinclair NR, Panoskaltsis A. Immunoregulation by Fc signals. *Immunol. Today* 1987; **8**:76–79.
25 Lüdin C, Hofstetter H, Sarfati M, Levy CA, Suter U, Alaimo D, Kilchherr E, Frost H, Delespesse G. Cloning and expression of the cDNA coding for a human lymphocyte IgE receptor. *EMBO J.* 1987; **6**:109–114.
26 Prenner BM. Double-blind placebo-controlled trial of intranasal IgE pentapeptide. *Ann. Allergy* 1987; **58**:332–335.
27 Kenten J, Helm B, Ishizaka T, Cattirin P, Gould H. Properties of a human immunoglobulin ε-chain fragment synthesized in *Escherichia coli*. *Proc. Natl. Acad. Sci. USA* 1984; **81**: 2955–2959.

7

New ideas on the prevention of allergy

B. BJÖRKSTÉN

Introduction

There appears to be an increase in the prevalence of atopic diseases in industrialized as well as developing countries [1, 2]. Symptomatic treatment has improved dramatically over the past 15 years and as a consequence the quality of life for the allergic individuals has improved. Despite this, however, severe asthma is still by far the most common chronic debilitating disease in childhood and the mortality in asthma has not declined in children or adults. The ideal treatment of all diseases is prevention rather than symptomatic treatment. In the sensitized allergic individual, the most important step towards prevention is obviously to avoid any contact with the offending allergen. While this may be feasible for certain uncommon allergens or for allergens occurring only in defined places, it is not practically feasible for ubiquitous allergens.

A better way to prevent allergy would be to prevent sensitization. Since sensitization usually occurs during childhood and since most allergic diseases begin in this age group, preventive measures should be instituted early in life. However, there are close similarities between the sensitization process in childhood and sensitization occurring later in life when an adult is exposed to a new allergen in a new job. Thus, it is reasonable to assume that many of the preventive measures recommended in childhood would also be relevant in the prevention of occupational allergy. In this review the current possibilities for allergy prevention, particularly in childhood, will be discussed. The chapter will be limited to avoidance of sensitization rather than prevention of symptoms in the allergic individual. The various methods to predict allergy will also be briefly discussed. The effect of the environment on the development of allergy will be analysed, focusing on the interplay between individual susceptibility and environmental influence. Finally, the practical consequences of this as related to prevention of allergy will be discussed.

Prediction

Allergic disease is strongly influenced by genetic factors but no single dominant or recessive gene predisposed to allergy has yet been identified. As reviewed by Marsh *et al.*, atopic diseases are multifactorial, affected by a polygenic inheritance as well as by various environmental influences [3]. There appears to exist a primary abnormality in the regulation of the IgE antibody formation. In principle, there are a number of tests available for identification of individuals at risk for developing allergy (Table 7.1). From a practical point of view, however, there are only a few methods that are simple enough to allow a widespread use for screening purposes. They include family history and determination of total IgE and IgE against defined allergens.

Table 7.1 Identification of individuals at risk of allergic disease

	Age group		
	At birth	Infants	Children and adults
Family history	+	+	+
IgE antibody determination			
Total serum IgE	+	+	+
Specific IgE antibodies in serum or skin	–	+	+
Lymphocyte tests	?	+	?
Leucocyte phosphodiesterase levels	+	?	?
Cell membrane lipids	?	?	?
Blood eosinophils	?	+	?

Family history

A good family history has a high sensitivity for identification of allergy-risk individuals. The specificity is, however, low, since a positive family history is commonly encountered and many individuals with allergic relatives do not develop allergy themselves. To improve the accuracy, family history should be limited to immediate relatives, i.e. parents, brothers and sisters. The predictive capacity of a positive family history may be further improved by registration not only of the number of relatives with allergic disease, but also which organs are affected in them. Thus, Kjellman [4] has shown that the risk of allergy in a child increases with the number of affected relatives and that the risk was up to 75% if both parents were allergic with symptoms from the same organ. In a continuing study of 1700 newborn babies [5] in which the results so far have been compiled for the first 10 years of life, a sensitivity of 49% and a specificity of 73% for family history as predictor of allergy has been found [Croner and Kjellman, unpublished observations]. Kjellman [6] has recommended the use of a 'family allergy score' as a simple tool to identify risk babies; obvious allergy in a family member giving a score of 2 for each affected member and possible disease giving a score of 1. A total score of 4 is considered to indicate a high risk of allergy. In children with a score of 2 or 3, further predictive tools should be employed, while individuals with lower scores would be regarded as low-risk infants.

So far, the use of various genetic markers has not improved the predictive capacity of family history, i.e. blood groups and α_1-antitrypsin phenotypes.

Determination of IgE

Formation of IgE is a hallmark of atopic disease. Excessive IgE formation is not synonymous with atopic disease but may appear in response to parasitic infection, as part of immune deficiency, and in several uncommon disorders. Further, all children with allergic disease, even some with severe illness, may not exhibit elevated serum IgE levels or demonstrable IgE antibodies against any known allergen.

The IgE formation has certain characteristic features which are summarized in Table 7.1. Most of the knowledge regarding this has been gathered in animal studies, but the results obtained so far appear to be relevant for humans [7]. As originally shown in rats, the IgE formation is part of the early normal immune response. Similarly, an early appearance of IgE antibodies against respiratory syncytial virus has been demonstrated in infected infants [8]. In infectious mononucleosis there is an early pronounced increase in IgE levels which is apparent after about 10 days. This increase is then followed after about 1 month by a period during which the IgE levels are lower than at the onset of disease, indicating an active suppression of IgE formation. Similarly, low concentration of IgE against egg white and cow's milk are found in almost one-third of healthy infants at a time when these foods are introduced into the diet [9]. But the levels are higher in infants who later develop atopic disease, indicating that there is a primary defect, expressing itself as lack of suppression of excessive IgE formation in atopic children. Further, there are now several studies demonstrating that many infants, who later in life will develop atopic disease, already have elevated serum IgE levels at birth [5]. These antibodies are not of maternal origin, since antibodies of the IgE isotype do not cross over the placental barrier. As it is quite rare to demonstrate IgE antibodies in cord blood against a particular antigen, the elevated levels indicate a spontaneous antibody formation which is not efficiently suppressed. All these clinical observations lend support to data obtained in animal studies, that lack of suppression of IgE synthesis may be a characteristic trait in atopy.

It is well known that extremely low doses of antigen, in the picogram range, are particularly effective in inducing IgE formation in rats. This is apparently also true for humans. Thus, it is relatively common for infants to become sensitized to food antigens like cow's milk and hen's egg which are present in extremely low amounts in the mother's breast milk [9, 10].

Animal experiments have indicated that the early IgE antibody response may be primarily local. In rats, it has been shown that subcutaneous (s.c.) immunization in the neck region or via aerosol leads to detectable IgE formation after about 2 weeks in the draining axillary and mediastinal lymph nodes, while s.c. immunization in the tail root leads to similar early synthesis in the inguinal lymph nodes [11]. Only later does IgE appear in the systemic circulation. Less data are available in humans to lend support for the local synthesis of IgE but a dichotomy between systemic and local immune response may at least in part explain the fact that IgE antibodies in the circulation, in the absence of a present or previous clinical history of allergy, are a strong indicator of subsequent appearance of atopic disease [9, 12].

Other predictive markers

Several observations indicate that the increased IgE levels in atopic individuals could be a consequence of a primary defect in the regulation of the IgE formation. There are at least two prospective studies showing that abnormal T-cell counts and function are primary defects in many atopic children and therefore in principle can be utilized as predictive tests. These include reduced numbers of lymphocyte-forming rosettes with sheep erythrocytes *in vitro* [13], lower levels of T-cells positive for the surface antigen marker OKT8 and less capacity to concanavalin A-induced suppression *in vitro* [14]. These studies support the hypothesis that defective induction of T-cell mediated suppression of IgE formation is a primary defect in atopy [reviewed in 15]. Such T-cell studies are, however, too complicated from a technical point of view to be used for screening purposes.

Elevated levels of phosphodiesterase (PDE) in blood leucocytes have been reported in patients with atopic dermatitis (and in newborn infants of atopic parents [16]). This enzyme is involved in degradation of cAMP, and increased activity could result in hyperreactivity of the cell but also possibly in increased IgE levels. The relationship with atopic disease has not yet been demonstrated in a prospective study but we have preliminary data indicating that infants with high levels of leucocyte PDE in the cord blood cells develop atopic disease during infancy. If confirmed, this may give new insight into the pathogenetic mechanisms behind the development of allergy.

It has also been reported that the lipid composition of the cell membrane is altered in atopic individuals. Thus, an increased ratio of linoleic acid over α-linolenic acid and decreased levels of arachidonic acid have been found in white blood cells [17, 18].

In addition, there are some other unspecific findings that may precede clinical allergy. They include eosinophilia and infantile colics but these findings are too unspecific to be of practical use in allergy prediction.

Exposure to allergens

An individual must be exposed to the allergen in order to be sensitized and to develop allergy against it. This sensitization is influenced by the amount and time of exposure. However, the age at which the exposure takes place also seems to play a role.

There appears to be a period in early life during which the individual is more easily sensitized than later in life. As a consequence the season of birth seems to play a role. A significantly increased risk of subsequent development of allergy to birch and grass pollen in children born in the spring has been reported from Finland [19]. Later it was shown that the risk of subsequent pollen allergy was affected, not only by the month of birth, but also by the intensity of the first pollen season, underlining the importance of exposure to allergens as a risk factor to sensitization [20]. Similar results have been reported for grass allergy in the UK [21] and for ragweed allergy in individuals born approximately 3 months before the ragweed pollen season, culminating in those individuals that were born during the peak pollen season [22]. Recently it was shown from our group [23] that the relationship between season of birth and allergy development was limited to children with a congenital propensity for allergy as indicated by elevated IgE levels in the cord blood. Thus, the environmental triggers of allergy were particularly important in high-risk babies.

Similar to the apparent effects of early exposure to pollen, early contacts with animal epithelia and house dust appear to influence the incidence of allergy. It has been reported that among a population of adolescents with various allergies, those who were exposed to cats in the homes for at least 3 months during the first year of life more often had positive skin reactions to cat epithelium than those who were not exposed to cats early in life [20]. There was no significant difference in the proportion of positive skin tests to cats between those who had and those who did not have a cat in the house at the time of testing. British studies have shown that birth during the period of May to October is associated with a significantly increased risk of asthma to house-dust mite [21]. These observations in pollen-, dander- and mite-sensitive individuals all indicate that early exposure to an allergen increases the risk of allergy. Intensive exposure to an allergen plays a role not only in children but also in adults. This may have practical consequences for occupational allergy and it has been shown that reduction of exposure, e.g. by introduction of effective air cleaning devices, reduces the risk of occupational allergy.

Environmental triggers of atopic disease

In addition to allergenic exposure there are a number of environmental factors that facilitate sensitization to almost any allergen that the individual is exposed to.

Infections

There are now several studies suggesting an association between upper respiratory infections and appearance of allergic disease. There are several possible explanations for this apparent relation. Firstly, infections may trigger wheezing in a hyperreactive individual who is already sensitized. A low-grade allergic allergy may be clinically manifested when the extra burden of an infection is added to the respiratory tract. Secondly, a local inflammation caused by an infection may facilitate sensitization to an allergen to which the individual is simultaneously exposed. Thirdly, an infection may temporarily alter the normal immune regulation of the IgE antibody response, inducing a temporary dampened control, thus allowing an 'allergy break-through' [24]. There are no convincing epidemiological data to indicate that allergic individuals have an increased propensity for infections although there is little doubt that infections often trigger even severe symptoms in asthmatic individuals. There are several indications that infections may facilitate sensitization to environmental allergens. In a prospective study of children born into allergic families it was found that sensitization was often associated with upper respiratory infections occurring 2–6 weeks earlier [25]. In another series of studies, it has been shown that respiratory syncytial virus (RSV) infections may be associated with appearance of allergy [8]. Infants with RSV who develop wheezing manifest higher IgE responses to RSV than infected children who do not wheeze. The wheezing infants also have lower numbers of suppressor/toxic OKT8-positive cells. It is not, however, known whether this finding is a consequence of the infection or whether an infant with this imbalance in the T-cell population is more prone to develop wheezing in response to the infection.

Another possible effect of infections has been suggested in a study of antibodies to Epstein–Barr virus (EBV) in children in which it was found that the levels of IgE against EBV were significantly higher in children with atopic disease than in healthy controls [26]. This could possibly indicate that an infection with a polyclonal B-cell activator like EBV can result in hyper-production of IgE and development of allergy at least in individuals who are genetically predisposed towards atopic disease.

It is well known from animal studies that pertussis bacteria are adjuvants for induction of IgE formation against various antigens [27]. There are also indications that this can be true in humans, suggesting the possibility that whooping cough may trigger allergy. We have also very recently found that IgE antibodies against pertussis toxin are commonly encountered in children and adults during whooping cough and in the secondary response to pertussis immunizations [unpublished observations]. The possible adjuvant effect of pertussis and the fact that immunization often results in IgE responses, could possibly cast some doubt on the routine immunization of infants against pertussis, particularly if done at a time when the baby is heavily exposed to environmental allergens, e.g. during the pollen season.

Tobacco smoke

Tobacco smoke is the major indoor air pollutant. In fact, the concentration of smoke in a room when somebody is smoking far exceeds that of any emission by a polluting industry. There is now a confirmed clear relation between tobacco smoke and allergy. This is true for smoking adults in whom increased IgE levels and an increased prevalence of positive skin tests towards occupational allergens have been shown in multiple studies. It has also been shown that smokers more easily develop asthma triggered by occupational allergens than non-smokers who are exposed to the allergens to a similar degree [28].

The effect of tobacco smoke is, however, not limited to active smoking. In children of parents who smoke at home, a significantly earlier onset of allergy has been reported and wheezy bronchitis is about five times higher than in children of non-smoking parents [29]. The effect of tobacco smoke on sensitization to allergens may be explained by a local effect on the airways or by a direct effect on the immune system. The former notion is supported by our finding that smoking rats exposed to antigen in aerosol have higher IgE responses than subcutaneously immunized animals and non-smoking aerosol-immunized controls [30].

Air pollution

Other air pollution such as ozone, SO_2 and NO_2 may all increase serum IgE levels, at least in experimental animals [31]. Data are less clear cut in humans, but in a recent epidemiological survey of 5300 children in Sweden, we found that bronchial hyperreactivity and pollen allergy were both more common in children living near a moderately air polluting paper factory than among children living in a forested unindustrialized area about 40 km away from the factory [32]. If the parents smoked at home, then the prevalence of bronchial hyperreactivity and of pollen allergy were further increased showing a synergistic effect between the two pollutants.

Housing

Modern housing may be part of the explanation why allergy seems to be increasing in industrialized countries. Several studies have demonstrated the importance of exposure to house-dust mites in children and adults. The epidemiological studies by Turner and coworkers in Papua New Guinea clearly demonstrate the importance of environmental factors for the development of asthma [33]. Asthma was previously unknown in the population. Over the past 15 years they have been increasingly exposed to Western lifestyle. As a consequence, blankets are increasingly used and new food items, particularly various proteins, have been introduced into the diet. The blankets contain very high numbers of house-dust mites. The prevalence of asthma has increased among adults from 0 to more than 7%, but so far not among the children. Reasons for the increase are unknown but the study strongly indicates that environmental factors can profoundly influence the development of allergic disease.

Modern, well-insulated buildings with poor ventilation may also represent a risk factor for allergic sensitization. In temperate climates, the energy crises in 1973 resulted in improved building standards with regard to insulation. As a consequence of this, various forms of 'sick buildings' characterized by damage due to dampness and indoor mould growth have become increasingly common. In the previously cited epidemiological survey, we found that homes with damage due to dampness were associated with both a higher incidence of atopic disease and/or bronchial hyperreactivity in the children [32] (Table 7.2). In children living in houses with damage by dampness whose parents smoked at home there was a marked increase in allergic asthma and bronchial hyperreactivity as compared with children exposed to only one of these factors. The effect of the living conditions were most marked for children having a family history of asthma, supporting the notion that the environmental influences only play an important role in individuals with a genetic susceptibility to allergic disease.

Psychological factors

Animal studies and clinical observations in recent years have indicated that stress alters the immune response [34, 35]. It is also clinically well established that psychological factors affect the severity of asthmatic symptoms and allergic reactions. Very recently it has been shown from our group that families with a severely affected asthmatic child more often demonstrate a rigid interaction pattern than families of diabetic children or healthy children [36]. Family therapy to such families was shown to improve the severity of the asthma. It is not known whether the particular family interaction pattern in the families of asthmatic children is a consequence of the disease or represents a primary risk factor, increasing the risk of asthma in an atopic individual. The former possibility is contradicted by the fact that families in which a child had diabetes which is also a severe, chronic, potentially life-threatening disease, did not differ from control families with healthy children. Further research is clearly indicated in this area.

Pre- and perinatal medication and stress

Drugs administered to women during pregnancy may have some immune regulatory activities. Progesterone has been shown to significantly increase the cord blood IgE levels but does not seem to increase the onset of allergic disorders later in infancy and childhood [37]. In a double-blind placebo control study of the β-adrenergic receptor-blocking agent, metoprolol, during the treatment of moderate toxicosis of pregnancy it was found that children exposed *in utero* to the drug more often had elevated IgE levels in the cord blood and/or developed clinical allergy during the first 4 years of life than the children of placebo-treated control mothers [38].

Stress during the neonatal period may also possibly increase the risk of development of allergy later in life as indicated by a retrospective study [39]. It has also been reported that general anaesthesia in infants may be associated with later development of respiratory tract allergy [40]. Since both these studies were retrospective and did not control for a number of possible compounding factors, they should be confirmed by other studies before any definite conclusions can be drawn.

Sensitization via the mother

It has also been suggested that intra-uterine sensitization or priming of allergy can occur. The possible role of maternal diet during pregnancy for the development of allergy was recently evaluated in two Swedish studies [41, 42]. In the two prospective randomized studies of pregnancies in families with a history of allergy, four diets of the pregnant women during the last trimester were compared, i.e. (1) total avoidance of cow's milk protein and eggs; (2) avoidance of visible amounts of these foods; (3) a normal unrestricted diet; and (4) an unrestricted diet with an ensured extra ingestion of at least 1 litre of milk and one egg daily. The babies were then followed for 18 months and a blinded evaluation was done. There was no difference between the groups with regard to allergic disease or sensitization as determined by skin-prick tests. The IgE levels in serum were also similar at birth and at 6–18 months of age. These findings in the two studies, comprising more than 400 children, strongly indicate that maternal diet during pregnancy does not significantly contribute to the development of allergy.

In contrast to the situation *in utero*, breast-fed babies are influenced by their mothers' diet. It is thus now well known that nursing infants may be sensitized by minute amounts of foreign proteins that are present in the maternal milk [9, 10]. As a consequence of this observation we have tried to avoid sensitization in early infancy by restricting the diet of breast-feeding mothers by total avoidance of cow's milk, egg and fish during the first 3 months of lactation. The study in 115 babies showed that atopic dermatitis was significantly lower in the babies of mothers adhering to the diet during the first 6 months of life [unpublished observations]. At 9 and 18 months of age, however, the differences between the groups were not significant in this respect although the severity of disease was significantly less in the babies of mothers on the diet at 9 months.

Prevention (Table 7.2)

Even if genetic factors play a major role in the development of allergic disease, the atopic diseases definitely do not belong to the groups of diseases where genetic counselling to avoid child birth is warranted. From the previous discussion it seems reasonable, however, to strongly encourage a woman to stop smoking as early as possible during pregnancy. It is also reasonable to consider the possibility that a medication during pregnancy could affect the risk of allergy in the baby. On the other hand, this notion should not prevent adequate treatment of diseases during pregnancy. Immunotherapy to a pregnant woman should be continued since it may even be beneficial to the baby [43], but obviously the treatment should be given under careful avoidance of severe anaphylactic reactions.

Although dietary manipulation of the maternal diet during pregnancy has been suggested, this practice should be discouraged. Studies indicating an effect of such manipulation have not been limited to pregnancy but also included the first months of lactation. Two prospective studies that were limited to studying the effect of dietary manipulation during pregnancy both failed to reveal any effect. There are also definite risks involved with dietary manipulation. Thus, such manipulation resulted in lower maternal weight gain and lower birth weight of the babies, particularly if the mothers also smoked [41, 42]. This was found despite the fact that all the mothers in the

Table 7.2 Preventive measures in individuals at risk of allergic disease

At-risk individuals
Avoid exposure to tobacco smoke
Reduce exposure to strong allergens:
No household pets
Reduce house-dust mite numbers
Effective evacuation of allergens at work
Ensure good ventilation of house
Additional measures for infants
Exclusive breast feeding for at least 3 months
Total avoidance of other foods
Possibly controlled reduction of allergens in diet of breast-feeding mother

study had repeated individual and group counselling on a proper diet and avoidance of milk and egg was carefully compensated for.

In early infancy, all forms of artificial foods should be avoided for the first months and breast feeding should be strongly encouraged. Human milk is the normal optimal food for babies and it provides the infant with immune defence and immune regulatory components [44].

Although season of birth has been shown to affect the risk of allergy to pollen in temperate zones, this is not sufficient for any recommendations. Although it is undoubtedly true for pollens it is also true that birth during the autumn or winter is associated with increased risk of allergy to house-dust mites and furry pets. It is, however, reasonable to discourage families with a history of allergy to keep any pets at home and the exposure to house-dust mites should be reduced if possible, e.g. by avoiding wall-to-wall carpets and by reducing humidity indoors. Regular weekly vacuum cleaning of the house is reasonable but meticulous cleaning beyond that does not reduce the mite counts to an appreciable degree.

Avoidance of tobacco smoking is probably the single most important allergy preventive measure. This is true in adults to avoid occupational allergy as well as in infants and children where the parents can reduce allergy in their offspring by restraining from smoking at home.

Similarly, reduction of industrial emission may in the long term reduce the risk of sensitization. The effect of outdoor air pollution is, however, probably limited since it has not been clearly shown that the incidence of allergic disease is lower in unpolluted rural areas than in industrialized urban zones.

Concluding remarks

Several factors play a role in the apparent increase in the incidence of atopic diseases in industrialized countries, i.e. the diseases are a consequence of an interplay between a genetic propensity for allergy, time and amount of exposure to sensitizing allergens, and exposure to a number of non-specific triggers that facilitate sensitization. However, even if all known and hypothetical environmental influences are taken into account, this would not fully explain why allergic diseases are becoming more common. Since the genetic set-up of humans has not changed appreciably over the last century, unknown environmental triggers must be looked for, including lifestyle patterns, housing conditions, food and environmental pollution. There seems to be a period in early infancy during which the individual is particularly vulnerable to sensitization. It is important to identify individuals with an increased risk of allergic disease since they seem to be the ones that suffer from the various environmental influences. Such individuals may act as 'biological indicators' of risk factors in the environment. Although much remains to be learned about the development of allergy, several preventive measures can already be taken now, including reduction of allergen exposure and avoidance of smoking and exposure to tobacco smoke.

References

1 Nelson HS. The atopic diseases. *Ann. Allergy* 1985; **55**:441–447.
2 Taylor B, Wadsworth J *et al.* Changes in the reported prevalence of childhood eczema since the 1939–45 war. *Lancet* 1984; **i**:1255.
3 Marsh DH, Hsu SH *et al.* Genetics of human immune responses to allergen. *J. Allergy Clin. Immunol.* 1980; **65**:322–332.
4 Kjellman N-IM . Atopic disease in seven-year old children. Incidence in relation to family history. *Acta Paediatr. Scand.* 1977; **66**:456–471.
5 Kjellman N-IM, Croner S. Cord blood IgE determination for allergy prediction. A follow-up to seven years of age in 1651 children. *Ann. Allergy* 1984; **53**:167–171.
6 Kjellman N-IM. Serum IgE and the predictive value of IgE determination. In *Advances in Pediatric Allergy* (Ed. Businco L), Elsevier, Amsterdam 1983, pp. 69–82.
7 Björkstén B, Ahlstedt S. The relevance of animal models for studies of immune regulation of atopic allergy. *Allergy* 1984; **39**:317–327.
8 Welliver RC, Ogra L. The role of IgE in pathogenesis of mucosal viral infections. *Ann. NY Acad. Sci.* 1983; 321–332.
9 Hattevig G, Kjellman B, Johansson SGO, Björkstén B. Clinical symptoms and IgE responses to common food proteins in atopic and healthy children. *Clin. Allergy* 1984; **14**:551–559.
10 Gerrard JW. Allergy in breast fed babies to ingredients in breast milk. *Ann. Allergy* 1979; **42**:69–71.
11 Ahlstedt S, Olaisson E, Thellin J, Björkstén B. Appearance of mast cells in bone marrow, peripheral blood and spleen of immunized rats. *Int. Arch. Allergy Appl. Immunol.* 1986; **80**:122–126.
12 Hagy JW, Settipane GA. Prognosis of positive allergy skin tests in an asymptomatic population. A three year follow-up of college students. *J. Allergy Clin. Immunol.* 1971; **48**:200–211.
13 Björkstén B, Juto P. Immunoglobulin E and T cells in infants. In *Proceedings of the XI International Congress of Allergy and Clinical Immunology* (Eds Kern JW, Ganderson MA), Macmillan Press, London 1983, pp. 144–148.
14 Chandra RK, Baker M. Numerical and functional deficiency of suppressor T cells precedes development of atopic eczema. *Lancet* 1983; **i**:1393–1394.

15 Björkstén B. Atopic allergy in relation to cell-mediated immunity. *Clin. Rev. Allergy* 1984; **2**:95–106.
16 Grewe SR, Chan SC, Hanifin JM. Elevated leukocyte cyclic AMP-phosphodiesterase in atopic disease: a possible mechanism for cyclic AMP-agonist hyporesponsiveness. *J. Allergy Clin. Immunol.* 1982; **70**:452–457.
17 Manku MS, Horrobin DF, Morse NL, Wright S, Burton JL. Essential fatty acids in the plasma phospholipids of patients with atopic eczema. *Br. J. Dermatol.* 1984; **110**:643–648.
18 Rocklin R, Thistle L, Gallant L, Manku MS, Horrobin D. Altered arachidonic acid content in polymorphonuclear and mononuclear cells from patients with allergic rhinitis and/or asthma. *Lipids* 1986; **21**:17–20.
19 Björkstén F, Suoniemi I, Koski V. Neonatal birch-pollen contact and subsequent allergy to birch pollen. *Clin. Allergy* 1980; **10**:581–591.
20 Björkstén F, Suoniemi I. Early allergen contacts, adjuvant factors and subsequent allergy. In: *Proceedings of the XI Congress of Allergy and Clinical Immunology* (Eds Kern JW, Ganderson MA), Macmillan Press, London 1983, pp. 144–148.
21 Morrison Smith J, Springet VH. Atopic disease and month of birth. *Clin. Allergy* 1979; **9**:153–157.
22 Settipane RJ, Hagy GW. Effect of atmospheric pollen on the newborn. *Rhode Island Med. J.* 1979; **62**:477–482.
23 Croner S, Kjellman N-IM. Predictors of atopic disease: cord blood IgE and month of birth. *Allergy* 1986; **41**:68–70.
24 Katz DH, Bargatze RF, Bogowitz CA, Katz LR. Regulation of IgE antibody production by serum molecules. V. Evidence that coincidental sensitization and imbalance in the normal damping mechanism results in 'allergic breakthrough'. *J. Immunol.* 1979; **122**:2191–2197.
25 Frick OL, German DF *et al.* Development of allergy in children. *J. Allergy Clin. Immunol.* 1979; **63**:228–241.
26 Strannegård I-L, Strannegård O. Epstein–Barr virus antibody in children with atopic disease. *Int. Arch. Allergy Appl. Immunol.* 1981; **64**:314–319.
27 Pauwels R, Van der Straeten M, Platteau B, Bazin H. The non-specific enhancement of allergy. I. *In vitro* effects of *Bordetella pertussis* vaccine on IgE synthesis. *Allergy* 1983; **38**:239–246.
28 Zetterström O, Osterman K, Machado L, Johansson SGO. Another smoking hazard: raised serum IgE concentration and increased risk of occupational allergy. *Br. Med. J.* 1982; **70**:199–204.
29 Liard R, Perdrizet S, Reiner P. Wheezy bronchitis in infants and patients 'smoking' habits. *Lancet* 1982; **i**:334–335.
30 Zetterström O, Nordvall SL, Björkstén B, Ahlstedt S, Stelander M. Increased IgE antibody responses in rats exposed to tobacco smoke. *J. Allergy Clin. Immunol.* 1985; **75**:594–598.
31 Gershwin J, Osebold JE, Zee YC. Immunoglobulin E-containing cells in mouse lung following allergen inhalation and ozone exposure. *Int. Arch. Allergy Appl. Immunol.* 1985; **65**:266–277.
32 Andrae S, Axelson O, Björkstén B, Fredriksson M, Kjellman N-IM. Symptoms of bronchial hyperreactivity and asthma in relation to environmental factors. *Arch. Dis. Child.* 1988; **87**:59–62.
33 Turner KJ, Dowse GK, Stewart GA, Alpers MP, Woolcock AJ. Prevalence of asthma in the South Fore people in the Okapa district of Papua New Guinea. *Int. Arch. Allergy Appl. Immunol.* 1985; **10**:581–591.
34 Editorial. Depression, stress and immunity. *Lancet* 1987; **i**:1467–1468.
35 Husband AJ, King MG, Brown R. Behaviourally conditioned modification of T cell subset ratios in rats. *Immunol. Lett.* 1987; **14**:91–94.
36 Gustafsson PA, Kjellman N-IM, Ludvigsson J, Cederblad M. Asthma and family interaction. *Arch. Dis. Child.* 1987; **62**:258–263.
37 Michel J, Bousquet J, Coulomb Y, Robinet-Lévy M. Prediction of the 'high-allergic-risk' newborn. In *Diagnosis and Treatment of IgE-mediated Diseases* (Ed. Johansson SGO), Excerpta Medica, Amsterdam 1981, pp. 35–47.

38 Björkstén B, Finnström O, Wichman K. Intrauterine exposure to beta-adrenergic receptor blocking agent metoprolol and allergy. *Int. Arch. Allergy Appl. Immunol.* 1988; **87**:59–62.
39 Salk L, Grellong BA. Perinatal complications in the history of asthmatic children. *Am. J. Dis. Child.* 1974; **127**:30–33.
40 Johnstone DE, Roghmann KJ, Pless IB. Factors associated with the development of asthma and hay fever in children: the possible risks of hospitalization, surgery, and anaesthesia. *Pediatrics* 1975; **77**:158–162.
41 Fälth-Magnusson K, Kjellman N-IM. Development of atopic disease in babies whose mothers were on exclusion diet during pregnancy—a randomized study. *J. Allergy Clin. Immunol.* 1987; **80**:968–975.
42 Antibodies IgG, IgA and IgM to food antigens during the first 18 months of life in relation to feeding and development of atopic disease. *J. Allergy Clin. Immunol.* 1988; **81**:743–749.
43 Glovsky MM, Rejzek E, Ghekiere L. Can tolerance be induced *in utero* in children by immunotherapy of atopic pregnant mothers? *J. Allergy Clin. Immunol.* 1982; **69**:100.
44 Chandra RK. Immunological aspects of human milk. *Nutr. Rev.* 1978; **36**:265–272.

Discussion session

DEWDNEY: do we know anything about chemicals which might induce IgE production? You mentioned factory smoke and other environmental factors.

BJÖRKSTÉN: TDI hypersensitivity has been shown to be IgE-mediated and it is almost exclusively smokers who become sensitized. Also, nurses working with the powdered laxative, ispaghula, may be sensitized to the compound, and again it is almost exclusively smoking nurses who are sensitive. We have shown that we can sensitize smoking rats to an inhaled antigen much more easily than non-smoking rats.

DEWDNEY: do you know what it is in smoking that has this effect?

BJÖRKSTÉN: perhaps the inflammation induced by the smoke opens up tight junctions thus allowing the antigen to pass through more easily.

8

Suppression of the IgE response with human proteins developed by biotechnology

D.H. KATZ

Introduction

Because of the depth of studics performed on it during the past 10 years, more is known about the diversity of components of the IgE system than about any other known antibody isotype. Thus, in both rodent and human species we know much about:

1 The physicochemical properties of the ε heavy chain, and the corresponding cDNAs for human and rodent ε chains have been cloned.

2 The fine physiology of regulatory mechanisms controlling IgE synthesis, including the existence of multiple IgE-selective regulatory proteins/peptides (some of which have been cloned).

3 The consequences of target cell triggering by IgE which initiates a cascade of inflammatory mediator activities in the surrounding tissues.

Quite significantly, we have come to realize that IgE—perhaps uniquely—has evolved with *two* distinct corresponding ε-specific Fc receptors (FcεR): the originally discovered, high-affinity FcεR present on mast cells and basophils—now termed FcεRI—and the more recently discovered, moderate-affinity FcεR which is displayed (in surprising numbers in certain circumstances) on lymphocytes—now termed FcεRII. Understanding all of the mechanisms involved in regulation of the IgE system is crucial for us to succeed in controlling many immediate hypersensitivity diseases.

In recent years, we have focused mainly on the IgE system of both humans and rodents because we feel that it provides a particularly illuminating window through which to view general mechanisms of immune system regulation. Thus, studies in animal systems from our own laboratory and from those of others have indicated that the IgE system is regulated by a complex cellular network and IgE-selective soluble factors which exert either suppressive or enhancing effects on IgE synthesis [1–5]. In our laboratory, the original factors were discovered in the circulating body fluids of rodents and were shown to exert IgE-selective biological effects following passive transfer to recipient animals. The first two were denoted suppressive factor of allergy (SFA) and enhancing factor of allergy (EFA) because of their respective suppressive and enhancing effects on IgE synthesis [6, 7]. Subsequently, we found comparable factors produced by human lymphocytes following appropriate stimulation *in vitro* [8–10].

FcεR$^+$ lymphoid cells are important regulatory components in the IgE system

Another important discovery in the IgE system has been that certain lymphocytes express Fc receptors for IgE [11–13]. The discovery of FcεR$^+$ lymphocytes took on added importance when it was demonstrated that: (1) the degree of FcεR expression, both in quantities of receptors per cell and the frequencies of cells expressing FcεR, is inducible (in both animal and human systems) by exposure to IgE itself [14–18]; and (2) IgE-induced FcεR expression can be selectively modulated by factors such as SFA and EFA in both animal [15, 16, 19, 20] and human [19–21] experimental systems.

Various regulatory roles of FcR$^+$ lymphocytes in immune responses have been suggested [22], and significant increases of FcR$^+$ lymphocytes have been found in some disease states which involve elevated production of the particular Ig isotype for which such FcR$^+$ cells have affinity [20, 23, 24]. For example, allergic humans often display concomitantly increased levels of both IgE and circulating FcεR$^+$ lymphoid cells [25]. Moreover, exposure of lymphocytes to sufficient amounts of a particular Ig not only induces the expression of FcR of corresponding isotype specificity, but also the production by these FcR$^+$ cells of factors that regulate Ig synthesis in an isotype-selective manner [17, 26–29].

In the IgE system, FcεR$^+$ lymphocytes and isotype-selective regulatory factors are interrelated participants in the regulatory events controlling IgE synthesis [16, 30, 31]. Among the isotype-selective soluble factors, some have been shown to be devoid of binding affinity for IgE (such as SFA and EFA), while others have been shown to be IgE-binding factors (IgE-BF), such as those described by Ishizaka and colleagues [26, 28, 32–35].

FcεR$^+$ lymphocytes produce regulatory molecules

The striking inducibility of FcεR expression by rodent [15, 17] or human [19, 20, 36] lymphoid cells following short-term *in vitro* exposure to IgE, the selective inhibition of FcεR expression by factors such as SFA and EFA [15, 16], and the IgE-induced production of both potentiating and suppressive species of IgE-BF [17] emphasized the need for a detailed analysis of the events involved in IgE-induced FcεR expression by lymphocytes and the role of soluble mediators in such processes. Such studies revealed that exposure of lymphoid cells to IgE *in vitro* induces both FcεR expression and the secretion of a family of soluble factors that, independently of IgE, can regulate FcεR expression by other lymphoid cells. We have termed such soluble factors IgE-induced regulants, or EIR [1–5, 14, 19, 20]. In the mouse system, at least six distinct soluble mediators have been identified as participants in this complex, and balanced, negative and positive control process for regulating production of IgE. Table 8.1 summarizes our current nomenclature and state of knowledge about these six factors.

Table 8.1 Summary of biologically active factors in the IgE cascade

Factor	Molecular weight (kD)	Biological effects on:	
		FcεR expression	IgE synthesis
Suppressive factor of allergy (SFA)	25	Inhibits	Suppresses
Enhancing factor of allergy (EFA)	25	Inhibits	Enhances
IgE-induced regulant/B-cells (EIR_B)	15, 30	Stimulates	Suppresses
IgE-induced regulant/T-cells (EIR_T)	45–60	Stimulates	Enhances
Suppressive effector molecule (SEM)*	15	Inhibits	Not active
Enhancing effector molecule (EEM)*	15	Inhibits	Not active

* SEM and EEM are IgE-binding factors.

Suppressive factor of allergy is a factor of around 25 kD which inhibits IgE-induced FcεR expression and, more importantly, suppresses IgE biosynthesis both *in vivo* in rodents and *in vitro* by human lymphoid cells. Enhancing factor of allergy is a 25 kD factor that also inhibits FcεR expression but, in contrast, exerts enhancing effects on IgE synthesis, both *in vivo* and *in vitro*. IgE-induced regulants derived from B-cells (EIR_B) consist of one species of 15 kD and a second of 30 kD which directly stimulate FcεR expression by B-cells but, like SFA, suppress IgE synthesis *in vivo*. IgE-induced regulant from T-cells (EIR_T) is a 45–60 kD factor that also directly stimulates FcεR expression but by T-cells rather than B-cells, and this factor enhances *in vivo* IgE synthesis. Suppressive effector molecule (SEM) and enhancing effector molecule (EEM) are both 15 kD in size and inhibit IgE-induced FcεR expression by B-cells, and both display IgE-binding properties. To date, however, we have failed to observe any biological activity of either of these latter factors on *in vivo* IgE biosynthesis.

Features of the IgE regulatory network

To briefly summarize what we know about how these molecules and cells interact; when lymphocytes are exposed to suitable levels of IgE, a subset of B-cells becomes activated as indicated by two events: (1) production of EIR_B; and (2) enhanced expression of FcεR. An example of this is shown in Fig. 8.1.

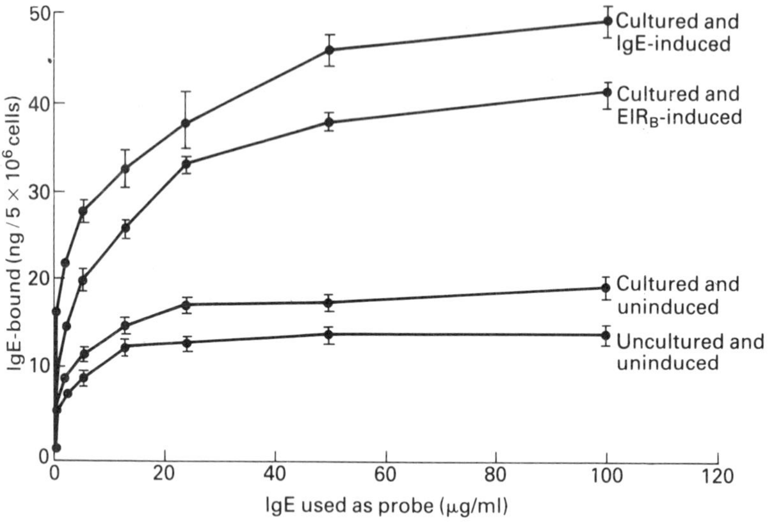

Fig. 8.1 Lymphocytes induced with IgE or monoclonal EIR_B exhibit enhanced expression of FcεR. BALB/c B-cells (anti-θ + C and G-10 passed) were cultured 18 h in either medium alone, medium containing 15 μg/ml monomeric IgE, or in medium containing 25% 99E9 B hybridoma-conditioned medium (i.e. EIR_B), or were uncultured. The cells were washed with saline and pH 4.0 buffer and resuspended in 0.1 ml MEM containing 5% FCS, 0.05% azide, 200 ng/ml ^{125}I-IgE (approximately 10^6 c.p.m.), and varied concentrations of cold IgE. After 5 h incubation at 4 °C, the cells were pelleted through oil and the amount of IgE bound determined as a function of tracer IgE present in the cellular fraction. Non-specific binding was determined in control tubes for each cellular population by pre-incubation with 300 μg/ml cold IgE prior to addition of ^{125}I-IgE; non-specific c.p.m. binding levels were 3755 ± 149, 3677 ± 394, 4004 ± 216 and 4058 ± 119 for uncultured/uninduced, cultured/uninduced, cultured/IgE-induced and cultured EIR_B-induced cellular populations respectively.

Using radiolabelled monoclonal IgE as a probe, one can measure levels of FcεR expression by mouse B-cells either taken directly from the animal and assayed as shown by the lower curve on Fig. 8.1, or following 18 h of culture either in medium alone or medium containing either 15 μg/ml of monoclonal IgE or EIR_B isolated from a B-cell hybridoma line producing this factor. What is clear is that exposure of B-cells to either IgE or EIR_B results in up-regulation of FcεR expression.

We have documented two functional activities for EIR_B [2, 3], one of which is the up-regulation of FcεR by B-cells, in the absence of IgE, as presented in Fig. 8.1. The second activity is EIR_B-induced activation of Ly-1$^+$ T-cells to synthesize and secrete SFA, the protein which selectively suppresses *in vivo* antibody responses when administered to intact animals. In addition to its *in vivo* effects on IgE synthesis, SFA also suppresses inducible FcεR expression by B-cells. Although the precise mechanism is still undefined, SFA appears to modulate B-cell FcεR expression by first acting on Ly-1$^+$ cells and inducing release of SEM which, in turn, inhibits EIR_B-induced B-cell FcεR expression. Suppression of *in vivo* IgE synthesis by SFA may or may not involve the participation of SEM [2–4].

A third molecule of central importance to this network is the T-cell derived EIR_T. EIR_T is produced by Ly-2$^+$ T-cells as yet another consequence of IgE- and EIR_B-induced enhancement of B-cell FcεR expression. The mechanism by which this occurs has yet to be defined, but we are certain that EIR_B does not directly stimulate T-cells to produce EIR_T. EIR_T *then* induces Ly-2$^+$ T-cells to express FcεR and to release EEM [5].

Besides the aforementioned regulatory effects on FcεR expression *in vitro*, EIR_B and EIR_T have been shown to selectively suppress and enhance, respectively, IgE biosynthesis in immunized mice. Interestingly, quite recently it has been shown that interleukin-4 (IL-4) up-regulates expression of FcεRII by B-cells *in vitro* [37] and enhances *in vivo* IgE responses of mice [38]. Interleukin-4 thus appears to behave like EIR_B in terms of effects on FcεRII expression (B-cells), but like EIR_T (and *not* EIR_B) in terms of enhancing IgE synthesis *in vivo*. There are several clear points of distinction between IL-4 and EIR_B and EIR_T. First, while IL-4 and EIR_B both up-regulate FcεR expression on B-cells, IL-4 enhances while EIR_B suppresses *in vivo* IgE synthesis. Second, while IL-4 and EIR_T both enhance IgE synthesis, EIR_T induces FcεR expression by T-cells, while IL-4 lacks this property. Finally, the three entities can be distinguished on the basis of size—IL-4 is approximately 20 kD, EIR_B is 15 kD and 30 kD and EIR_T is approximately 45–50 kD.

Application of biotechnology to regulation of the IgE system

We are presently at the exciting point of translating purely biological phenomena into molecular terms. It therefore becomes crucial to purify and characterize the molecules responsible for these biological events. Recently, we constructed a human T-cell hybridoma line (ICI-1D6) that produces large human SFA which selectively suppresses both *in vitro* IgE biosynthesis by cultured human mononuclear cells [19, 20] and *in vivo* IgE antibody responses of sensitized living rodents. This makes it possible to compare *in vivo* efficacy of human SFA with its effects on defined target-cell populations in the *in vitro* antibody secretion assay system. The fact that the very same human lymphocyte products exert isotype-selective biological activity *in vivo* in rodents as well as *in vitro* on human lymphocytes also argues for the physiological significance of such soluble factors and the cells from which they originate.

Constitutive production of human SFA by a T-cell hybridoma

Activated human peripheral T-cells were fused with a lymphoma line and the resulting hybridomata were screened for SFA activity. This process yielded a monoclonal line, denoted ICI-1D6, which displayed stable, constitutive production of human SFA.

Figure 8.2 illustrates the suppressive activity of ICI-1D6 hybridoma culture fluid on *in vivo* IgE synthesis. CAF_1 mice developed high primary IgE responses to dinitrophenol–keyhole limpet haemocyanin (DNP–KLH) (group I), which was suppressed by treatment with conventional mouse SFA (group II). As shown by groups III–VI, comparable suppression (approximately 50% of the untreated control response) was obtained with unfractionated, concentrated ICI-1D6 supernatant fluid at doses as low as 50 μl per injection (the inhibitory activity titrated out at lower doses of SFA and was undetectable below 20 μl; not shown). This suppressive effect was selective for IgE synthesis, since IgG responses were unaffected (not shown). Also, the activity was clearly due to biologically active SFA produced by the hybridoma cells, since treatment of mice with the same doses of fresh culture medium alone (groups VI–VIII) had no detectable effect on IgE synthesis.

Stepwise purification of native human SFA from hybridoma ICI-1D6

A four-step purification process was devised to purify the native protein to homogeneity. Figure 8.3 assembles five densitometer scans of silver-stained SDS-PAGE gels performed at each step of the purification process, beginning with the concentrated crude supernatant fluid of 30 l of ICI-1D6 (lane a). The initial ammonium sulphate precipitation step removed much of the higher molecular weight contaminants (cf. lanes a and b), and the gel permeation chromatography step (lane c) removed essentially all of the remaining serum albumin, transferrin and insulin. However, the complex protein composition of these fractions clearly indicated that further resolution was required to obtain a homogeneous preparation of SFA. After high-performance liquid chromatography (HPLC) in trifluoroacetic acid (TFA): acetonitrile solvent, the resulting material consisted of only two to three discernible polypeptides (lane d), and final resolution was achieved by HPLC in heptafluorobutyric acid (HFBA): acetonitrile (lane e). As shown in lane e, the SFA-containing fractions display mainly a single polypeptide band of around 25 kD.

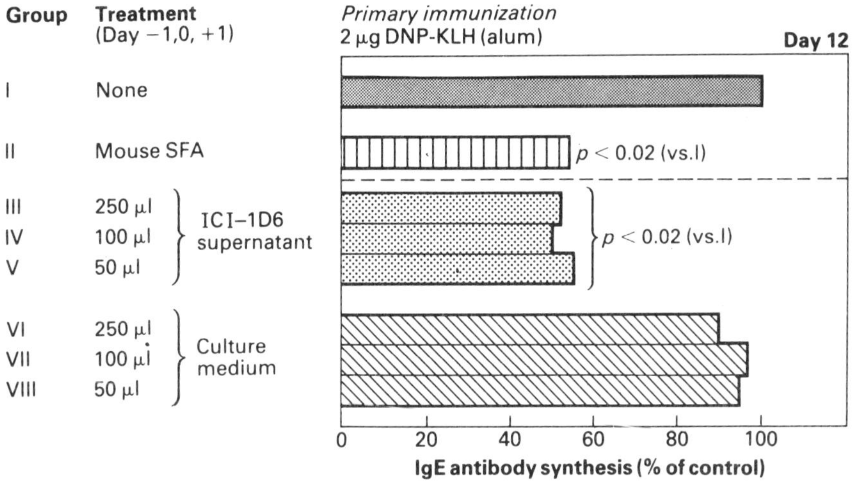

Fig. 8.2 Human SFA produced by T-cell hybridoma, ICI-1D6, suppresses *in vivo* IgE synthesis in high-responder CAF_1 mice. Groups of CAF_1 mice (six mice per group) were either not treated (but given saline as control injection) or treated with conventional ascites-derived mouse SFA, fresh culture medium or culture supernatant fluids from the SFA-secreting ICI-1D6 human hybridoma. In the treatment regimen, mice were injected twice a day (at 7 h intervals) on day −1, day 0 and day 1 with the materials indicated, and in the dose indicated per injection (mouse SFA injections were at 400 μl per injection). The injections were alternately made intravenously and intraperitoneally. On day 0, all mice were immunized with 2 μg of DNP–KLH in 2 mg of alum. The day 12 primary IgE and IgG responses were measured by PCFIA analysis of individual serum samples. The data graphically presented are expressed in terms of the per cent IgE response of the untreated control group (group I: 232 ng/ml). Statistically significant differences in levels of response in experimental groups versus control are indicated beside the corresponding bars. There were no statistically significant differences in IgG responses among the various groups (not shown).

In vivo activity of purified native human SFA in mice

SELECTIVE SUPPRESSION OF PRIMARY IgE ANTIBODY RESPONSES

The experiment presented in Fig. 8.4 demonstrates: (1) IgE-selectivity of the suppressive effects of human SFA; and (2) efficacy of its action even when administered as a single dose. Thus, as shown by the stippled bars, treatment of mice in groups II and III with purified human SFA resulted in 50–60% suppression of IgE synthesis relative to responses of controls in group I. As indicated by the hatched bars, the same treatment had no inhibitory effects on IgG synthesis. Interestingly, with the purified human protein we found that a single dose administered on the same day as antigen sensitization was equally effective (group III) as the routinely employed six-dose treatment regimen administered over 3 consecutive days (group II).

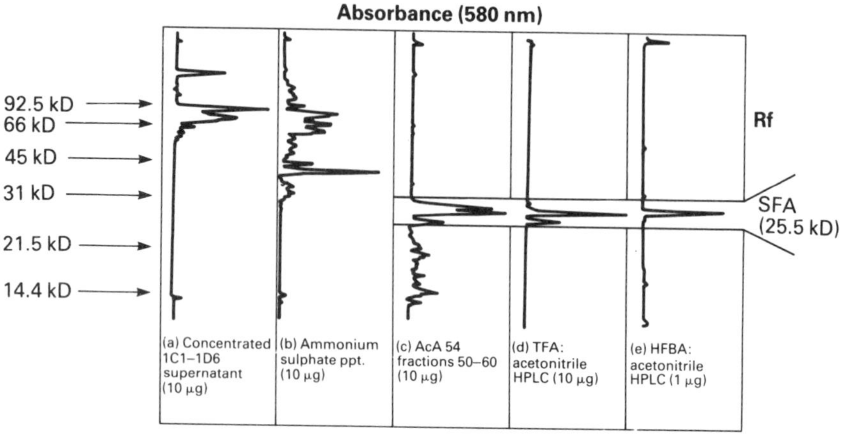

Fig. 8.3 Stepwise purification of human SFA: SDS-PAGE (12.5%) analysis. Concentrated supernatant fluid from the ICI-1D6 hybridoma (starting material) and samples from each step of the four-step purification procedure were electrophoresed on 12.5% polyacrylamide slab gels containing SDS. Lanes (a) to (d) were loaded with 10 μg of total protein, while lane (e) was loaded with approximately 1 μg of protein. After electrophoresis, the gels were silver-stained and then scanned with a Bio-Rad model 1650 scanning densitometer interfaced to a DEC Pro 350 computer using a Waters' system interface module as an analogue to digital converter. The digitized scans were scaled and plotted using Waters' Expert chromatography software, which normalized the respective scans to the highest quantity of absorbance detected in each lane.

The arrows to the left of lane (a) denote the migration positions of the molecular weight standards, which are phosphorylase B (92.5 kD), bovine serum albumin (66 kD), ovalbumin (45 kD), carbonic anhydrase (31 kD), soybean trypsin inhibitor (21.5 kD) and lysozyme (14.4 kD). The material contained in each of the five lanes (a) to (e) is labelled on the figure. Lane (e) contains the product of the second reversed-phase HPLC step, demonstrating that the protein is essentially homogeneous after elution with heptafluorobutyric acid.

SUPPRESSION OF SECONDARY IgE RESPONSES

Since human SFA can be viewed as a potential therapy for human allergy, it is important to determine whether the purified protein is effective in inhibiting IgE responses of previously sensitized animals. In our previous work with mouse SFA, we documented that secondary IgE responses were susceptible to the inhibitory effects of mouse SFA [6]. The experiment summarized in Fig. 8.5 confirms this to be true with the human protein.

Mice immunized 10 months previously were either not treated (group I) or treated with a 5-day course of human SFA (groups II and III). Seven days after the treatment, all animals were boosted. Two days later, mice in group III were subjected to another 5-day course of human SFA. All groups of mice were bled on day 16 after secondary challenge and analysed for levels of serum IgE and IgG. Clearly, as shown by the hatched bars, the IgE levels of

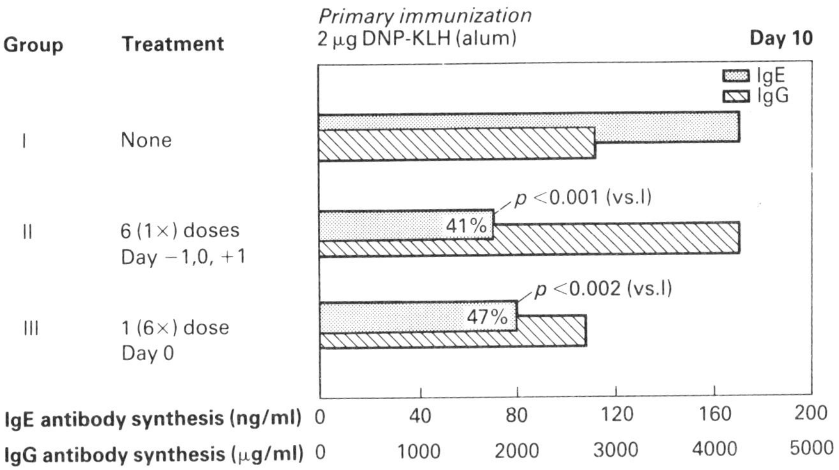

Fig. 8.4 A single dose of purified native SFA selectively suppresses *in vivo* IgE synthesis by high-responder CAF_1 mice. Groups of CAF_1 mice (six mice per group) were either not treated (but injected with saline: group I) or treated with HPLC-purified human SFA (groups II and III). The mode of immunization and bleeding and antibody determinations (on day 10 after immunization) were identical to those described in Fig. 8.2. Mice in group II were injected with SFA (1 μl per injection) according to the same 3-day regimen (six doses) as described in Fig. 8.2. Mice in group III, on the other hand, received a single injection of 6 μl of SFA, administered intravenously, on the same day as primary immunization. The data are presented as the mean IgE (stippled bars) and IgG (hatched bars) serum levels determined from individual serum samples of mice in each group on day 10 after primary immunization. Statistically significant differences in levels of response relative to group I are indicated beside the corresponding bars. The per cent values enclosed within the horizontal IgE bars of groups II and III denote the per cent response of each of these groups in comparison to control group I.

mice in group III were significantly depressed relative to controls in group I. The responses of group II mice were also lower by 38%, but this was not statistically significant. In contrast, IgG responses were of comparable magnitudes in all three groups, as depicted by the stippled bars.

Development of human SFA as a therapeutic product

This brings us to the next phase of this story—namely, the development of human SFA as a therapeutic product to use in human patients. Work currently underway is aimed at cloning the cDNA for human SFA, incorporating the cDNA into an appropriate bacterial expression vector and ultimately purifying the resulting recombinant protein.

Once this has been achieved we will have taken a significant step forward in developing the technology for production of large-scale quantities of human SFA in order to begin human clinical trials someday in the near future.

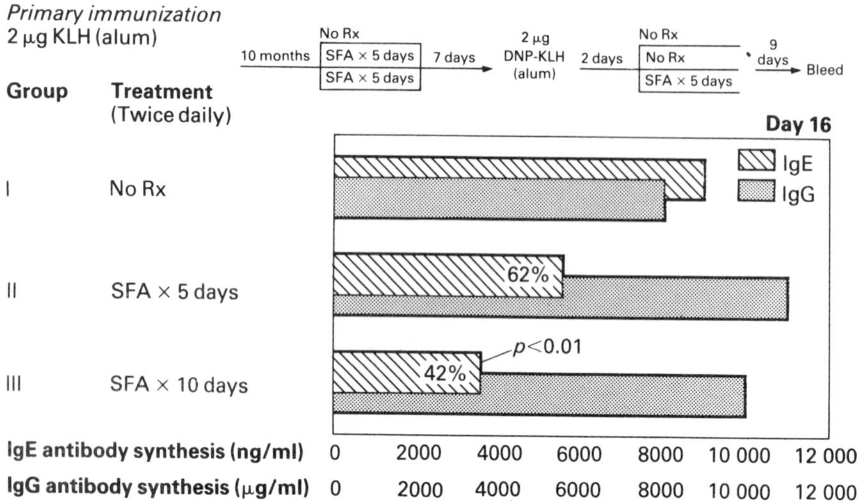

Fig. 8.5 Secondary IgE responses of high-responder CAF_1 mice can be selectively suppressed by treatment with native human SFA. CAF_1 mice were immunized with 2 μg of KLH in alum (4.0 mg) and then rested for 10 months. At that time, groups of such mice (10 mice per group) were either not treated (group I) or treated with native human SFA for 5 consecutive days (groups II and III). In this treatment regimen, injections were given twice per day (1 μl per injection), alternately administered intravenously and intraperitoneally. At the conclusion of this 5-day treatment course, mice were rested for 7 days and then all groups were challenged with 2 μg of DNP–KLH adsorbed on 2 μg of alum intraperitoneally. Two days after this immunization, mice in group III were given a second 5-day course of human SFA according to the same treatment regimen administered during the first course. At the completion of that 5-day treatment, all mice were rested for an additional 9 days and then bled. The data are presented as mean IgE (hatched bars) and IgG (stippled bars) levels determined from individual serum samples of mice in each group on day 16 after immunization with DNP–KLH. Statistically significant differences in the levels of the IgE responses in group III relative to control group I are indicated next to the corresponding data bar. The per cent values enclosed within the horizontal IgE bars of groups II and III denote the per cent response of each of these groups in comparison to control group I.

What is presented herein reflects the cumulative accomplishments of many dedicated colleagues in our group who have brought their expertise from various scientific disciplines together in a targeted approach aimed at resolving the unanswered questions about this important system.

In conclusion, it is now clear that the IgE system is regulated by an intricate series of cellular and molecular interactions. We now have a better understanding of how finely tuned the system is, and have a grasp on many of the cellular and molecular participants in the regulatory network. Moreover, we are becoming more and more knowledgeable about the types of external disturbances that can upset the balance of such fine tuning. This allows us to move closer to the time that new therapies will be developed for treatment of IgE-mediated disorders. Now that the cloning and expression of the gene coding for human SFA is foreseeable in the near future, perhaps the product of this genetic engineering effort will become part of the new generation of therapeutic approaches to this major disease process.

Acknowledgements

I am deeply indebted to all of my colleagues who have participated in various aspects of the studies presented herein. My sincere appreciation goes to Janet Czarnecki and Debbie Bosinger for superb assistance in the preparation of the manuscript.

This is publication number 125 from the Medical Biology Institute and publication number 8 from QUIDEL, La Jolla, California, USA. Portions of the research cited were supported by NIH grants AI-19476 and AI-19477, while the work relating to human SFA-secreting hybridoma cells and use of material therefrom, was funded by QUIDEL, La Jolla, California, USA.

References

1 Marcelletti JF, Katz DH. $FcR\varepsilon^+$ lymphocytes and regulation of the IgE antibody system. I. A new class of molecules, termed IgE-induced regulants (EIR), which modulate FcRε expression by lymphocytes. *J. Immunol.* 1984; **133**:2821–2827.

2 Marcelletti JF, Katz DH. $FcR\varepsilon^+$ lymphocytes and regulation of the IgE antibody system. II. $FcR\varepsilon^+$ B lymphocytes initiate a cascade of cellular and molecular interactions that control FcRε expression and IgE production. *J. Immunol.* 1984; **133**:2829–2836.

3 Marcelletti JF, Katz DH. $FcR\varepsilon^+$ lymphocytes and regulation of the IgE antibody system. III. Suppressive factor of allergy (SFA) is produced during the *in vitro* FcRε expression cascade and displays corollary physiologic activity *in vivo*. *J. Immunol.* 1984; **133**:2837–2844.

4 Marcelletti JF, Katz DH. $FcR\varepsilon^+$ lymphocytes and regulation of the IgE antibody system. IV. Delineation of target cells and mechanisms of action of SFA and EFA in inhibiting *in vitro* induction of FcRε expression. *J. Immunol.* 1984; **133**:2845–2851.

5 Marcelletti JF, Katz DH. $FcR\varepsilon^+$ lymphocytes and regulation of the IgE antibody system. V. Preliminary physicochemical characterization of the T cell-selective IgE-induced regulant EIR_T. *J. Immunol.* 1986; **137**:2599–2610.

6 Katz DH, Bargatze RF, Bogowitz CA, Katz LR. Regulation of IgE antibody production by serum molecules. IV. Complete Freund's adjuvant induces both enhancing and suppressive activities detectable in the serum of low and high responder mice. *J. Immunol.* 1979; **122**:2184–2190.

7 Katz DH, Bargatze RF, Bogowitz CA, Katz LR. Regulation of IgE antibody production by serum molecules. V. Evidence that coincidental sensitization and imbalance in the normal damping mechanism results in 'allergic breakthrough'. *J. Immunol.* 1979; **122**:2191–2197.

8 Zuraw BL, Nonaka M, O'Hair CH, Katz DH. Human IgE antibody synthesis *in vitro*: stimulation of IgE responses by pokeweed mitogen and selective inhibition of such responses by human suppressive factor of allergy (SFA). *J. Immunol.* 1981; **127**:1169–1177.

9 Katz DH, Nonaka M, Zuraw BL, Cohen PA, O'Hair CH. Regulation of human IgE antibody synthesis *in vitro*. In *Human B-Lymphocyte Function: Activation and Immunoregulation* (Eds Fauci AS, Ballieux RE), Raven Press, New York 1982, pp. 181–194.

10 Chen PP, Nonaka M, O'Hair CH, Cohen PA, Zuraw BL, Katz DH. Human IgE synthesis *in vitro* by pokeweed mitogen-stimulated human lymphoid cells: verification with a re-confirmed ε-specific radioimmunoassay. *J. Immunol.* 1984; **133**:1909–1913.

11 Gonzalez-Molina A, Spiegelberg HL. A subpopulation of normal human peripheral B lymphocytes that bind IgE. *J. Clin. Invest.* 1977; **59**:616–624.

12 Fritsche R, Spiegelberg HL. Fc receptors for IgE on normal rat lymphocytes. *J. Immunol.* 1978; **121**:471–478.

13 Yodoi J, Ishizaka K. Lymphocytes bearing Fc receptors for IgE. I. Presence of human and rat T lymphocytes with Fcε receptors. *J. Immunol.* 1979; **122**:2577–2583.

14 Katz DH. Regulation of the IgE system: experimental and clinical aspects. *Allergy* 1984; **39**:81–106.

15 Chen S-S, Bohn JW, Liu F-T, Katz DH. Murine lymphocytes expressing Fc receptors for IgE (FcRε). I. Conditions for inducing FcRε$^+$ lymphocytes and inhibition of the inductive events by suppressive factor of allergy (SFA). *J. Immunol.* 1981; **127**:166–173.

16 Katz DH, Chen S-S, Liu F-T, Bogowitz CA, Katz LR. Biologically active molecules regulating the IgE antibody system: biochemical and biological comparisons of suppressive factor of allergy (SFA) and enhancing factor of allergy (EFA). *J. Mol. Cell. Immunol.* 1984; **1**:157–166.

17 Yodoi J, Ishizaka T, Ishizaka K. Lymphocytes bearing Fc receptors for IgE. II. Induction of Fcε receptor bearing rat lymphocytes by IgE. *J. Immunol.* 1979; **123**:455–462.

18 Hoover RG, Gebel HM, Dieckgraefe BK, Hickman S, Rebbe NF, Hirayama H, Ovary Z, Lynch RG. Occurrence and potential significance of increased numbers of T cells with Fc receptors in myeloma. *Immunol. Rev.* 1981; **56**:115–139.

19 Katz DH, Marcelletti JF. A cascade of regulatory factors controls IgE antibody synthesis *in vivo* and *in vitro*. In *Progress in Immunology VI* (Eds Cinader B, Miller RG), Academic Press, New York 1986, pp. 879–889.

20 Katz DH, Marcelletti JF. Regulation of the IgE antibody system in humans and experimental animals. In *Progress in Immunology V* (Eds Yamamura Y, Tada T), Academic Press, Orlando 1984, pp. 465–482.

21 Katz DH. The IgE antibody system is coordinately regulated by FcR epsilon-positive lymphoid cells and IgE-selective soluble factors. *Int. Arch. Allergy Appl. Immunol.* 1985; **77**:21–25.

22 Gupta S, Good RA. Subpopulations of human T lymphocytes: laboratory and clinical studies. *Immunol. Rev.* 1981; **56**:89–114.

23 Adachi M, Yodoi J, Masuda T, Takatsuki K, Uchino H. Altered expression of lymphocyte Fcα receptor in selective IgA deficiency and IgA nephropathy. *J. Immunol.* 1983; **131**:1246–1251.

24 Hoover RG, Hickman S, Gebel HM, Rebbe N, Lynch RG. Expression of Fc receptor-bearing T lymphocytes in patients with immunoglobulin G and immunoglobulin A myeloma. *J. Clin. Invest.* 1981; **67**:308–315.

25 Thompson LF, Mellon MH, Zeiger RS, Spiegelberg HL. Characterization with monoclonal antibodies of T lymphocytes bearing Fc receptors for IgE (Tε cells) and IgG (Tγ cells) in atopic patients. *J. Immunol.* 1983; **131**:2772–2776.

26 Ishizaka K, Yodoi J, Suemura M, Hirashima M. Isotype-specific regulation of the IgE response by IgE-binding factors. *Immunol. Today* 1983; **4**:192–195.

27 Adachi M, Yodoi J, Noro N, Masuda T, Uchino H. Murine IgA binding factors produced by FcαR(+) T cells: role of FcγR(+) cells for the induction of FcαR and formation of IgA-binding factor in Con A-activated cells. *J. Immunol.* 1984; **133**:65–71.

28 Uede T, Huff TF, Ishizaka K. Suppression of IgE synthesis in mouse plasma cells and B cells by rat IgE-suppressive factor. *J. Immunol.* 1984; **133**:803–810.

29 Kishimoto T, Hirai Y, Suemura M, Nakanishi K, Yamamura Y. Regulation of antibody response in different immunoglobulin classes. IV. Properties and functions of 'IgE class-specific' suppressor factor(s) released from DNP-mycobacterium-primed T cells. *J. Immunol.* 1978; **121**:2106–2112.

30 Spiegelberg HL. Lymphocytes bearing Fc receptors for IgE. *Immunol. Rev.* 1981; **56**:199–218.

31 Ishizaka K, Ishizaka T. Mechanisms of reagenic hypersensitivity and IgE antibody response. *Immunol. Rev.* 1978; **41**:109–148.

32 Hirashima M, Yodoi J, Ishizaka K. Regulatory role of IgE-binding factors from rat T lymphocytes. III. IgE-specific suppressive factor with IgE-binding activity. *J. Immunol.* 1980; **125**:1442–1448.

33 Suemura M, Yodoi J, Hirashima M, Ishizaka K. Regulatory role of IgE-binding factors from rat T lymphocytes. I. Mechanism of enhancement of IgE response by IgE-potentiating factor. *J. Immunol.* 1980; **125**:148–154.
34 Yodoi J, Hirashima M, Ishizaka K. Regulatory role of IgE-binding factors from rat T lymphocytes. II. Glycoprotein nature and source of IgE-potentiating factor. *J. Immunol.* 1980; **125**:1436–1444.
35 Hirashima M, Yodoi J, Ishizaka K. Regulatory role of IgE-binding factors from rat T lymphocytes. IV. Formation of IgE-binding factors in rats treated with complete Freund's adjuvant. *J. Immunol.* 1980; **125**:2154–2160.
36 Parker CW, Schechtel T, Falkenhein S, Huber M. Induction of IgE receptors on human lymphocytes. *Immunol. Communications* 1983; **12**:1–10.
37 Kikutani H, Inui S, Sato R, Barsumian EL, Owaki H, Yamasaki K, Kaisho T, Uchibayashi N, Hardy RR, Hirano T, Tsunasawa S, Sakiyama F, Suemura M, Kishimoto T. Molecular structure of human lymphocyte receptor for immunoglobulin E. *Cell* 1986; **47**:658–665.
38 Coffman RL, Ohara J, Bond MW, Carty J, Zlotnik A, Paul WE. B cell stimulatory factor-1 enhances the IgE response of lipopolysaccharide-activated B cells. *J. Immunol.* 1986; **136**:4538–4541.
39 Marquis DM, Smolec JM, Katz DH. Use of a portable ribosome-binding site for maximizing expression of a eukaryotic gene in *Escherichia coli. Gene* 1986; **42**:175–183.

Discussion session

BJÖRKSTÉN: suppression of the IgE response is often considered the goal of immunotherapy. Nevertheless, there is often no relationship between clinical efficacy, reduced sensitivity and reduction in IgE. Perhaps IgE is important in sensitization only and once it is triggered there are other factors working in causing the symptoms of allergy.

KATZ: I would like to comment on that because you are absolutely right. One of the things we have learnt is that the specific IgE response is crucial for ongoing sensitization. There is, in addition, a relationship between IgE levels and the number of FcεR-bearing cells. When patients are going through the off-season period of remission the levels of these cells in the circulation are drastically reduced. As soon as the season begins again, even without drastic increases in serum IgE, these cells return. I am concerned with what the regulatory proteins do to these cells. We are not exactly certain why the numbers of these cells correlate with the disease process.

DEWDNEY: if you give SFA in your phase 2 and 3 clinical trials will it be sufficient to show that one does get a reduction of IgE or will one have to show quite clearly that one has improved the allergic state?

KATZ: it only matters to me that the patient feels better and I am not so concerned if the IgE level goes down. Nevertheless, objective criteria other than simple diary scores are required. Measurements of FcR-positive cells might be useful and we do see fluctuations in those cells associated with symptomatology.

DEWDNEY: as you know, that sort of patient monitoring has made specific allergy treatment very difficult.

KATZ: I know.

KAY: why did you not elect to produce EIR_B?

KATZ: we are, but it is about 2–3 years behind. Also I am not certain that it does exactly the same thing as SFA or will work in the same indications as SFA will. There may be very different indications. I do not know.

KAY: a 30% suppression of IgE does not seem very much to me.

KATZ: no, the values you saw were the percentage of response relative to the control, so if it said 30% it meant that it was depressed 70%.

KAY: 70% might not be enough.

KATZ: yes, most of the time we see 60%, maybe at best 75% suppression relative to controls and I think that is important *vis á vis* Bengt Björkstén's concern about depressing IgE too greatly. The system has its own internal clock which prevents it from going any further down.

LEE: have you considered delivering SFA topically to mucosal surfaces instead of giving it parenterally? If you gave it topically to mucosal surfaces you might down-regulate IgE synthesis locally without depressing systemic IgE levels.

KATZ: I do not know quite how to do that other than by aerosol and we are still far too early in our research to be able to successfully give a protein this size by aerosolization.

WILHELMS (Mannheim): how long did the effects in the animal studies last and did you see any sign of toxicity by application of your suppressor factor?

KATZ: no, the material is due to go into full-course toxicity studies within the next month but in the type of experiment that we performed in the past with mice we have given it repeatedly numerous times over long periods of time to mice without any overt evidence of toxicity. The effects will last for many months when mice are given the proper dose and purified material.

QUESTIONER (unidentified): there has been recent work on the heterogeneity of the IgE molecule. Have you looked at heterogeneity of suppression?

KATZ: no, we have not.

QUESTIONER (unidentified): do you think that heterogeneity exists and is a valid entity?

KATZ: I think there is limited heterogeneity in IgE but there is no evidence to indicate that any heterogeneity puts one kind of IgE out of the normal regulatory network made up by the majority of IgE.

HOLGATE: can you give us a little more information about the bioavailability of the molecule because you have been injecting it into these animals, so presumably you have an idea of how long radiolabelled material may last or how it is degraded? Do you know?

KATZ: we are in the process of doing these pharmacological studies now and I have no evidence to present on radiolabelled material.

9

Immunotherapy of asthma

S. DREBORG

Introduction—the status of immunotherapy

When immunotherapy (IT) was introduced in 1911 by Noon [1], the documentation of pharmaceutical products was just beginning. Therefore, as was the case with other old treatment principles, the safety and efficacy of IT with allergenic extracts was not documented according to modern principles. This has been done mainly during the last decade.

A prerequisite for controlled IT trials and proper clinical use of IT is that allergens used are standardized with known, repeatable potency and composition, and are stable, i.e. freeze-dried and reconstituted with appropriate stabilizing agents [2, 3]. Furthermore, reliable, repeatable methods for symptom/medication scoring and challenge testing must be used and the repeatability of these tests in the hands of the investigator should be documented [4, 5]. Immunotherapy trials must be run double-blind or objective parameters used for evaluation [4, 5].

Clinical efficacy of immunotherapy

In a recent review [4] we reported on 26 controlled IT trials in asthmatics. Nineteen of these showed clinical efficacy as measured by symptom/medication scoring, subjective evaluation and/or allergen challenge. They were performed with house dust (1/3 efficacious), house-dust mite (8/11), ragweed (1/2), grass (3/4), cat/dog (4/4) and *Cladosporium* (2/2) allergenic preparations. The non-specific bronchial hyperreactivity (BHR) decreased in treated compared with untreated patients in two of four studies [6–9]. The late asthmatic reaction (LAR) was reduced in 3/3 trials [10–12]. Furthermore, in one trial, eosinophil chemotactic activity (ECA) [13] and eosinophil cationic protein (ECP) [14] in serum decreased during the pollen season in patients given IT with birch-pollen allergen preparation compared with untreated patients. A decrease in BHR was noted in parallel, which may indicate an influence of IT on allergic inflammation in the bronchi. However, influence on this parameter has been inconsistent [6–9], probably due to the fact that BHR decreases slowly during several months also when patients are subject to total elimination [15]. The reason for the slow recovery is probably the destruction of bronchial mucosa which is a dominating feature even in patients with mild asthma [16]. A long-term decrease in BHR can be expected in all patients with decreased bronchial sensitivity to allergen [17] as the change in BHR is related to the immediate rather than the late-phase bronchial reaction [18, 19].

Adverse drug reactions induced by immunotherapy

As recently reviewed [4], no major adverse drug reactions to allergen IT have been reported. However, the incidence of general allergic reactions, even deaths, during IT, has been high [5, 20, 21]. Such accidents can be avoided, provided patient information, and personnel and patient supervision is adequate [5, 20]. Instructions should be given to patients/parents regarding self-supervision including reporting of late reactions at home to previous injections, status of the patient's allergy/asthma between injections, etc. Patients must be symptom-free with or without anti-inflammatory medication for at least 3 days prior to each injection. The doctor should supervise patients with proper clinical investigations, including lung function tests, before each injection and 30 min after when leaving the office.

Mechanisms of action

How IT of asthma works is not clearly understood. In a recent review [4] we summarized the present status. An updated summary of our conclusions is given in Table 9.1. There are conflicting data on the role of circulating allergen-specific IgG, IgG1 and IgG4 antibodies, indicating that the blocking-antibody theory is probably not a relevant mechanism. Furthermore, circulating allergen-specific IgE antibodies increase during the first months of IT. During the same time, the sensitivity as measured by symptom/medication scoring, skin-prick test and shock organ-provocation tests decreases. Therefore, changes in specific IgE antibodies are probably not a reflection of changes in clinical sensitivity. Based on these clinical findings, the most probable mechanisms are changes in the affinity of allergen-specific IgE antibodies and/or changes in allergen-specific histamine-releasing factors (Table 9.1). Of special interest is the decrease of serum ECA and ECP found by Rak *et al.* [13, 14], which was parallel to decrease in BHR during season. However, the mechanisms behind these findings are as yet not known.

Indications

Modern treatment of allergic asthma is based on the diagnosis of bronchial hyperreactivity and allergy of clinical relevance, and avoidance of the

Table 9.1 Possible mechanisms of action of immunotherapy in asthma (from reference 4 by permission of the publishers)

Proposed mechanism	Comments
Increase of allergen specific IgG-blocking antibodies	Decreases despite maintained efficacy [17, 22]
Decrease in allergen-specific IgE	A negative correlation between changes in allergen specific IgE and symptoms, skin and conjunctival sensitivity after 3 months of immunotherapy [23]
Increase in allergen-specific IgA and IgG in secretions	Has not been proven to correlate to clinical outcome [24]
Decrease of the allergen-induced basophil releasability	Does correlate to clinical outcome in some but not all studies (see further under IgE and histamine-releasing factors) [24]
Decrease in lymphocyte proliferation and lymphokine production by lymphocytes at exposure to allergen	Has not been proven to correlate to clinical outcome [24]
Increase in T-suppressor cells	Has not been proven to correlate to clinical outcome [25]
Switch from IgG1 to IgG4	Increase of IgG4 and ratio IgG4/IgG1 correlated to bad outcome [26] or to efficacy [27]
Change in affinity of IgE antibodies	Might be a clinically important observation [28, 29] (see IgE)
Decrease of allergen-specific histamine releasing factor(s) influencing basophil releasability [30, 31]	Fits best to available clinical data (see IgE versus clinical parameters, ECP, ECA and BHR changes)
Increase in anti-idiotypic antibodies	A negative correlation to changes in specific IgE and IgG antibodies has been found [32] but these changes were correlated to clinical outcome
Decrease in serum ECP and ECA with decrease of BHR	These findings [13, 14] can explain clinical improvement, but the mechanism inducing these changes has not been revealed (see under histamine-releasing factors). ECA, ECP and BHR changes are a result of the immediate allergic reaction

offending allergens and irritating stimuli, including proper supervision of the allergen content in the environment [33]. If these measures are not sufficient to reduce asthmatic symptoms, medication with 'mast-cell stabilizing agents' and inhalant steroids [34, 35] should be administered. At exacerbations, β_2 stimulants and theophyllines should be used. When total avoidance cannot be achieved, it is possible to reduce the ongoing allergic inflammation in the bronchi by IT. Such treatment should only be used by doctors with extensive knowledge and experience of IT, willing to supervise the treatment [4, 5]. Thus, the main problem with this type of treatment is not lack of efficacy, but the risk of adverse reactions if not properly supervised [36]. Many different approaches have been used to reduce the risk of allergic reactions during IT. One of these is to reduce the allergenicity of allergens by coupling to polyethylene glycol.

Monomethoxy polyethylene glycol modified allergens

Introduction

Monomethoxy polyethylene glycol (mPEG)-modified allergens have been developed with two main goals: to suppress the specific IgE production [37]

Table 9.2 Reagents used for modification of allergens

Reagent	Mol. wt	Structure
Formaldehyde (allergoid)	30	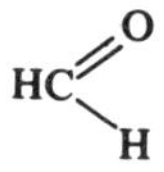
Glutaraldehyde	100	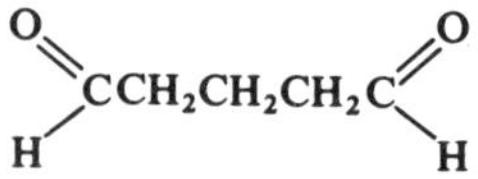
Monomethoxy polyethylene glycol (mPEG)	3000 5000	$CH_3O(CH_2CH_2O)_nCH_2CH_2OH$
Alginate	18 000	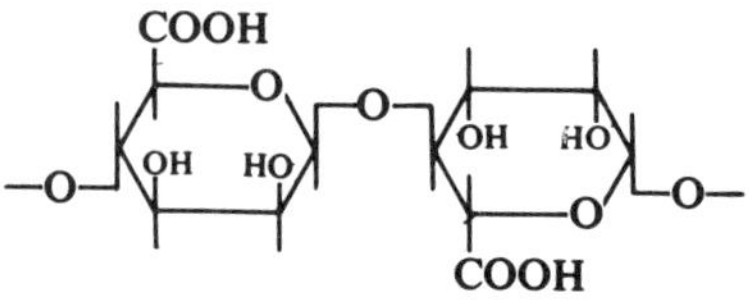
Poly-D-glutamic acid D-lysine	50 000	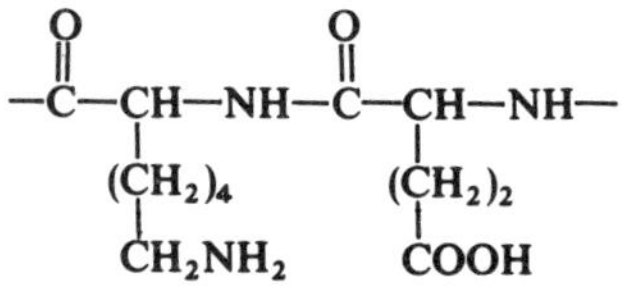

and to reduce the allergenicity in order to lessen the risk of severe general reactions [38]. Several other product lines have been developed with these aims (Table 9.2). However, our group favoured development of mPEG-modified allergens for several reasons, which will be described in more detail below.

Rationale for development of mPEG-modified allergens

Monomethoxy polyethylene glycol-modified allergens were initially developed to suppress the production of specific IgE [37]. The initial experiments were performed in animals prior to or in connection with primary immunization. When mPEG-modified allergens were first given to humans, no IgE antibody suppression was found [39, 40] probably due to the fact that the human IgE response was long standing in contrast to the initial response studied in mice. This finding has been confirmed in animal models [41]. Since then, the main objective of allergen modification with mPEG has been to reduce the allergenicity to lessen the risk for severe general reactions without losing the clinical effect in relation to the corresponding unmodified material.

Monomethoxy polyethylene glycol modification of allergens

Monomethoxy polyethylene glycol-modified allergens are prepared by reacting a mixed anhydride derivative of monomethoxy polyethylene glycol succinate (mol. wt 3000 or 5000) with the primary amino groups of allergens [38]. Standardization of the modified product in relation to the source material, i.e. a well-standardized and immunochemically characterized intermediate product (IMP) [42], has been made by determination of protein content by amino acid analysis. Furthermore, the modified material has been characterized regarding the number of mPEG molecules per antigen molecule [38]. The allergenicity of the modified product in relation to the unmodified source material is normally determined by radioallergosorbent test (RAST) inhibition, but has been documented in several experiments by histamine release and skin tests [38, 43, 44]. The unmodified allergen, the mPEG-modified allergen and PEG have undergone prolonged toxicity studies in two animal species according to the Nordic Council on Medicines' guidelines for registration of allergen preparations [45] as described by Johnsson [46].

Monomethoxy polyethylene glycol-modified allergens used in clinical trials have been modified to a high degree. The allergenicity has been reduced to less than 1% as measured by RAST inhibition and the allergenicity as measured by skin tests to a varying degree depending upon the patient's sensitivity to the many different allergenic components present in the allergen source material (five and 500 times) [38, 43, 44]. Only one mite preparation has been used in clinical trials. The allergenicity of that preparation was reduced only to about 40–60% by RAST inhibition and to 10% relative to the unmodified allergen preparation [44] by parallel-line bioassay using the skin-prick test method [47].

PHARMACOLOGY

According to unpublished experimental data, the release of mPEG-modified birch-pollen extract from the injection site is slower than that of the corresponding unmodified preparation. Furthermore, the circulation time of the modified preparation is markedly prolonged from hours to days [48].

Clinical experience with mPEG-modified allergens

To date, about 300 patients have been treated in trials comparing the effect of mPEG-modified ragweed pollen, grass pollen, deciduous tree pollen, *Dermatophagoides pteronyssinus* (*Der. p.*) mite allergen and honey-bee venom (HBV) allergen preparations with that of the corresponding unmodified material or placebo [38, 43, 44, 49–55].

SUPPRESSION OF LONG-STANDING IgE PRODUCTION

When it was first proposed to develop mPEG-modified allergens for human use, this was done on the basis of a proposal by Lee and Sehon [37]. However, when the first trial in humans performed with mPEG-modified ragweed was evaluated after 1 year of IT [Norman, personal communication], it was obvious that no IgE suppression could be achieved in humans with long-standing IgE production. This negative result has later been confirmed in humans [39, 40] and in animal models [41]. On the basis of this experience, the aim with development of mPEG-modified allergens was changed to developing a product with less allergenicity, but maintained efficacy in relation to the unmodified source material.

STARTING DOSE AND TOP DOSE

In the first trials with mPEG-modified short-ragweed allergen preparations [Norman, personal communication, 40] the allergenicity was reduced to less than 1% as measured by RAST inhibition. Nevertheless, the reduction in allergenicity as measured by skin tests differed from five to 500 times. This was probably due to the fact that allergenic components were modified to different degrees in combination with differences in sensitization pattern between patients. To ensure safety, the same starting dose was used in patients treated with mPEG-modified as with unmodified ragweed [Norman, personal communication, 36]. To investigate the highest possible initial dose, daily increasing doses of mPEG-modified allergens were given [49]. This approach induced a number of late cutaneous reactions in contrast to when the same initial dose was given followed by weekly injections. The latter group also reached higher top doses [49]. By daily injections an accumulation of mPEG-modified ragweed, inducing an ongoing allergic inflammation, probably took place. This can be avoided by giving a lower dose once a week or once a fortnight [50, 51]. However, systematic investigations of the possible highest initial dose are necessary to optimize the use of these and other allergens.

In most trials, the same top dose as with unmodified allergens, normally 100 000 biological units (BU), has been used. However, in the initial trial by Norman [personal communication], a much higher dose than normally used was aimed at (150 μg of antigen E (AgE) 3.7 million BU or 50 000 allergen units (AU)) and reached by patients given mPEG-modified ragweed, but not by patients on unmodified extract.

INTERVAL

As documented in animals, the release of allergen from the injection site as well as the circulation time of allergen was much increased. The interval between injections has in most trials been 2 weeks. As mentioned above, the interval was decreased to one day in one group of patients [49] in order to document the possible maximum-tolerated initial dose [49]. This regimen induced many reactions. Hypothetically, an ongoing allergic inflammation was built up in the subjects increasing the reactivity of their bronchi. Also in a study by Björkander *et al.* [50], where mPEG-modified grass-pollen extract was given at weekly intervals, such reactions were observed. No such complications were seen in this or other trials when the interval was increased to 2 weeks [40, 51–55].

EFFECT IN RELATION TO PLACEBO/CONTROLS

Juniper *et al.* [40] documented the efficacy of mPEG-modified short ragweed in relation to placebo. The patients were given a short course of injections just prior to the season. There was a clinically documented efficacy, which was not, however, as impressive as in many other trials investigating the efficacy of IT in hay fever. Therefore, more prolonged injection schedules have been used in subsequent studies. Mosbech *et al.* [44, 54, 55] investigated the efficacy of mPEG-modified *Der. p.* allergen in relation to the unmodified preparation and controls. During the first year of IT they found a significant decrease in conjunctival, nasal and bronchial sensitivity within the group receiving mPEG-modified *Der. p.* allergen and also a significant decrease in non-specific BHR as measured by histamine challenge tests. However, there was a significant difference only between the mPEG-modified group and controls regarding conjunctival sensitivity. According to symptom scores there was a better effect of the unmodified allergen preparation.

EFFECT IN RELATION TO THE CORRESPONDING UNMODIFIED ALLERGEN PREPARATION

Norman [personal communication] found a similar or slightly more pronounced decrease in conjunctival sensitivity and symptom scores in the group given mPEG-modified short-ragweed allergen compared with the one given the corresponding unmodified short-ragweed pollen allergen preparation. In a subsequent study by Ring *et al.* [51] comparing mPEG-modified with unmodified grass-allergen preparation, a similar influence on symptom scores, skin sensitivity and conjunctival sensitivity at corresponding dose levels was found. Basomba *et al.* [52] compared an mPEG-modified *Der. p.* allergen preparation to the native allergen preparation and found a similar decrease in skin and conjunctival sensitivity which was highly significant in relation to the pretreatment sensitivity. Furthermore, a slight decrease in the bronchial sensitivity of mPEG-treated patients was found (not significant). In a pilot study Öhman *et al.* [43] found a good protection to spontaneous bee stings in patients given mPEG-modified HBV. The therapy also produced a pronounced decrease in skin sensitivity. However, no intentional stings were performed. Therefore, Müller *et al.* [56] performed a double-blind study comparing mPEG-modified HBV to ordinary HBV. A top dose of 200 μg venom protein was given. A slightly more pronounced reactivity at bee-sting challenge was found in patients given the mPEG-modified allergen material, but the patient material was limited.

IMMUNOLOGICAL RESPONSE TO mPEG-MODIFIED ALLERGENS

In the first trials with mPEG-modified ragweed-pollen allergen preparation, there was a slightly less pronounced AgE-specific IgE and IgG response with the mPEG-modified than with the unmodified material [Norman, personal communication]. In a trial with a less modified mPEG-modified *Der. p.* allergen preparation performed by Basomba *et al.* [53], the *Der. p.*-specific IgE and IgG response was slightly more pronounced with the modified than with the unmodified preparation.

The response to individual components of allergenic preparations was investigated in two trials with mPEG-modified HBV [57]. There was a pronounced IgE and IgG response to all identified allergens. This indicates that determinants with a capacity to induce an immune response were available on these modified preparations.

THE DOSE–EFFECT RELATIONSHIP

In two trials, various top doses have been investigated [50, 51].

In the trial performed by Ring *et al.* [51], the effect of mPEG-modified grass mix and the corresponding unmodified allergen preparation was investigated at three different dose levels (20 000, 120 000, 200 000 BU) reached prior to three subsequent seasons. The effect was similar for the two preparations at all three dose levels as measured by symptom scores during the subsequent season and changes in conjunctival and skin sensitivity. Furthermore, Björkander *et al.* [50] found a similar effect using mPEG-modified and the corresponding unmodified grass-mix pollen allergen preparation in a 'low dose' (top dose 100 000 BU), but a more pronounced decrease in the conjunctival and skin sensitivity when higher doses of mPEG-modified grass mix were given (top dose 1 million BU).

ADVERSE DRUG REACTIONS (ALLERGIC SIDE EFFECTS)

In the first published trials with mPEG-modified ragweed-pollen allergen preparation [40, 49] there were some late general reactions. Injections were given at weekly or less than weekly intervals. In subsequent studies, injections have been given at 2 week intervals. As a result, there have been fewer reactions than with the corresponding unmodified preparations [50–52]. Only one serious systemic reaction has been reported [51].

INDUCTION OF IgE-MEDIATED SENSITIVITY TO mPEG-RELATED DETERMINANTS

In the first publication by Juniper *et al.* [40] an increased sensitivity to the mPEG-modified preparation was reported in some patients given mPEG-modified but not in those given placebo. In a second trial by Juniper *et al.* [49], five patients reacted with small weals to PEG. Further studies are needed to disclose whether IT with mPEG-modified (and other modified) products may cause sensitization to determinants related to the modifying agent.

It has also been reported that some patients exhibit an increase in IgM anti-PEG antibodies during the initial phase of IT with mPEG-modified allergens [58]. No clinical correlate to this type of sensitization has been found.

Possible advantages in relation to unmodified allergen preparations

Monomethoxy polyethylene glycol-modified allergens are slowly released from the injection site and the circulation time is prolonged in relation to that of unmodified allergen preparations. Theoretically, this fact decreases the possibility of severe general reactions. Only one severe reaction has been reported where about 300 patients have been treated in three trials with mPEG-modified ragweed, two with HBV, one with tree pollen, three with grass pollen and three with *Der. p.* mPEG-modified allergen preparations. About 10 000 injections have been given in total.

The clinical response seems to be similar to that obtained with the corresponding amount of unmodified allergen preparation. An immune response is obtained to all relevant components as measured by crossed immunoelectrophoresis/crossed radioimmunoelectrophoresis (CIE/CRIE). Thus, the effect is similar to that obtained with unmodified allergen preparations, but with fewer side effects. This means that higher doses [50] can be achieved, i.e. the efficacy can be improved [50, 51]. From this follows that the indications for IT can be more liberal provided strict diagnostic criteria and clinical supervision [4, 5] are used.

Possible advantages in relation to other types of modified allergen preparations

The major advantage in relation to other types of preparations, e.g. those obtained by modification by low molecular weight substances, such as glutaraldehyde or formaldehyde, is that the degree of modification can be supervised and the batch-to-batch consistency can be controlled by simple chemical and immunochemical methods.

Immunotherapy in the future

Allergen preparations

It has been clearly shown that the effect in IT is allergen specific [59, 60]. Therefore, allergen preparations used for IT in the future should contain allergens of clinical relevance to the patient. At present the best preparations available are those which are partly purified containing all relevant allergens in a composition reflecting the source material [42]. However, it can be assumed that immunological stimulation is achieved on a molar basis. This means that about the same amount of relevant allergen molecules from each allergenic component should be present in preparations used for IT. This can be achieved by purifying single components and then mixing these components in relevant quantities. Another approach is that proposed by Weeke

[61, 62]. IgE or IgG from individual patient's sera is bound to columns by anti-IgE or protein A respectively. Allergen is then bound to the specific IgE or IgG molecules and finally patient-specific allergens are eluted. Thus, patients can be treated with allergens corresponding to the specific IgE molecules present in their respective sera. By this initial treatment, specific IgG antibodies are induced. These antibodies are directed only against molecules allergenic in that patient. Allergen-specific IgG produced by the patient can then be used for further purification of individually important allergens. Such sophisticated treatment will not be available for several years. However, it includes a possibility to give a very specific treatment. By elution of IgG antibodies together with allergen and injection of the complexes, it will probably be possible to reduce allergic reactions in connection with IT.

Antigen-presenting cells use fragments of antigens, often six to eight amino acids. Such fragments are antigenically active, but cannot induce allergic reactions. It has therefore been proposed to use fragments of allergens with the relevant determinants for IT. However, no experimental data supporting this possibility are available.

In the short term, modified allergens with reduced allergenicity, and proven efficacy, such as mPEG-modified allergens will be the standard.

STANDARDIZATION OF ALLERGENS FOR IT

For many years the same standardization systems have been used for diagnostic and therapeutic preparations. Methods have been developed in Scandinavia [42, 45, 63] and in the USA [64] for biological standardization of allergenic preparations. These methods aim at minimizing the difference in potency between allergen preparation produced from different allergen source materials. The standardization methods are based on skin tests. It has been suggested [5] that biological units based on skin tests are not relevant for therapeutic allergen preparations. The intensity of allergic reactions, e.g. the weal and flare reaction in the skin, is related to the potency of the allergen preparation used, i.e. to the number of allergenic molecules (determinants) in the solution used. The clinical and immunological effect of IT is probably also related to the number of allergenic molecules available for antigen-presenting cells. Thus, biological units based on diagnostic tests ought to be relevant also for therapeutic allergenic preparations.

There are other possibilities to biologically standardize (equilibrate) allergenic preparations intended for therapeutic use. The most constant finding during IT is a decrease in skin sensitivity [47]. The skin reactivity is less influenced by exposure to allergens and irritants. Therefore, the change in skin sensitivity better reflects the releasability of effector cells. The change in skin sensitivity can be expressed as the change in allergen concentration needed to induce a certain skin response, e.g. that of histamine dihydrochloride 10 mg/ml or similar reference substance, or as the change in concentration eliciting the same weal reaction before and after IT as measured by parallel-line bioassay [47]. To be able to assign the same unit to standardized allergenic preparations of different species, I propose that changes in skin sensitivity obtained at different dose levels reached within defined time limits are used for estimating by regression analysis the dose inducing a change in skin sensitivity of 10 times in a representative sample of patients. This calculated dose is assigned 10 therapeutic units (TU). Probably such TUs will correlate to BUs used at present for diagnostic allergenic preparations.

Mode of administration

At present, subcutaneous IT dominates. It includes the risk of severe allergic reactions. Therefore, in the future other possibilities will probably be exploited.

Sublingual allergen drops have been used in central Europe for many decades. However, there is no scientific documentation supporting its efficacy in rhinitis or asthma [65–70].

Oral administration of allergen extract in enteric coated capsules has been tested without success using mite and grass-pollen extracts [71, 72]. In a recent pilot study [73] and three double-blind, placebo-controlled studies [74–76], birch-pollen extract in high doses administered in enteric coated capsules have been shown to reduce seasonal symptoms, conjunctival and skin sensitivity. This type of treatment will never be economically justified because of the high doses required and because it does not work in grass pollinosis or mite allergy which are much greater problems than birch pollinosis. The reason for this failure is not clear, but is probably that the dose of grass and mite allergen that reaches immunocompetent cells in the gut mucosa is too low. Furthermore, the influence on asthma, although theoretically possible, has not been systematically investigated.

Nasal administration of aqueous allergenic extracts reduces nasal symptoms during the pollen season, but induces symptoms prior to this season [77]. This type of treatment should be further explored and its effects on asthma should be investigated.

Indications

Indications vary between regions. In northern Europe only patients with severe hay fever and a few patients with asthma are treated. In southern Europe, IT is on the increase and in central Europe and the USA it is popular, but not increasing. If allergic side effects can be avoided, by measures proposed by the subcommittee on IT within the European Academy of Allergology and Clinical Immunology (EAACI) [5], there will be a broader indication for IT as it has now been shown that the effect lasts for at least 8 years [78].

Combination of IT with other measures

As mentioned above, a combination of allergen and specific IgG antibodies might reduce allergic side effects, but this type of treatment has not been widely used and the scientific documentation is scarce [79]. To inject human monoclonal antibodies against major allergens prior to or in combination with active allergenic material is at present just a theoretical possibility.

Furthermore, stimulation of the immune system with appropriate allergens in combination with immunopharmacologically active agents such as interleukin (IL)-4 or Cyclosporin has not been tried, but might be a possibility that should be investigated.

At present, there is strong evidence that during IT as well as when patients outgrow their allergy, specific IgE is present in the circulation often in higher concentrations than when the patient had symptoms. Thus, measures aimed at reducing mediator release from IgE-receptor bearing cells should be explored.

Concluding remarks

Inhalant allergen IT, when used as prescribed [4, 5] at present, is effective in a dose–effect related manner for the treatment of asthma. The major problem with IT, allergic reactions to injected allergen, can be avoided by mPEG-modification of allergens. Such allergens have a similar effect as unmodified allergens, but induce fewer allergic reactions when given in comparable doses. Future research should be concentrated on allergen-specific factors decreasing the releasability of effector cells, rather than on IgE regulation.

References

1 Noon S. Prophylactic inoculation against hay fever. *Lancet* 1911; 1572–1573.

2 Foucard T, Bennich H, Johansson SGO. Studies on the stability of diluted allergen extracts using the radioallergosorbent test (RAST). *Clin. Allergy* 1973; **3**:91–102.

3 Nelson HS. The effect of preservatives and dilution on the deterioration of Russian Thistle (*Salsola pestifer*), a pollen extract. *J. Allergy Clin. Immunol.* 1979; **63**:417–425.

4 Dreborg S, Mosbech H, Weeke B. Immunotherapy (hyposensitization) and bronchial asthma. In *Ballière's Clinical Immunology and Allergy* (Ed. Kay AB), vol. 2, No. 1, Ballière Tindall, London 1988, pp. 245–258.

5 Immunotherapy Sub-committee of The European Academy of Allergology and Clinical Immunology. Immunotherapy. *Allergy* **43** (Suppl. 6).

6 Taylor WW, Ohman JL, Lowell FC. Immunotherapy in cat-induced asthma. *J. Allergy Clin. Immunol.* 1978; **61**:283–287.

7 Ohman JL, Findlay SR, Leitermann KM. Immunotherapy in cat-induced asthma. Double-blind trial with evaluation of *in vivo* and *in vitro* responses. *J. Allergy Clin. Immunol.* 1984; **74**:231–239.

8 Sundin B, Lilja G, Graff-Lonnevig V, Hedlin G, Heilborn H, Norrlind K, Pegelow K-O, Løwenstein H. Immunotherapy with partially purified and standardized animal dander extracts. I. Clinical results from a double-blind study on patients with animal dander asthma. *J. Allergy Clin. Immunol.* 1986; **77**:478–487.

9 Bousquet J, Caluzel AM, Chanal I, Dhivert H, Hejjaoui A, Michel FB. Non specific bronchial hyperreactivity in asthmatic subjects after immunotherapy with a standardized mite extract. *Am. Rev. Respir. Dis.* 1987; **135**:A315.

10 Price JF, Warner JO, Hey EN, Turner MW, Soothill JF. A controlled trial of hyposensitization with tyrosine adsorbed *Dermatophagoides pteronyssinus* antigen in childhood asthma: *in vivo* aspects. *Clin. Allergy* 1984; **14**:209–219.

11 Formgren H, Olofsson E, Dreborg S, Lanner Å. Bronchial sensitivity during immunotherapy (IT) with Pharmalgen *D. farinae* (*D.f.*). *Proceedings of the 13th Congress of the European Academy of Allergology and Clinical Immunology*, Budapest 1986, p. 363.

12 Metzger WJ, Donnelly A, Richerson HB. Modification of late asthmatic responses (LAR) during immunotherapy for *Alternaria*-induced asthma. *J. Allergy Clin. Immunol.* 1983; **71**:119.

13 Rak S, Håkansson L, Venge P. Eosinophil chemotactic activity in allergic patients during the birch pollen season: the effect of immunotherapy. *Int. Arch. Allergy Appl. Immunol.* 1987; **82**:349–350.

14 Rak S, Löwhagen O, Venge P. The effect of immunotherapy on bronchial hyperresponsiveness and eosinophil cationic protein in pollen allergic patients. *J. Allergy Clin. Immunol.* 1988; **82**:470–480.

15 Platts-Mills TAE, Mitchell EB, Nock P, Tovey ER, Moszoro H, Wilkins SR. Reduction of bronchial hyperreactivity during prolonged allergen avoidance. *Lancet* 1982; **ii**:675–678.

16 Laitinen LA, Heino M, Laitinen A, Kava T, Haahtela T. Damage of the airway epithelium and bronchial reactivity in patients with asthma. *Am. Rev. Respir. Dis.* 1985; **131**:599–606.

17 Dreborg S, Agrell B, Foucard T, Kjellman N-IM, Koivikko A, Nilsson S. A double-blind, multicenter immunotherapy trial in children, using a purified and standardized *Cladosporium herbarum* preparation. I. Clinical results. *Allergy* 1986; **41**:131–140.

18 Durham SR, Graneek BJ, Hawkins R, Newman-Taylor AJ. The temporal relationship between increases in airway responsiveness to histamine and late asthmatic responses induced by occupational agents. *J. Allergy Clin. Immunol.* 1987; **79**:398–406.

19 Durham SR, Cookson WOCM, Craddock CF, Benson MK. Falls in eosinophils parallel the late asthmatic response and associated increases in airway responsiveness. *J. Allergy Clin. Immunol.* 1988; **80**:249.

20 Dreborg S. New aspects on immunotherapy. Safety aspects. *Proceedings of the 14th Congress of the European Academy of Allergology and Clinical Immunology*, Mallorca 1987, pp. 123–126.

21 Lockey RF, Benedict LM, Turkeltaub PC, Bukantz SC. Fatalities from immunotherapy (IT) and skin testing (ST). *J. Allergy Clin. Immunol.* 1987; **79**:660–677.

22 Karlsson R, Agrell B, Dreborg S, Foucard T, Kjellman N-IM, Koivikko A, Einarsson R. A double-blind, multicenter immunotherapy trial in children, using a purified and standardized *Cladosporium herbarum* preparation. II. *In vitro* results. *Allergy* 1986; **41**:141–150.

23 Urbanek R, Kuhn W, Holgersson M, Dreborg S. Changes in conjunctival provocation test (CPT) and skin prick test (SPT) specific IgE and IgG during immunotherapy (IT) with grass pollen preparations (PP). *Ann. Allergy* 1985; **55**:259.

24 Djurup R. The subclass nature and clinical significance of the IgG antibody response in patients undergoing allergen-specific immunotherapy. *Allergy* 1985; **40**:469–486.

25 Rocklin RE, Sheffer AL, Greinder DK, Melmon KL. Generation of antigen-specific suppressor cells during allergy desensitization. *N. Engl. J. Med.* 1980; **302**:1213–1219.
26 Djurup R, Malling H-J. High IgG4 antibody level is associated with failure of immunotherapy with inhalant allergens. *Clin. Allergy* 1987; **17**:459–468.
27 Löfkvist T, Svensson G, Agrell B, Dreborg S, Einarsson R. Monitoring of allergen specific IgG and IgG subclasses in patients undergoing immunotherapy (IT). *J. Allergy Clin. Immunol.* 1988; **80**:293.
28 Ahlstedt S, Eriksson NE. Immunotherapy in atopic allergy—antibody titres and avidities during hyposensitization with birch and timothy pollen allergens. *Int. Arch. Allergy Appl. Immunol.* 1977; **55**:400–411.
29 Warner JA, Pienkowski MM, Plaut M, Norman PS, Lichtenstein LM. Identification of histamine releasing factor(s) in the late phase of cutaneous IgE-mediated reactions. *J. Immunol.* 1986; **136**:2583–2587.
30 Conroy MC, Adkinson NF, Lichtenstein LH. Measurement of IgE on human basophils: relation to serum IgE and anti-IgE induced histamine release. *J. Immunol.* 1977; **118**:1317–1321.
31 MacDonald SM, Lichtenstein LM, Proud D, Plaut M, Nacleiro RM, MacGlashan DW, Kagey-Sobotka A. Studies of IgE-dependent histamine releasing factors: heterogeneity of IgE. *J. Immunol.* 1987; **139**:506–512.
32 Bose R, Marsh DG, Delespesse G. Anti-idiotypic to anti-Lolp I (Rye) antibodies in allergic and non-allergic individuals. Influence of immunotherapy. *Clin. Exp. Immunol.* 1986; **66**:231–240.
33 Platts-Mills TAE. Dust mite avoidance in the treatment of asthma. *Ann. Allergy* 1985; **55**:419–420.
34 Cockcroft DW. Airway hyperresponsiveness: therapeutic implications. *Ann. Allergy* 1987; **59**:405–414.
35 Woolcock AJ, Yan K, Salome CM. Effect of therapy on bronchial hyperresponsiveness in the long-term management of asthma. *Clin. Allergy* 1988; **18**:165–176.
36 Committee on Safety of Medicines. CMS update. Desensitizing vaccines. *Br. Med. J.* 1986; **293**:94.
37 Lee WY, Sehon AH. Suppression of reaginic antibodies with modified allergens. I. Reduction in allergenicity of protein allergens by conjugation to polyethylene glycol. *Int. Arch. Allergy Appl. Immunol.* 1978; **56**:159–170.
38 Åkerblom E. Monomethoxypolyethylene glycol modified allergens. *Arbeiten aus dem Paul-Ehrlich-Institut (Bundesamt für Sera und Impfstoffe) dem Georg-Speyer-Haus und dem Ferdin-and-Blum-Institut*, Heft 78, Fischer Verlag, Stuttgart, New York 1983, pp. 231–239.
39 Norman PS, King TP, Alexander JF, Kagey-Sobotka A, Lichtenstein LM. Immunologic responses to conjugates of antigen E in patients with ragweed hay fever. *J. Allergy Clin. Immunol.* 1984; **73**:782–789.
40 Juniper EF, Roberts RS, Kennedy LK *et al.* Polyethylene glycol-modified ragweed pollen extract in rhinoconjunctivitis. *J. Allergy Clin. Immunol.* 1985; **75**:578–585.
41 Ahlstedt S, Björkstén B, Åkerblom E. Antibody responses to honey bee venom and monomethoxypolyethylene glycol modified honey bee venom in mice. *Int. Arch. Allergy Appl. Immunol.* 1983; **71**:228–232.
42 Dreborg S, Einarsson R, Longbottom J. The chemistry and standardization of allergens. In *Handbook of Experimental Immunology*, vol. I, Immunochemistry, 4th edn. (Ed. Weir DM), Blackwell Scientific Publications, Edinburgh 1986, pp. 10.1–10.28.
43 Öhman S, Björkander J, Dreborg S, Lanner Å, Malling H-J, Weeke B. A preliminary study of immunotherapy with a monomethoxy polyethylene glycol modified honey bee venom preparation. *Allergy* 1986; **41**:81–88.
44 Mosbech H, Dreborg S, Påhlman I, Stahl Skov P, Steringer I, Weeke B. Modification of house dust mite allergens by monomethoxypolyethylene glycol. Allergenicity measured by *in vitro* and *in vivo* methods. *Int. Arch. Allergy Appl. Immunol.* 1988; **84**:145–149.
45 Nordic Council on Medicines. Registration of allergen preparations. Nordic guidelines. 2nd edn. *NLN Publication* No. 23, Uppsala, 1989.

46 Jonsson GBJ. Toxicity studies with allergens—requirements, guidelines. *Arbeiten aus dem Paul-Ehrlich-Institut (Bundesamt für Sera und Impfstoffe) dem Georg-Speyer-Haus und dem Ferdinand-Blum-Institut*, Heft 78, Fischer Verlag, Stuttgart, New York 1983, pp. 191–193.
47 Dreborg S. The skin prick test. Methodological studies and clinical applications. *Linköping University Medical Dissertation*, No. 239, Linköping 1987.
48 Dreborg S. mPEG-modified allergen preparations. In *New Trends in Allergy II* (Eds Ring J, Burg G) Springer-Verlag, Heidelberg 1986, pp. 312–316.
49 Juniper EF, O'Connor J, Roberts RS, Evans S, Hargreave FE, Dolovich J. Polyethylene glycol-modified ragweed extract: comparison of two treatment regimens. *J. Allergy Clin. Immunol.* 1986; **78**:851–856.
50 Björkander J, Sundberg R, Ljungstedt–Påhlman I, Dreborg S. Immunotherapy with a monomethoxy polyethyleneglycol (mPEG-gm) modified and unmodified (gm) partly purified grass mix allergen preparation. *Proceedings of the 13th Congress of the European Academy of Allergology and Clinical Immunology*, Budapest 1986, p. 244.
51 Ring J, Pryzbilla B, Frey C, Galosi A, Burow G, Ljungstedt-Påhlman I, Dreborg S. A two years double blind study with a monomethoxy polyethylene glycol modified and an unmodified, partly purified grass mix allergen preparation. Clinical evaluation at different dose levels. *Proceedings of the 13th Congress of the European Academy of Allergology and Clinical Immunology*, Budapest 1986, p. 243.
52 Basomba A, Almodovar A, Campos A, Garcia Villalmanzo A, Peleaz A, Ljungstedt-Påhlman I, Dreborg S. One year double-blind immunotherapy (IT) study in adult mite asthmatics. Comparison between mPEG-modified (mPEG) and the corresponding unmodified (Der.p.) mite allergen preparations. I. Clinical results. *Proceedings of the 14th Congress of the European Academy of Allergology and Clinical Immunology*, Mallorca 1987, p. 164.
53 Basomba A, Almodovar A, Campos A, Garcia Villalmanzo A, Peleaz A, Ljungstedt-Påhlman I, Dreborg S. One year double-blind immunotherapy (IT) study in adult mite asthmatics. Comparison between mPEG-modified (mPEG) and the corresponding unmodified (*Der. p.*) mite allergen preparations. II. *In-vitro* results. *Proceedings of the 14th Congress of the European Academy of Allergology and Clinical Immunology*, Mallorca 1987, p. 170.
54 Mosbech H, Dreborg S, Frölund L, Påhlman I, Svendsen UG, Søborg M, Taudorf E, Weeke B. Immunotherapy with mPEG modified and unmodified *D. pteronyssinus* extract. Clinical results in a one year double blind study. *Ann. Allergy* 1985; **55**:389.
55 Mosbech H, Dreborg S, Poulsen L, Påhlman I, Stahl Skov P, Steringer I, Søborg M, Weeke B. Immunotherapy with mPEG modified and unmodified *D. pteronyssinus* extract. Paraclinical results of a two-year double-blind study. *Proceedings of the 14th Congress of the European Academy of Allergology and Clinical Immunology*, Mallorca 1987, p. 271.
56 Müller U, Rabson AR, Bischof M, Lomnitzer R, Dreborg S, Lanner Å. A double-blind study comparing monomethoxypolyethylene glycol-modified honey bee venom and unmodified honey bee venom for immunotherapy. I. Clinical results. *J. Allergy Clin. Immunol.* 1987; **80**:252–260.
57 Nordvall SL, Uhlin T, Öhman S, Björkander J, Malling H-J, Weeke B, Dreborg S, Lanner Å, Einarsson R. IgG and IgE antibody patterns after immunotherapy with monomethoxy polyethyleneglycol modified honey bee venom. *Allergy* 1986; **41**: 89–94.
58 Richter AW, Åkerblom E. Polyethylene glycol reactive antibodies in man: titer distribution in allergic patients treated with monomethoxy polyethylene glycol modified allergens or placebo and in healthy blood donors. *Int. Arch. Allergy Appl. Immunol.* 1984; **74**:336–339.
59 Norman PS, Lichtenstein LM. The clinical and immunological specificity of immunotherapy. *J. Allergy Clin. Immunol.* 1978; **61**:370–377.
60 Østerballe O, Löwenstein H, Pral P, Skov P, Weeke B. Immunotherapy in hay fever with two major allergens 19, 25 and partially purified extracts of timothy grass pollen. *Allergy* 1981; **36**:183–189.
61 Søndergaard I, Weeke B. Purification of patient related allergens by means of bioaffinity chromatography on a Sepharose anti-IgE patient IgE immunosorbent. *Allergy* 1984; **39**:473–479.
62 Søndergaard I, Weeke B. Isolation of patient-related antigens from allergen extracts by means

of protein A-sepharose-patient IgG1,2,4 bio-affinity chromatography. *Allergy* 1984; **39**:551–559.

63 Dreborg S, Basomba A, Belin L, Durham S, Einarsson R, Eriksson NE, Frostad AB, Grimmer Ø, Halvorsen R, Holgersson M, Kay AB, Nilsson G, Malling H-J, Sjögren I, Weeke B, Våla I-J, Zetterström O. Biological equilibration of allergen preparations: methodological aspects and reproducibility. *Clin. Allergy* 1987; **17**:537–550.

64 Turkletaub PC. Biological standardization based on quantitative skin testing. The ID_{50}EAL method. *Arbeiten aus dem Paul-Ehrlich-Institut (Bundesamt für Sera und Impfstoffe) dem Georg-Speyer-Haus und dem Ferdinand-Blum-Institut*, Heft 80, Fischer Verlag, Stuttgart, New York 1987, pp. 169–182.

65 Rebien W, Wahn U, Puttonen E, Maasch JH. Comparative study of immunological and clinical efficacy of oral and subcutaneous hyposensitization. *Allergologie* 1980; **3**:101–109.

66 Urbanek R, Gehl R. Efficacy of oral hyposensitization treatment of house dust mite allergy. *Monatsschr. Kinderheilkd.* 1982; **130**:150–152.

67 Reinert M, Reinert U. Oral hyposensitization with pollen solutions and placebos. *Prax. Klin. Pneumol.* 1983; **37**:228–234.

68 Urbanek R, Kuhn W, Binder U. Efficacy of oral and parenteral hyposensitization with pollen extracts. *Deutsch Med. Wochenschrift* 1983; **108**:1433–1437.

69 Cooper PJ, Darbyshire J, Nunn AJ, Warner JO. A controlled trial of oral hyposensitization in pollen asthma and rhinitis in children. *Clin. Allergy* 1984; **14**:541–550.

70 van Niekerk CH, de Wet JI. Efficacy of grass-maize pollen oral immunotherapy in patients with seasonal hay fever: a double-blind study. *Clin. Allergy* 1987; **17**:507–514.

71 Mosbech H, Dreborg S, Madsen F, Ohlsson H, Stahl-Skov P, Taudorf E, Weeke B. High dose grass pollen tablets used for hyposensitization in hay fever patients. *Allergy* 1987; **42**:451–455.

72 Taudorf E, Weeke B. Orally administered grass pollen. *Allergy* 1983; **38**:561–564.

73 Björkstén B, Möller C, Broberger U, Ahlstedt S, Dreborg S, Johansson SGO, Juto P, Lanner Å. Clinical and immunological effects of oral immunotherapy with a standardized birch pollen extract. *Allergy* 1986; **41**:290–295.

74 Möller C, Dreborg S, Lanner Å, Björkstén B. Oral immunotherapy of children with rhinoconjunctivitis due to birch pollen allergy. A double blind study. *Allergy* 1986; **41**:271–279.

75 Taudorf E, Laursen L, Lanner Å, Björkstén B, Dreborg S, Søborg M, Weeke B. Oral immunotherapy in birch pollen hay fever. *J. Allergy Clin. Immunol.* 1987; **80**:129–132.

76 Björkstén B, Croner S, Dreborg S, Lanner Å. A double blind study of oral immunotherapy (OIT) in children allergic to birch pollen using high doses of allergen. *J. Allergy Clin. Immunol.* 1986; **77**:214.

77 Nickelsen J *et al.* Local intranasal immunotherapy for ragweed allergy rhinitis. I. Clinical response. *J. Allergy Clin. Immunol.* 1981; **68**:33–40.

78 Mosbech H, Østerballe O. Does the effect of immunotherapy last after termination of treatment? *Allergy* 1988; In press.

79 Szégli G, Negut E, Peligrad I, Matache C, Muresan D, Radu R, Chirila M, Dinca L. Alergim®—a new antiallergic product which down-regulates B-lymphocyte functions. *Arch. Roum. Path. Exp. Microbiol.* 1987; **46**:73–82.

Discussion session

WHEELER (Beecham): why does the basophil or mast cell get less sensitive to allergen challenge when the serum IgE measurements go up?

DREBORG: there are several explanations. There may be a difference in affinity of IgE antibodies to the Fc receptor of mast cells. There may be an interaction by other antibodies, anti-idiotypic antibodies. Such antibodies have been shown to be induced during immunotherapy. Histamine releasing factors might be involved altering mast-cell/basophil histamine releasability.

LEE: could you just summarize for us evidence that immunotherapy reduces mast cell releasability. Are you able to separate reduction in mast cell releasability from changes in target organ sensitivity *in vivo*?

DREBORG: no. I think at present we cannot differentiate between the effect on mast cells and the total organ.

QUESTIONER (unidentified): you said that the basophil *in vitro* release went down?

DREBORG: yes, provided whole blood is used. There are several trials using washed lymphocytes showing no influence at all.

KAY: I would like to ask you about the effect of immunotherapy on the late-phase reaction because I think that this is one of the particularly interesting features of successful immunotherapy as I understand it. Firstly, do you know whether mPEG-modified allergens will also ablate the late-phase reaction, I imagine they would, but have you in fact looked at this? Secondly, do you think that this effect on late-phase reactions could be entirely explained on the basis of a mast-cell mediated phenomenon or do you have any thoughts on any other cells that might be involved?

DREBORG: the first question, there are no data available as yet. I have no data on effector cells.

DEWDNEY: one of your tables said that PEG allergens might modulate the effect of eosinophils. What data had you in mind? I really wondered what you felt the role of the eosinophil was? Do you see it as contributing to the pathology in the lung?

DREBORG: yes indeed.

DEWDNEY: and having a very special role over and above that of the neutrophil or other cells—I just wondered what the data were?

DREBORG: I think the eosinophil has been discussed as being more important because of its content of specific proteins capable of destroying bronchial epithelium. Others can give more evidence than I can. My evidence that they were influenced was the reduction of eosinophil chemotactic activity in serum during the season and concomitant reduction of the bronchial sensitivity to histamine in asthmatics during the season.

QUESTIONER (unidentified): I cannot agree with you on the point concerning the skin tests after immunotherapy. We have performed skin tests before and after immunotherapy in a large group of patients suffering from hay fever and have not found any changes in the skin tests.

DREBORG: I think it depends on how you perform them and whether doses high enough to influence the skin sensitivity are used. I know about at least 20 trials which show a reduction of skin sensitivity. These data came from parallel-line bioassays determining the dose necessary to induce the same skin response. If the skin-prick test technique is good, parallel assay is used for evaluation, then change in concentration eliciting the same response, i.e. the change in concentration can be used in the same way as is done, for example, in bronchial challenge.

General discussion (Chapters 6–9)

TERHO (Glaxo, Finland): first I have a critical comment to Dr Dewdney's term 'vaccine'. I think vaccination means prevention of a disease and hyposensitization means that you try to lessen allergic symptoms and so I would like to suggest not to use vaccine when you talk about hyposensitization.

It was interesting to hear about Swedish studies on the role of environment and genetic background to the development of allergy. It seems to me that the most important is really the genetic background and the environmental role is secondary. I think that increasing evidence has gathered that by manipulating the environment we can cause some delay in the development of allergy but we cannot prevent the development of allergy. If a person is genetically predisposed to the development of allergy they will develop it sooner or later whatever we do to the environment. In Finland we have some data recently published which supports this idea. The National Health Institute in Helsinki compared the prevalence of allergic symptoms among children in industrially polluted areas with less polluted areas and found that during the early years children had more allergic symptoms and infections in the polluted areas. I think that role of environment is really secondary and the amount of allergic symptoms in the community may be quite constant. If some increase is happening it is due more to genetic background and so I think we have more possibilities in the treatment of allergy. In this context I would anticipate that Dr Katz's approach would be successful.

DEWDNEY: I think your point is very well taken on the use of the word vaccine. It is a battle that has been fought and lost and my own view is that it is now so widely used that we will not go back to the purist use of it, but you are right.

BJÖRKSTÉN: concerning your points on environment and genetic influences, I think Dr Terho is right and wrong! Genetics has not changed over the past millenium, allergy apparently has and so it has to be an environmental factor. You are right in the sense that the environmental factors we have so far been looking for have only minor effects. Pollutants seem to be important for the sensitization process. In Papua New Guinea where 30 years ago asthma or breathing difficulties were unheard of, the adult population now have allergic asthma.

KATZ: the situation in New Guinea was man made. We gave them blankets containing the house-dust mite.

ALLEN (Glasgow): adult asthmatics often have multiple sensitivities to different allergens. These patients are very difficult to treat. On the other hand, patients with mild single sensitivity asthma are easy to control. Is there any experience of hyposensitization in the patients with multiple allergies?

DREBORG: no. The results of such studies are not available as yet.

QUESTIONER (unidentified): we have had many years experience of immunotherapy with Beecham's allergens and we have shown that after 2–3 years of immunotherapy with, for example, Pollinex (depot allergen vaccine) there was a decrease in skin-test reactivity and histamine release from basophils. This was not found in the first year nor after the first season, but after 2–3 years of immunotherapy.

DREBORG: maybe you are right. I know Pollinex and it is widely used. On the other hand, with potent allergens the same effect is noticed within 3 months, when reaching the top dose. Did you perform a blind trial?

KAY: what is the effect of SFA on allergen-specific reactions, bearing in mind that it is not an allergen-specific protein? For instance, what does it do to immediate and late-phase reactions in those animals to which it has been administered? Do you have information in that area?

KATZ: we have information that it does not, for example, directly interfere with mast cell triggering or sensitized mast cells. We know that it has no effect on traditional cell-mediated immune reactions but not necessarily of IgE-related delayed reactions. We know that it decreases allergen-specific triggering of an individual or of an animal that has been depressed in terms of IgE responsiveness by previous treatment (as you expect because the serum load is decreased). So you either use passive or active cutaneous anaphylaxis as a measurement of IgE circulating in the blood. You see a corollary decrease in these reactivities.

BJÖRKSTÉN: is there a danger of immune deficiency if one of the isotypes is switched off totally.

KATZ: may I re-emphasize, it does not switch it off. I think that is a very important point because I agree we would be treading in very dangerous waters to do that. One of the advantages of this system is that if you get it down to a certain level the system is turned back on endogenously.

BJÖRKSTÉN: could your serum measurements reflect preformed IgE, as Keven Turner has shown in macrophages and monocytes?

KATZ: we have nothing directly to contradict that possibility.

DEWDNEY: what we are discussing is isotypic selective suppression of IgE, or induced immunodeficiency. Do any of you have views on this?

QUESTIONER (unidentified): is there any evidence that animals or humans who lack IgE or who have extremely low IgE are susceptible to infection?

KATZ: I would like to say that I do not believe that there is such a thing as a human being who lacks IgE. This is such a vital primary defence mechanism that its total absence is incompatible with life. To my knowledge an ε-agammaglobulinaemic human has not been described.

BJÖRKSTÉN: there are no diseases associated with extremely low IgE levels.

III

Agents which Suppress Inflammation

10

Anti-inflammatory agents in the treatment of bronchial asthma

A.B. KAY

Introduction

In the early part of this century, experimentalists were greatly impressed with the asthma-like symptoms produced in animals by drugs such as muscarine, pilocarpine and physostigmine. Atropine, hyoscyamine and chloroform abolished these effects and for many years were used in the treatment of asthma. Thus, the asthma attack was thought to be associated with excessive vagal discharges triggered by peripheral irritation, or centrally from cerebral influences. In addition, the classical work on histamine by Sir Henry Dale and others suggested a clear interrelationship between chemical mediators, anaphylaxis, allergy, smooth muscle contraction and asthma. The asthmatic paroxysm was regarded essentially as spasm of bronchial muscle, although 'vascular turgescence' of the bronchial mucus membrane and increased secretion of mucus glands were also recognized.

Concepts on the pathogenesis of asthma have changed considerably over the past few years. Basically, we no longer think of the asthma attack as bronchospasm in isolation but bronchoconstriction superimposed on a background of inflammation. This major modification in our thinking has important implications for therapy. Up until the 1950s, the mainstay of treatment for asthma were the antispasmodic agents, i.e. adrenaline, atropine and theophylline. Corticosteroids were introduced very cautiously into the treatment of asthma. At first they were reserved only for the extreme, life-threatening variants of the disease. Indeed, initially, there was considerable controversy as to the precise indications for cortisone-like drugs in 'prevention' of day-to-day asthma. Several years of clinical experience with corticosteroids raised two important questions. The first was whether the response to prednisolone, which was often dramatic, was consequent to the anti-inflammatory properties of the drug. The second regarded the state of the airways *between* attacks since in these situations it was clear that corticosteroids could often 'open up' the airways and appreciably increase lung function. Thus, many cases of asthma were recognized as *chronic disease* with acute exacerbations. In other words, spasm and paroxysm appeared to be superimposed on airways that were already chronically inflamed and, as a consequence, obstructed.

This chapter attempts to shed light on a number of questions. For instance, how inflamed is the bronchial mucosa of the mild asthmatic, that is to say, the patient who is reasonably controlled with bronchodilators alone? There have now been a number of studies in asthma using the technique of bronchoalveolar lavage which have established that mild asthmatic volunteers have increases in the numbers of inflammatory cells, particularly eosinophils. Not only were there increases in eosinophils in these lung washes but also of shed epithelial cells, and the numbers correlated with a degree of non-specific bronchial hyperresponsiveness [1]. In addition, the eosinophils were 'activated' as assessed by the concentrations of eosinophil granule-derived major basic protein in the lung lavage fluid. A recent study of bronchial biopsies in mild asthmatics clearly demonstrated shedding, or desquamation, of epithelial cells [2]. Furthermore, when bronchoalveolar lavage was performed before, and 3 weeks after, a course of sodium cromoglycate (SCG), patients who responded clinically to SCG had a highly significant decrease in the percentage of lung eosinophils [3]. Sodium cromoglycate and the new chemical entity nedocromil sodium (Tilade) have quite marked effects on inflammatory cells *in vitro* [4]. For instance, at nanomolar concentrations these drugs inhibited cell activation *in vitro*. Thus, SCG (which seems to be particularly effective in children) and Tilade (which has encouraging corticosteroid-sparing effects in adult asthmatics) might exert part of their effect through their ability to dampen infiltration of eosinophils and possibly other inflammatory cells. In a recent preliminary study, Tilade was shown to be equally effective as inhaled topical corticosteroids in treating asthma symptoms.

What about other inflammatory cells such as the neutrophil, macrophage and platelet? An early infiltration of neutrophils is observed after either allergen challenge or inhalation of ozone. This, in turn, is associated with increased hyperresponsiveness. Neutrophils are prominent in the various models of allergen-induced asthma in experimental animals. It is difficult, if not impossible, to say whether the neutrophil is more or less important than the eosinophil in bronchial asthma since the inflammatory process is extremely complex and the dynamics of cell accumulation and activation vary considerably from one individual to another.

Macrophages and platelets are also suspected of playing a role in asthma because, like the eosinophil, they contain type 2 IgE receptors. These are of lower affinity than the type 1 receptors on mast cells and basophils but seem to be functionally active in that stimulation of these cells via an IgE-dependent trigger leads to the release of granule contents and the elaboration of lipid mediators [5].

It must be borne in mind that mucus hypersecretion, like epithelial destruction, is a typical accompaniment of inflammation at mucosal surfaces. Mucus hypersecretion is an important component of asthma and contributes greatly to the plugging of small airways (a characteristic feature of asthma deaths).

In this review, evidence is cited which supports the following scenario. In between attacks, the asthmatic has an abnormal bronchial mucosa. To a greater or lesser degree it is inflamed in the sense that there is mucosal oedema together with infiltrating inflammatory cells. The eosinophil is prominent and it is believed that, in particular, the major basic protein and other cationic proteins from eosinophil granules lead to destruction of epithelial cells with exposure of irritant receptors on the basement membrane. This, in turn, leads to an increase in bronchial responsiveness. Thus, the paroxysms of asthma, i.e. bronchospasm, are superimposed on an underlying inflammatory process. Present evidence suggests that the inflammatory process lowers the threshold for the development of increased bronchial responsiveness after triggers such as allergens, exercise, fumes, irritants, etc. In theory, an inhaled, non-steroidal agent, free from appreciable side effects would be an ideal drug to prevent or even reverse this process. A number of other compounds such as nedocromil sodium and SCG appear to go a long way to fulfilling these criteria, although it must always be borne in mind that corticosteroids should never be withheld in chronic, or acute, severe asthma.

Bronchial hyperresponsiveness

Bronchial hyperresponsiveness is the increased reactivity of the airways to a wide variety of pharmacological and physical agents. Although hyperresponsiveness underlies much of the symptomatology of asthma, i.e. wheeziness provoked by exercise, exposure to cold air, fumes, smokes and sprays, as well as nocturnal and early-morning symptoms, it is also observed in other conditions, albeit to a lesser extent. For instance, there is a transient increase in bronchial hyperresponsiveness in normal individuals after upper respiratory tract infections, as well as in allergic rhinitis sufferers at the height of the pollen season. Hyperresponsiveness is also a feature of some patients with chronic obstructive airway disease and cystic fibrosis. The majority of cases of bronchial asthma of any severity have appreciable bronchial hyperresponsiveness. In fact, for the purposes of this article it will be assumed that in studies of asthma, where the clinical features are well documented, hyperresponsiveness is present even though a formal methacholine or histamine provocation concentration $(PC)_{20}$ may not have been performed.

What is inflammation?

Inflammation is the response of vascularized tissue to injury and serves to resolve and repair the effect of damage. The causes of inflammation, like those of cell injury, are diverse and include infectious agents (bacteria, viruses and parasites), physical agents (burns, radiation and trauma), chemical agents (drugs, toxins and industrial agents) and immunological reactions such as allergy and autoimmunity. The histopathological features of inflammation consist of changes in vascular blood flow and calibre of small blood vessels followed by alterations in vascular permeability leading to a series of white cell events. Acute inflammation is of short duration and characterized by exudation of fluid and plasma proteins (oedema) and leucocyte emigration, with neutrophils being prominent. In contrast, chronic inflammation is of longer duration with a dense infiltration of lymphocytes and macrophages together with proliferation of blood vessels and connective tissue. In certain specialized circumstances, such as allergy and asthma, the eosinophil is a prominent cell. Both acute and chronic inflammation are associated with some degree of fibrin deposition with platelet adherence and the release of platelet products. Basophils are also inflammatory cells and are prominent in certain forms of delayed-type hypersensitivity and are also found in the upper airways in allergic rhinitis. At the present time there is no conclusive evidence that the basophil participates in pathological processes in the lung. Thus, the cells migrating from the blood vessels which cause airway inflammation include the neutrophil, eosinophil, lymphocyte, macrophage and platelet with the role of the basophil remaining uncertain. Certain fixed tissue cells such as mast cells and epithelial cells also participate as does the fibroblast. Inflammation at mucosal surfaces has two additional important features—mucus hypersecretion and shedding or denudation of the airway surface. It is important at the outset, to make the point that it is very unlikely that one cell, or one mediator, will explain totally the mechanisms of hyperreactivity. It is more likely that the *combined* effects of cells and mediators are required for the observed effects.

We know from the pathology of asthma that there are certain features which are very typical of the disease. These include intense infiltration of eosinophils and deposition of eosinophil products in and around the bronchial epithelium [6, 7]. There are also large numbers of lymphocytes and macrophages. Neutrophils are sometimes present but not invariably so. There is marked hyperaemia and dilatation of blood vessels and considerable mucus hypersecretion with plugging of the small airways. Shedding of epithelial cells is a common finding and appears as clumps in the sputum (Creola bodies). Other important features include thickening of the basement membrane and goblet-cell hyperplasia.

Microvascular leakage

It is virtually certain that microvascular leakage occurs in asthma of any severity not only because of the histopathological findings but because sputum [8, 9] and bronchoalveolar lavage (BAL) fluid from asthmatics contain elevated concentrations of albumin [10, 11]. Many of the mediators implicated in asthma are known to cause microvascular leakage at postcapillary venules. These include histamine, bradykinin, sulphidopeptide leukotrienes and platelet-activating factor (PAF-acether) [12–15]. In addition, stimulation of the vagus nerve and capsaicin, via release of sensory neuropeptides such as substance P, cause microvascular leakage in rodents [16]. Mediators which increase bronchial blood flow might be expected to exaggerate leakage in asthmatic airways. Oedema, as a result of increased capillary permeability, may have several sequelae relevant to asthma. These include a contribution to narrowing of small airways, epithelial shedding, the formation of mucus plugs, inhibition of mucociliary clearance and, by providing a rich source of plasma proteins, substrate for complement-derived anaphylatoxins and kinins. In the guinea pig, adrenaline is very effective in reversing microvascular leakage [17] and is a traditional drug in the treatment of asthma, particularly acute severe asthma. In certain circumstances, adrenalin might have advantages over selective β-agonists and this possibility should be explored.

Neutrophils

The evidence that neutrophils by themselves play an important role in bronchial hyperresponsiveness associated with ongoing clinical asthma is weak. On the one hand, a number of studies in experimental animals, as well as control models of asthma in humans, indicate that neutrophils may play a part early on in the asthma process. Neutrophils appear to be 'normal' residents of larger airways both in normoresponsive and hyperresponsive individuals. For instance, Wardlaw *et al.* [1] found a large percentage of neutrophils in the bronchial wash of non-atopic controls, hay fever sufferers as well as mild asthmatics. The numbers were approximately equal in all groups and considerably higher than those observed in bronchoalveolar wash. In fact, neutrophils accounted for almost 50% of the total cell count in bronchial wash in normal subjects. It is relevant that scrapings from the nasal mucosa of normal individuals also contained large numbers of neutrophils [Wardlaw, unpublished observations] as did normal conjunctiva.

Numerous studies indicate that peripheral blood neutrophils become 'activated' after allergen- and exercise-induced early- and late-phase asthmatic responses (EPR, LPR) [18–21]. Activation was assessed by increased membrane expression of complement receptors and enhanced cytotoxicity for complement-coated targets.

Significant increases in the percentage of neutrophils in BAL and bronchial mucus have been observed in subjects experiencing LPR after bronchial challenge [22]. Comparable observations were made in BAL during LPR in sensitized subjects challenged with toluene diisocyanate, in which it was found that increases in neutrophils, as well as eosinophils and lymphocytes, were inhibited by prior administration of oral prednisolone [11]. In contrast, BAL neutrophilia was not a feature of late reactions elicited by plicatic acid in red-cedar asthma [23].

The elaboration of a high molecular weight neutrophil chemotactic activity (HMW-NCA) into the circulation of patients after allergen- or exercise-induced EPR and LPR is well documented [24–27]. High molecular weight neutrophil chemotactic activity was associated with molecules having a molecular size of approximately 600 kD and a near neutral isoelectric point. This activity has recently been identified in 'real asthma', i.e. from the serum of asthmatics admitted to hospital with acute severe disease ('status asthmaticus') [28]. The molecular size of HMW-NCA in acute severe asthma was heterogeneous, i.e. 800, 600 and <20 kD. When peripheral blood mononuclear cells from patients with acute severe asthma were cultured in serum-free medium, HMW-NCA could also be detected in the supernatant [29]. Present evidence suggests that HMW-NCA may be derived from lymphocytes and/or monocytes and that it is related to the 10 kD neutrophil chemotactic factor (now fully characterized and sequenced by a number of groups [30–34]). The high molecular weight of the serum factor might be an artefact of heating to 56 °C for 30 min and this explanation is currently being explored.

Neutrophils have the potential for producing considerable airway damage and consequent hyperresponsiveness by virtue of their capacity to generate potent lipid mediators such as prostaglandins (PG), thromboxanes (Tx), leukotriene (LT)B_4 and PAF-acether. The cell has been implicated in ozone-induced and antigen-induced hyperreactivity in dogs [35, 36] and rabbits [37]. Furthermore, supernatants from phagocytosing neutrophils *in vitro* induced hyperreactivity when nebulized into the airways of rabbits [38]. The active agents in this model are yet to be identified. There was also a seven-fold increase at 6 h and 17-fold increase at 17 h in neutrophils in BAL in a guinea pig model of late-phase and 'late late'-phase bronchoconstriction [39]. On the other hand, whereas nedocromil sodium blocked the late reaction and subsequent eosinophil infiltration in BAL it did not affect the neutrophil infiltration, suggesting that in the late-phase and the 'late late'-phase neutrophil infiltration is less critical for the development of airway obstruction [40].

Eosinophils

There is now very persuasive evidence which suggests that the eosinophil is perhaps the single, most important pro-inflammatory cell in the asthma process. It is well known that a blood and sputum eosinophilia is often, but not invariably, found in association with mild or severe, acute or chronic asthma. A blood eosinophilia accompanied late-phase but not single early asthmatic responses and there was an inverse correlation between the blood eosinophil count and the degree of non-specific bronchial hyperreactivity as measured by the methacholine PC_{20} [41]. The accumulation of eosinophils and eosinophil products (major basic protein, MBP; eosinophil cationic protein, ECP; eosinophil-derived neurotoxin, EDN) have been observed in BAL during allergen-induced LPR [22, 42]. Similar observations were made in red-cedar asthma in which it was observed that plicatic acid inhalation elicited a BAL eosinophilia together with sloughing of bronchial epithelial cells [23]. In a placebo-controlled, double-blind study, SCG was shown to suppress the local accumulation of eosinophils in bronchial mucus and BAL fluid, and these reductions in lung eosinophils were related to clinical improvement [3].

Eosinophils are also very prominent cells in many of the histopathological sections obtained from asthma deaths [6, 43]. Major basic protein was prominent in the bronchial wall in the mucus plugs of virtually all of these patients, even though only a few intact eosinophils were observed by routine light microscopy [44]. Major basic protein concentrations are also elevated in the sputum from asthmatics. In fact the characteristic pathology has been termed 'chronic eosinophilic desquamative bronchitis'.

Eosinophil cationic protein and MBP are both cytotoxic to the respiratory epithelium [45] and both may account for the denudation of the epithelium as seen in asthma. Asthmatics with airway hyperreactivity ($PC_{20} < 4$ mg/ml) had significant elevations in the eosinophil count and concentrations of MBP in BAL fluid [1]. Furthermore, there were significant correlations between the amount of MBP recovered and the percentage of eosinophils. These changes were even more marked when asthmatics with airway hyperreactivity were compared with subjects with normoreactive airways (symptomatic asthmatics above). There were inverse correlations between the PC_{20} and the percentage of eosinophils and epithelial cells and the amount of MBP in BAL. This study, and that by Lam *et al.* [23] clearly supported the hypothesis that bronchial hyperresponsiveness is secondary to epithelial cell damage mediated through eosinophil-derived granule products.

In addition to the basic proteins of the eosinophil granule, membrane-derived agents may also play a role in the pathogenesis of asthma. Eosinophils produced considerable quantities of LTC_4 after ionophore- [46, 47], IgG-

[48] and IgE-dependent stimuli [5], and small elevations in LTC_4 were noted in BAL in the late-phase reactions when fluid from diluent challenge was compared with that obtained from allergen challenge [22]. Leukotriene D_4, or its 20-OH-LTB_4 metabolite, were also observed in BAL from patients with symptomatic asthma [49, 50].

Eosinophils have the capacity to generate considerable quantities of PAF-acether [51, 52]. Platelet-activating factor may be of particular relevance to asthma because of its ability to cause vasoconstriction and increased vascular permeability, to be chemotactic for eosinophils [53], to enhance mucus secretion and to increase bronchial hyperresponsiveness after inhalation in humans [54]. There are no convincing studies of PAF-acether elaboration associated with clinical asthma.

The mechanism of recruitment of the eosinophil, in preference to the neutrophil, to the human asthmatic airway still needs to be explained. Platelet activating factor is a potent chemotactic factor for eosinophils but is equally effective in evoking directional neutrophil migration. It seems likely that *in vitro* chemotaxis is not a true model of cell accumulation *in vivo* since it does not take into account the special requirements of adhesion to endothelial cells. Furthermore it is now appreciated that T-cell derived products which play a vital role in eosinophil maturation also affect the mature cell. For instance, both granulocyte/macrophage-colony stimulating factor (GM-CSF) and interleukin (IL) 5 activate mature eosinophils in terms of increased cytotoxicity and oxidative metabolism and prolong the life of eosinophils *in vitro*. In preliminary experiments it was shown that IL-5 causes eosinophil hyperadhesiveness to serum-coated glass. Thus T-cell products and/or PAF-acether acting either alone, in combination or in sequence might alter vascular endothelial cells in such a way to cause selective attachment of eosinophils.

Recruitment and activation of eosinophils are strongly inhibited by corticosteroids, an effect that could also explain the efficacy of these drugs in modifying late-phase bronchoconstriction and to a lesser extent increased bronchial hyperresponsiveness.

Mast cells and basophils

In humans, mast cells are located in the lumen (where they can be recovered by BAL), bronchial epithelium and submucosa (lung parenchyma). Basophils have not been identified in bronchial pathology or in any situation associated with hyperresponsiveness, although they probably exist as they do in most other organs.

Current evidence suggests that the early allergic asthmatic reaction is predominantly mast-cell mediated. The immediate response to inhaled allergens in atopic subjects (and the accompanying elevations in plasma histamine) is rapid in onset and easily reversed by inhaled β_2-adrenoreceptor agonists (i.e. albuterol) and SCG [55]. Thus, in these situations these drugs are thought to act primarily on the mast cell. Corticosteroids given over a period of time also attenuate the immediate reaction [56] possibly by depletion of mast cells in the mucosa [57, 58]. Following allergen challenge in atopic individuals there was an elevation in plasma histamine [59] and in BAL fluid [60]. This occurred within minutes of challenge and over the following few hours increased urinary secretion of a major catabolite—*N*-methylhistamine—was identified [61]. Mast cells recovered from the airways by lavage within the first 15 min of allergen challenge have all the morphological features of non-cytotoxic degranulation [60, 62]. Human lung mast cells also elaborate LTC_4, PGD_2, PAF-acether, various chemotactic peptides, proteolytic enzymes and proteoglycans. The pathophysiological role of these lipid mediators, proteolytic enzymes and proteoglycans is unknown.

In ongoing, day-to-day asthma there is an inverse correlation between the methacholine PC_{20} and the percentage of mast cells in BAL [1]. Furthermore, asthmatics with airway hyperreactivity have significant increases in spontaneous histamine release. On the other hand, increases in the number of BAL mast cells are also found in sarcoidosis, fibrosing alveolitis and, in particular, hypersensitivity pneumonitis where they comprise as many as 4–8% of the differential cell count [63]. Thus, there does not seem anything in particular about the mast cell in asthma to believe that this cell plays a special role except in the early bronchospasm observed in susceptible individuals after the inhalation of aeroallergens. Even in this circumstance the role of cells bearing the other lower-affinity FcεRII (i.e. eosinophils, platelets and macrophages) has to be reckoned with.

What is the role of the mast cell in LPR and ongoing asthma? Almost a decade ago it was shown that $F(ab)'_2$ anti-IgE, when injected into the skin, produced an LPR which had many of the histopathological features of allergen-induced LPR. For this reason it was hypothesized that mast cells were essential for LPR and that mast cell-derived chemotactic factors accounted for the subsequent infiltration of eosinophils, neutrophils and basophils. However, the anti-IgE may also have interacted with macrophages or lymphocytes through their FcεRII, and these cells in turn may have contributed to the LPR. In fact, evidence is now accumulating that LPR (mainly from the skin, and to a lesser extent in the lung) might be a form of delayed-type hypersensitivity and the evidence for this is discussed below under 'Lymphocytes'.

Activation of mast cells as a pathogenetic mechanism of immediate bronchoconstriction is not limited to allergen exposure since it is also an important component of asthma provoked by exercise, cold air and hyperventilation [64].

Thus, the evidence that mast cells are involved in immediate bronchoconstriction to immunological and non-immunological stimuli is still persuasive. The evidence that mast cells play a role in the LPR is debatable and there is little evidence to suggest that the cell plays a pivotal role in ongoing, day-to-day, chronic asthma. Mast cells are a feature of inflammation in general and are found in association with a variety of pulmonary pathologies. Indeed the most dramatic increases in mast-cell numbers are found in interstitial lung disease.

Monocytes and macrophages

The majority of cells recovered from BAL in normal as well as asthmatic subjects are macrophages. These cells contain functional IgE receptors (FcεRII) [65, 66] and the numbers of these IgE-bearing cells are substantially greater in atopic asthmatics than normal controls [67]. Macrophages have the capacity to produce a wide range of eicosanoids including PAF-acether. Macrophages clearly participate in airway responses to inhaled allergens. Appreciable amounts of β-glucuronidase were recovered in cell-free supernatants of BAL [68] and there was a substantial increase in the number of macrophages after allergen challenge [60]. This increase appears to be due to the migration of monocytes into the lung because most of the increase can be accounted for by peroxidase-positive cells (a characteristic feature of monocytes compared with alveolar macrophages).

Lung macrophages are also activated in LPR after allergen challenge as shown by an increase in the number of complement rosettes [22]. Chronic severe asthmatics who are relatively unresponsive to corticosteroids have increased numbers of circulating activated macrophages [69]. At one time this was thought to be a primary macrophage defect but this now appears to be secondary to T-cell activation [70].

Platelets

Initial experiments in the rabbit and in the guinea pig showed that platelet depletion prevented PAF-acether-induced airway hyperreactivity [71]. Platelets have been recovered in the lavage fluid following allergen challenge [72] and platelet factor 4 was identified in the plasma of atopic subjects [73] (although in a further study this was not confirmed [74]). Platelets bear the second IgE Fc receptor which is functionally active as shown by IgE-dependent oxidative metabolism and cytotoxicity [75]. By and large, the evidence that platelets play a pivotal role in human asthma, and are directly concerned in airway hyperresponsiveness, is unconvincing. Microvascular leakage associated with fibrin deposition and platelet trapping would be expected to involve elements of the coagulation pathways. More discerning experiments of nature, or precise antiplatelet drugs, might resolve this issue.

Lymphocytes

An area of considerable current interest is the possible role of the T-cell in the regulation and expression of the inflammation associated with allergy and asthma. The T-cell derived lymphokines, IL-4, IL-5 and interferon (IFN)-γ, are intimately involved in the regulation of IgE production [76]. Some lymphokines are active in the control of eosinophil production by the bone marrow (IL-5, GM-CSF, IL-3) and in the regulation of mast-cell differentiation. Others have chemotactic activity for neutrophils, eosinophils and basophil granulocytes as well as monocytes and can activate or degranulate these non-specific effector cells. T-cells also play a general role in the regulation of specific immune responses and are a probable target cell for desensitization immunotherapy.

Direct evidence for T-cell changes in asthma and allergy come from a variety of sources. Postmortem examination of the airways of asthmatic patients revealed large numbers of lymphocytes [43]. Increased numbers of 'atypical intraepithelial lymphocytes' were found in an ultrastructural morphological study of bronchial biopsies taken during life with subjects with mild asthma [77]. These lymphocytes are probably activated T-cells but formal proof of this is not yet available. Increased natural killer (NK) activity has been described in the peripheral blood of asthmatic patients. Natural killer activity is an inducible property of T-cells and of non-T, non-B lymphocytes and is thus a non-specific indicator of lymphocyte activation.

Measurements in chronic asthmatics indicated that patients who were relatively refractory to treatment with corticosteroids had a relative decrease in the numbers of circulating T-suppressor (CD8) cells [78]. These patients also had an abnormality of T-cell growth *in vitro* (colony counts in soft agar) since, unlike T-cells from normal individuals and corticosteroid-responsive subjects, cell proliferation from the refractory patients was not inhibited by optimal concentrations of methyl prednisolone [79]. A defect in concanavalin A-induced suppressor cell function has been described in asthma [80–84] and successful immunotherapy was associated with both an increase in the relative number of T-suppressor cells (OKT8) [85] and an abrogation of allergen-induced LPR [86].

T-cell subset changes have also been studied using the model of bronchial allergen challenge in three separate studies [60, 87, 88]. The design of each of these investigations was different but taking them together it appears that T-cells bearing the CD4 marker (helper/inducer subset) were depleted in peripheral blood and selectively retained in the lung following challenge, but the kinetics and significance of this finding are as yet unclear. In addition, it appears that there is a difference in the profile of regulatory T-cell subsets present in BAL in single early responders when compared with dual asthmatic (early- and late-phase) responders.

Skin studies offer a useful alternative to inhalation challenge where the ethical and practical limitations on access to the airways make it difficult to answer detailed questions about the kinetics of allergic reactions. A certain amount of caution is obviously required in extrapolating results from the skin to the lung, but evidence from animal studies suggests that the pathological features of skin and pulmonary actions are in fact very similar. Monoclonal antibodies against lymphocyte surface antigens have made it possible to identify functional subsets of T-cells both in cell suspensions and in tissue sections and also to assess activation states using immunocytochemistry to enumerate stained leucocytes. In a recent study these techniques were used to make a detailed study of cell traffic and activation in human allergen-induced cutaneous LPR and have shown a lymphocyte infiltration which was almost exclusively CD4-positive (helper/inducer subset) [89]. Some of these T-cells were activated in that they stained positively for the presence of IL-2 receptors. Further evidence of T-cell activation was provided by the observation that endothelial cells in the allergen-challenged biopsies showed an increased density of class II histocompatibility antigen (HLA-DR) expression compared with control sites. This indicated local secretion of the T-cell derived soluble inflammatory mediator, IFN-γ. In the same study it was shown that eosinophil accumulation and activation were striking features of the skin LPR with numerous activated eosinophils (EG-2^{+}) present at 6 h after challenge and persisting in tissue for up to 48 h.

The role of T-cells was also studied in acute severe asthma. T-cell subsets and the expression of lymphocyte-activation markers were measured in the peripheral blood of patients admitted to hospital with acute severe asthma ('status asthmaticus') [90]. Measurements were made on admission, day 3 and day 7 (or on discharge from hospital if this was sooner). The results were compared with controls (mild asthma, chronic obstructive airway disease (COAD) and normal individuals). The percentages of CD4-positive and CD8-positive T-cells, and the CD4/CD8 ratios, in the acute severe asthma patients and the control groups were similar and within the accepted normal range. In contrast, patients with acute severe asthma had significant elevations of the expression of three surface proteins associated with T-cell activation (IL-2R, HLA-DR and 'very-late activation' antigen, VLA-1), compared with control subjects. Phenotypic analysis of the IL-2R-positive T-cells showed that these cells were exclusively of the CD4 'helper-inducer' phenotype. The percentages of IL-2R- and HLA-DR-positive (but not VLA-1-positive) lymphocytes tended to decrease as the patients were treated and improved clinically, although in the 7-day observation period these remained elevated above control values. In addition, the serum concentrations of IFN-γ and soluble IL-2R were significantly elevated in patients with acute severe asthma as compared with all the control groups [91]. Concentrations decreased as the patients improved clinically during the first 7-day period of hospital treatment. At various time points during this initial 7-day period, a highly significant correlation was observed between the degree of airway obstruction as measured by the peak expiratory flow (PEF) rate and: (1) the percentages of peripheral blood T-cells expressing IL-2R; and (2) the serum concentrations of soluble IL-2R. Thus, taken together, all of these observations provide further evidence that CD4 T-cell activation is important in the pathogenesis of acute severe asthma.

Epithelial cells

The involvement of the airway epithelium in the pathogenesis of asthma is suggested by the finding of abnormal epithelium in airways of asthmatics. Thus, during asthma attacks, epithelial cells are desquamated and are found in the sputum as Creola bodies or Curschman's spirals. Postmortem examinations of asthmatics who die show extensive shedding of epithelial cells and severe epithelial inflammation. Even subjects with stable asthma show evidence of airway epithelial desquamation.

Until recently, airway epithelial cells were considered to form simply an inert, physical lining that covers the airways. It is now apparent that these cells, with their key location at the interface between the external environment and the internal milieu, play an important role in the defence of the airways.

Airway epithelial cells are capable of generating potent lipoxygenase products of arachidonic acid [92, 93]. Human airway epithelial cells produce 15-hydroxy-5,8,11,13-eicosatetraenoic acid (15-HETE) and 8,15-di-HETEs [94] and dog epithelial cells produce LTB_4 and selected HETEs [92]. Both LTB_4 and 8,15-di-HETEs [94, 95] are chemotactic for neutrophils, and this could explain the neutrophil infiltration of the airways that occurs with viral infection [96], ozone [97, 98] and other airway stimuli. The 15-HETEs are also of interest because they are reported to stimulate the release of mediators from mast cells [99]. Thus, epithelial cell stimulation by damage or by other mediators, produces lipoxygenase products which in turn stimulate other inflammatory cells.

Summary and concluding remarks

A diagrammatic representation of the interactions between mediators of hypersensitivity and leucocytes in early-, late-phase and ongoing asthma is shown in Fig. 10.1.

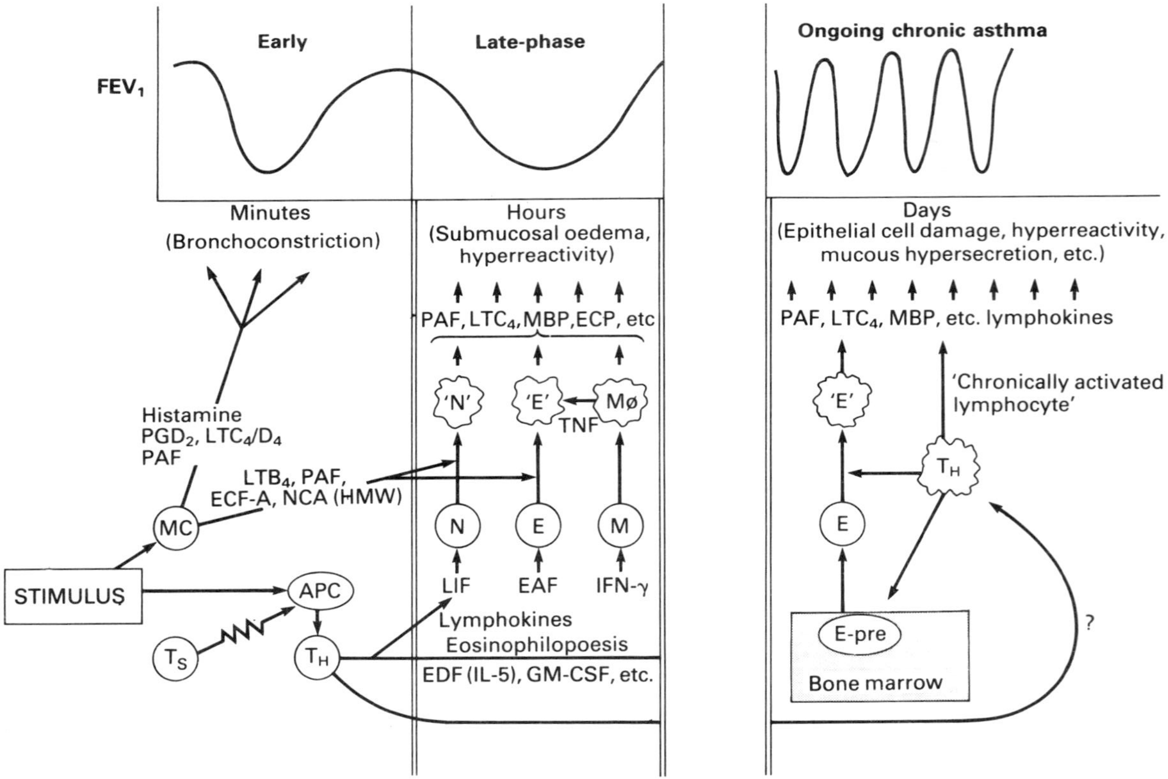

Fig. 10.1 Interactions between mediators of hypersensitivity and leucocytes in early-, late-phase and ongoing asthma. MC = mediator cell; N = neutrophils; E = eosinophils; Mø = macrophages; T_H = T-helper cells; T_S = T-suppressor cells; APC = antigen-processing cells; TNF = tumour necrosis factor; ECF-A = eosinophil chemotactic factor of anaphylaxis.

Early-phase, or immediate reactions are largely the result of bronchoconstriction consequent to the release of mediators such as histamine, PGD_2, LTC_4/D_4 and PAF-acether. The principal mediator cell is the mast cell (although other IgE receptor-bearing cells such as the macrophage, eosinophil and platelet might also be involved in this immediate response). The stimulus for mediator-cell activation may be either immunological (IgE dependent) or non-immunological (i.e. changes in osmolarity as a result of the respiratory water loss associated with exercise-induced asthma). Late-phase reactions appear to be a consequence of infiltration with neutrophils, eosinophils and macrophages. These cells are recruited and activated either by mast-cell associated chemotactic factors (such as LTB_4, PAF-acether, the eosinophil chemotactic factor of anaphylaxis or HMW-NCA) and/or 'lymphokines' derived from T-helper cells which have been stimulated by antigen processed by the antigen-processing cells. These mononuclear cell interactions are under the control of regulatory T-cells (T-suppressor cells) and it is speculated that the availability of these subsets may determine the magnitude of the LPR. Lymphokines and monokines which selectively activate neutrophils, eosinophils and monocytes include leucocyte inhibitory factor (LIF), eosinophil-activating factor (EAF) and IFN-γ respectively. Macrophage-derived tumour necrosis factor also amplifies the inflammatory response by its capacity to enhance eosinophil cytotoxicity. Eosinophil-derived agents such as PAF-acether, LTC_4, MBP and ECP might be responsible for submucosal oedema and non-specific bronchial hyperreactivity which are characteristic features of LPR. T-cell derived lymphokines such as EDF (IL-5), together with GM-CSF, might lead to eosinophilopoiesis and account for the prolonged eosinophilia of ongoing chronic asthma. The T-cell is prominent in the pathology of chronic asthma and is possibly 'chronically activated'. Thus, lymphocytes, driven by as yet undetermined 'antigens' (possibly viral) may perpetuate the inflammatory response in and around the bronchi. Interleukin-5-like products from these putative activated lymphocytes might perpetuate:

1 Eosinophil production by the bone marrow.
2 Its release into the circulation.
3 Its migration into bronchial tissue.
4 Activation to release PAF-acether, LTC_4, MBP, etc.

Lymphokines released directly from these T-helper cells might also influence bronchial pathology. The end result is epithelial cell damage, amplification of hyperreactivity, together with mucus hypersecretion and total or partial plugging of small to medium-size airways. The various aspects of this broad hypothesis are currently under investigation.

References

1 Wardlaw AJ, Dunnette S, Gleich GJ, Collins JV, Kay AB. Eosinophils and mast cells in bronchoalveolar lavage in mild asthma: relationship to bronchial hyperreactivity. *Am. Rev. Respir. Dis.* 1988;**137**:62–69.

2 Laitinen LA, Heino M, Laitinen A, Kave T, Haahtela T. Damage of the airway epithelium and bronchial reactivity in patients with asthma. *Am. Rev. Respir. Dis.* 1985; **131**:599–606.

3 Diaz P, Galleguillos FR, Gonzalez MC, Pantin C, Kay AB. Bronchoalveolar lavage in asthma: the effect of DSCG on leucocyte counts, immunoglobulins and complement. *J. Allergy Clin. Immunol.* 1984; **74**:41–48.

4 Moqbel R, Cromwell O, Walsh GM, Wardlaw AJ, Kurlak L, Kay AB. Effects of nedocromil sodium (Tilade) on the activation of human eosinophils and neutrophils and the histamine release from mast cells. *Allergy* 1988; **43**:268–276.

5 Moqbel R, MacDonald AJ, Kay AB. IgE-dependent release of leukotriene (LT) C_4 from human low density eosinophils. *J. Allergy Clin. Immunol.* 1988; **81**:208 (Abstract 158).

6 Dunnill MS. The pathology of asthma with special reference to changes in the bronchial mucosa. *J. Clin. Path.* 1960; **13**:27.

7 Dunnill MS, Massarella GR, Anderson JA. A comparison of the quantitative anatomy of the bronchi in normal subjects, in status asthmaticus, in chronic bronchitis and in emphysema. *Thorax* 1969; **24**:176.

8 Ryley HC, Brogan TD. Variation in the composition of sputum in chronic chest diseases. *Br. J. Exp. Pathol.* 1968; **49**:25.

9 Brogan TD, Ryley HC, Neale L, Yassa L. Soluble proteins of bronchopulmonary secretions from patients with cystic fibrosis, asthma and bronchitis. *Thorax* 1975; **30**:72.

10 Lam S, LeRiche JC, Kijek K, Phillips RT. Effect of bronchial lavage volume on cellular and protein recovery. *Chest* 1985; **88**:856.

11 Boschetto P, Fabbri LM, Zocca E, Milani G, Pivirotto F, Dal Vecchio A, Plevani M, Mapp CE. Prednisone inhibits late asthmatic reactions and airway inflammation induced by toluene diisocyanate in sensitized subjects. *J. Allergy Clin. Immunol.* 1987; **80**:261–267.

12 Saria A, Lundberg JM, Skofitsch G, Lembeck F. Vascular protein leakage in various tissues induced by substance P, capsaicin, bradykinin, serotonin, histamine and by antigen challenge. *Naunyn-Schmiedeberg's Arch. Pharmacol.* 1983; **324**:212.

13 Persson CGA. Leakage of macromolecules from the tracheobronchial circulation. *Am. Rev. Respir. Dis.* 1987; **135**:S71.

14 Hua X-Y, Dahlen S-E, Lundberg JM, Hammarstrom S, Hedqvist P. Leukotrienes C_4 and E_4 cause widespread and extensive plasma extravasation in the guinea pig. *Naunyn-Schmiedeberg's Arch. Pharmacol.* 1985; **330**:136.

15 Evans TW, Chung K, Rogers DF, Barnes PJ. Effect of platelet-activating factor on airway vascular permeability: possible mechanisms. *J. Appl. Physiol.* 1987; **63**:479.

16 Lundberg JM, Saria A, Lundblad L, Angaard A, Martling C-R, Theodorsson-Norheim E, Stjarne P, Hokfelt T. Bioactive peptides in capsaicin-sensitive C-fiber afferents of the airways: functional and pathophysiological implications. In *Neural Control in Health and Disease* (Eds Kaliner MA, Barnes PJ), Marcel Dekker, New York 1987, p. 417.

17 Boschetto P, Roberts NM, Rogers DF, Barnes PJ. Effect of anti-asthma drugs on microvascular leakage in guinea-pig airways. *Am. Rev. Respir. Dis.* 1988; In press.

18 Papageorgiou N, Carroll M, Durham SR, Lee TH, Walsh GM, Kay AB. Complement receptor enhancement as evidence of neutrophil activation after exercise-induced asthma. *Lancet* 1983; **ii**:1220–1223.

19 Moqbel R, Durham SR, Shaw RJ, Walsh GM, MacDonald AJ, Mackay JA, Carroll MP, Kay AB. Enhancement of leukocyte cytotoxicity after exercise-induced asthma. *Am. Rev. Respir. Dis.* 1986; **133**:609–613.

20 Carroll M, Durham SR, Walsh GM, Kay AB. Activation of neutrophils and monocytes after allergen- and histamine-induced bronchoconstriction. *J. Allergy Clin. Immunol.* 1985; **75**:290–296.

21 Durham SR, Carroll M, Walsh GM, Kay AB. Leucocyte activation in allergen-induced late-phase asthmatic reactions. *N. Engl. J. Med.* 1984; **311**:1398–1402.
22 Diaz P, Gonzalez MC, Galleguillos FR, Ancic P, Cromwell O, Kay AB. Leucocytes and mediators in bronchoalveolar lavage during allergen-induced late-phase asthmatic reactions. *J. Allergy Clin. Immunol.* 1987; **79**:256 (Abstract).
23 Lam S, LeRiche J, Phillips D, Chan-Yeung M. Cellular and protein changes in bronchial lavage fluid after late asthmatic reaction in patients with red cedar asthma. *J. Allergy Clin. Immunol.* 1987; **80**:44–50.
24 Atkins PC, Norman M, Weiner H, Zweiman B. Release of neutrophil chemotactic activity during immediate hypersensitivity reactions in humans. *Ann. Int. Med.* 1977; **86**:415–418.
25 Nagy L, Lee TH, Kay AB. Neutrophil chemotactic activity in antigen-induced late asthmatic reactions. *N. Engl. J. Med.* 1982; **306**:497–501.
26 Lee TH, Nagy L, Nagakura T, Walport MJ, Kay AB. The identification and partial characterisation of an exercise-induced neutrophil chemotactic factor in bronchial asthma. *J. Clin. Invest.* 1982; **69**:889–899.
27 Lee TH, Nagakura T, Papageorgiou N, Iikura Y, Kay AB. Exercise-induced late asthmatic reactions with neutrophil chemotactic activity. *N. Engl. J. Med.* 1983; **308**:1502–1505.
28 Buchanan DR, Cromwell O, Kay AB. Neutrophil chemotactic activity in acute severe asthma ('status asthmaticus'). *Am. Rev. Respir. Dis.* 1987; **136**:1397–1402.
29 Buchanan DR, Fitzharris P, Cromwell O, Kay AB. Neutrophil chemotactic activity from cultured blood mononuclear cells in acute severe asthma. *Thorax* 1987; **42**:749 (Abstract).
30 Yoshimura T, Matsushima K, Tanaka S, Robinson EA, Appella E, Oppenheim JJ, Leonard EJ. Purification of a human monocyte-derived neutrophil chemotactic factor that has peptide sequence similarity to other host defense cytokines. *Proc. Natl. Acad. Sci. USA* 1987; **84**:9233–9237.
31 Van Damme J, Van Beeumen J, Opdenakker G, Billiau A. A novel, NH_2-terminal sequence-characterized human monokine possessing neutrophil chemotactic, skin-reactive, and granulocytosis-promoting activity. *J. Exp. Med.* 1988; **167**:1364–1376.
32 Gregory H, Young J, Schröder J-M, Mrowietz U, Christophers E. Structure determination of a human lymphocyte derived neutrophil activating peptide (LYNAP). *Biochem. Biophys. Res. Commun.* 1988; **151**:883–890.
33 Maestrelli P, Tsai J-J, Cromwell O, Kay AB. The identification and partial characterization of a human mononuclear cell-derived neutrophil chemotactic factor apparently distinct from IL-1, IL-2, GM-CSF, TNF and IFN-gamma. *Immunology* 1988; **64**:219–225.
34 Maestrelli P, O'Hehir RE, Lamb JR, Tsai J-J, Cromwell O, Kay AB. Antigen-induced neutrophil chemotactic factor from cloned human T lymphocytes. *Immunology* 1988; **65**:605–609.
35 Lee L-Y, Bleecker ER, Nadel JA. Effect of ozone on bronchomotor response to inhaled histamine aerosol in dogs. *J. Appl. Physiol.* 1977; **43**:626–631.
36 Chung KF, Becker AB, Lazarus SC, Frick OL, Nadel JA, Gold WM. Antigen-induced airway hyperresponsiveness and pulmonary inflammation in allergic dogs. *J. Appl. Physiol.* 1985; **558**:1347–1353.
37 Marsh WR, Irvin CG, Murphy KR, Behrens BL, Larsen GL. Increases in airway reactivity to histamine and inflammatory cells in bronchoalveolar lavage after the late asthmatic response in an animal model. *Am. Rev. Respir. Dis.* 1985; **131**:875–879.
38 Irvin CG, Baltopoulos G, Henson P. Airways hyperreactivity produced by products from phagocytosing neutrophils. *Am. Rev. Respir. Dis.* 1985; **131**:A278 (Abstract).
39 Hutson PA, Church MK, Clay TP, Miller P, Holgate ST. Early and late-phase bronchoconstriction after allergen challenge of nonanesthetised guinea pigs. 1. The association of disordered airway physiology to leukocyte infiltration. *Am. Rev. Respir. Dis.* 1988; **137**:548–557.
40 Church MK, Hutson PA, Holgate ST. Comparison of nedocromil sodium and albuterol against late phase bronchoconstriction and cellular infiltration in guinea pigs. *Am. Rev. Respir. Dis.* 1988; **137**:136 (Abstract).

41 Durham SR, Kay AB. Eosinophils, bronchial hyperreactivity and late-phase asthmatic reactions. *Clin. Allergy* 1985; **15**:411–418.
42 De Monchy JG, Kauffman HF, Venge P, Koeter GH, Jansen HM, Sleuter HJ, DeVries K. Bronchoalveolar eosinophils during allergen-induced late asthmatic reactions. *Am. Rev. Respir. Dis.* 1985; **131**:373–376.
43 Dunnill MS. The pathology of asthma. In *Allergy, Principles and Practices* (Eds Middleton E Jr, Reed CE, Ellis EF), CV Mosby, St Louis, Missouri, 1978, pp. 678–686.
44 Filley WV, Holley KE, Kephart GM, Gleich GJ. Identification by immunofluorescence of eosinophil granule major basic protein in lung tissues of patients with bronchial asthma. *Lancet* 1982; **ii**:11–16.
45 Gleich GJ, Frigas E, Loegering DA, Wassom DL, Steinmuller D. Cytotoxic properties of the eosinophil major basic protein. *J. Immunol.* 1979; **123**:2925–2927.
46 Weller PF, Lee CW, Foster DW, Corey EJ, Austen KF, Lewis RA. Generation and metabolism of 5-lipoxygenase pathway leukotrienes by human eosinophils: predominant production of leukotriene C_4. *Proc. Natl. Acad. Sci. USA* 1983; **80**:7626–7630.
47 Shaw RJ, Cromwell O, Kay AB. Preferential generation of leukotriene C_4 by human eosinophils. *Clin. Exp. Immunol.* 1984; **56**:716–722.
48 Shaw RJ, Walsh GM, Cromwell O, Moqbel R, Spry CJF, Kay AB. Activated human eosinophils generate SRS-A leukotrienes following physiological (IgG-dependent) stimulation. *Nature* 1985; **316**:150–152.
49 Lam S, Chan H, LeRiche JC, Chan-Yeung M, Salari H. Release of leukotrienes in patients with bronchial asthma. *J. Allergy Clin. Immunol.* 1988; **81**:711–717.
50 Wardlaw AJ, Hay H, Cromwell O, Collins JV, Kay AB. Leukotrienes B_4 and C_4 in bronchoalveolar lavage in bronchial asthma and other respiratory diseases. *J. Allergy Clin. Immunol.* In press.
51 Lee TC, Lenihan DJ, Malone B, Ruddy LL, Wasserman SI. Increased biosynthesis of platelet-activating factor in activated human eosinophils. *J. Biol. Chem.* 1984; **259**:5520–5530.
52 Champion A, Wardlaw AJ, Moqbel R, Cromwell O, Shepherd D, Kay AB. IgG-dependent generation of platelet-activating factor by normal and 'low density' human eosinophils. *J. Allergy Clin. Immunol.* 1988; **81**:207 (Abstract 157).
53 Wardlaw AJ, Moqbel R, Cromwell O, Kay AB. Platelet activating factor: a potent chemotactic and chemokinetic factor for human eosinophils. *J. Clin. Invest.* 1986; **78**:1701–1706.
54 Cuss FM, Dixon CM, Barnes PJ. Effects of platelet activating factor on pulmonary function and bronchial responsiveness in man. *Lancet* 1986; **ii**:189.
55 Howarth PH, Durham SR, Lee TH, Kay AB, Church MK, Holgate ST. Influence of albuterol, cromolyn sodium and ipratropium bromide on the airway and circulating mediator responses to antigen bronchial provocation in asthma. *Am. Rev. Respir. Dis.* 1985; **132**:986.
56 Burge PS, Efthimiou J, Turner-Warwick M, Nelmes TJ. Double-blind trials of inhaled beclomethasone dipropionate and fluocortin butyl ester in allergen-induced immediate and late asthmatic reactions. *Clin. Allergy* 1982; **12**:523–531.
57 King SJ, Miller HRP, Newlands GFJ, Woodbury RG. Depletion of mucosal mast cell protease by corticosteroids: effect on intestinal anaphylaxis in the rat. *Proc. Natl. Acad. Sci. USA* 1985; **82**:1214–1218.
58 Otsuka H, Denburg JA, Befus AD, Hitch D, Lapp P, Rajan RS, Bienenstock J, Dolovich J. Effect of beclomethasone dipropionate on nasal metachromatic cell subpopulations. *Clin. Allergy* 1986; **16**:589–595.
59 Lee TH, Brown MJ, Nagy L, Causon R, Walport MJ, Kay AB. Exercise-induced release of histamine and neutrophil chemotactic factors in atopic asthmatics. *J. Allergy Clin. Immunol.* 1982; **70**:73–81.
60 Metzger WJ, Zavala D, Richerson HB, Moseley P, Iwamota P, Monick M, Sjoerdsma K, Hunninghake GW. Local allergen challenge and bronchoalveolar lavage of allergic asthmatic lungs. Description of the model and local airway inflammation. *Am. Rev. Respir. Dis.* 1987; **135**:433–440.

61 De Monchy JG, Keyzer JJ, Kauffman HF, Beaumont F, DeVries K. Histamine in late asthmatic reactions following house dust mite inhalation. *Agents Actions* 1985; **16**:252.
62 Metzger WJ, Richerson HB, Warden K, Monick M, Hunninghake GW. Bronchoalveolar lavage of allergic asthmatic patients following allergen provocation. *Chest* 1986; **89**:477–483.
63 Haslam PL, Dewar A, Butchers P, Primett ZS, Newman Taylor A, Turner-Warwick M. Mast cells, atypical lymphocytes in bronchoalveolar lavage in extrinsic allergic alveolitis. *Am. Rev. Respir. Dis.* 1987; **135**:35–47.
64 Lee TH, Assoufi BK, Kay AB. The link between exercise, respiratory heat exchange, and the mast cell in bronchial asthma. *Lancet* 1983; **i**:520–522.
65 Melewicz FM, Kline NE, Cohen AB, Spiegelberg HL. Characterization of Fc receptor for IgE on human alveolar macrophages. *Clin. Exp. Immunol.* 1982; **49**:364.
66 Joseph M, Tonnel AB, Torpier G, Capron A, Arnoux B, Benveniste J. Involvement of IgE in the secretory processes of alveolar macrophages from asthmatic patients. *J. Clin. Invest.* 1983; **71**:221.
67 Capron M, Jouault T, Prin C, Joseph M, Ameisen JC, Butterworth AE, Papin JP, Kusneirz JP, Capron A. Functional study of a monoclonal antibody to IgE-Fc receptor of eosinophils, platelets and macrophages (Fc R_2). *J. Exp. Med.* 1986; **164**:72.
68 Tonnel AB, Gosset P, Joseph M, Fournier E, Capron A. Stimulation of alveolar macrophages in asthmatic patients after local provocation test. *Lancet* 1983; **i**:1406.
69 Kay AB, Diaz P, Carmichael J, Grant IWB. Corticosteroid-resistant chronic asthma and monocyte complement receptors. *Clin. Exp. Immunol.* 1981; **44**:576–580.
70 Grant IWB, Wyllie AH, Poznansky MC, Gordon ACH, Douglas JG. Corticosteroid resistance in chronic asthma. In *Asthma. Physiology, Immunopharmacology, and Treatment* (Eds Kay AB, Austen KF, Lichtenstein LM), Academic Press, London 1984, pp. 359–374.
71 Mazzoni L, Morley J, Page CP, Sanjar S. Induction of airway hyperreactivity by platelet activating factor in the guinea-pig. *J. Physiol.* 1985; **365**:107.
72 Metzger WJ, Hunninghake GW, Richerson HB. Late asthmatic reactions: inquiry into mechanisms and significance. *Clin. Rev. Allergy* 1985; **3**:145–165.
73 Knauer KA, Lichtenstein LM, Adkinson NF Jr, Fish JE. Platelet activation during antigen-induced airway reactions in asthmatic subjects. *N. Engl. J. Med.* 1981; **304**:1404–1407.
74 Durham SR, Dawes J, Kay AB. Platelets in asthma. *Lancet* 1985; **ii**:36.
75 Joseph M, Auriault C, Capron A, Vorng H, Viens P. A new function for platelets: IgE-dependent killing of schistosomes. *Nature* 1983; **303**:310–312.
76 Leung DYM, Geha RS. Regulation of the human IgE antibody response. *Int. Rev. Immunol.* 1987; **2**:75–91.
77 Jeffery PK, Nelson FC, Wardlaw A, Kay AB. Quantitative analysis of bronchial biopsies in asthma. *Am. Rev. Respir. Dis.* 1987; **135**:A316.
78 Poznansky MC, Gordon ACH, Grant IWB, Wyllie AH. A cellular abnormality in glucocorticoid resistant asthma. *Clin. Exp. Immunol.* 1985; **61**:135–142.
79 Poznansky MC, Gordon ACH, Douglas JG, Krajewski AS, Wyllie AH, Grant IWB. Resistance to methylprednisolone in cultures of blood mononuclear cells from glucocorticoid-resistant asthmatic patients. *Clin. Sci.* 1984; **67**:639–645.
80 Harper TB, Gaumer HR, Waring W, Brannon RB, Salvaggio JE. A comparison of cell-mediated immunity and suppressor T cell function in asthmatic and normal children. *Clin. Allergy* 1980; **10**:555–563.
81 Rola-Pleszczynski M, Blanchard R. Suppressor cell function in respiratory allergy. *Int. Arch. Allergy Appl. Immunol.* 1981; **64**:361–370.
82 Rivlin J, Kuperman O, Freier S, Godfrey S. Suppressor T-lymphocyte activity in wheezy children with and without treatment by hyposensitization. *Clin. Allergy* 1981; **11**:353–356.
83 Hwang KC, Fikrig SM, Friedman HM, Gupta S. Deficient concanavalin A-induced suppressor-cell activity in patients with bronchial asthma, allergic rhinitis, and atopic dermatitis. *Clin. Allergy* 1985; **15**:67–72.
84 Ilfeld D, Kivity S, Feierman E, Topilsky M, Kuperman O. Effects of *in vitro* colchicine and oral theophylline on suppressor cell function of asthmatic patients. *Clin. Exp. Immunol.* 1985; **61**:360–367.

85 Rocklin RE, Sheffer AL, Greineder DR, Melmon KL. Generation of antigen-specific suppressor cells during allergy desensitization. *N. Engl. J. Med.* 1980; **302**:1213–1219.

86 Warner JO. Significance of late reactions after bronchial challenge with house dust mite. *Arch. Dis. Child.* 1976; **51**:297–301.

87 Gerblich AA, Campbell AE, Schuyler MR. Changes in T lymphocyte subpopulations after antigenic bronchial provocation in asthmatics. *N. Engl. J. Med.* 1984; **310**:1349–1352.

88 Gonzalez MC, Diaz P, Galleguillos FR, Ancic P, Cromwell O, Kay AB. Allergen-induced recruitment of bronchoalveolar (OKT4) and suppressor (OKT8) cells in asthma. Relative increases in OKT8 cells in single early responders compared with those in late-phase responders. *Am. Rev. Respir. Dis.* 1987; **136**:600–604.

89 Frew AJ, Kay AB. The relationship between infiltrating $CD4^+$ lymphocytes, activated eosinophils and the magnitude of the allergen-induced late phase cutaneous reaction. *J. Immunol.* 1988; **141**:4158–4164.

90 Corrigan CJ, Hartnell A, Kay AB. T lymphocyte activation in acute severe asthma. *Lancet* 1988; **i**:1129–1132.

91 Corrigan CJ, Kay AB. Activated ($CD4^+$) T lymphocyte activation in acute severe asthma is accompanied by an elevation of serum concentrations of gamma-interferon and the soluble interleukin-2 receptors. *Thorax* 1988; **43**:814p. (Abstract).

92 Holtzman MJ, Aizawa H, Nadel JA, Goetzl EJ. Selective generation of leukotriene B_4 by tracheal epithelial cells from dogs. *Biochem. Biophys. Res. Commun.* 1983a; **114**:1071–1076.

93 Hunter JA, Finkbeiner WE, Nadel JA, Goetzl EJ, Holtzman MJ. Predominant generation of 15-lipoxygenase metabolites of arachidonic acid by epithelial cells from human trachea. *Proc. Natl. Acad. Sci. USA* 1985; **82**:4633–4637.

94 Shak S, Perez HD, Goldstein IM. A novel dioxygenation product of arachidonic acid possesses potent chemotactic activity for human polymorphonuclear leukocytes. *J. Biol. Chem.* 1983; **258**:14948–14953.

95 Walsh JJ, Dietlin LF, Low FN, Burch GE, Mogabgab WJ. Bronchotracheal response in human influenza. *Arch. Intern. Med.* 1961; **108**:376–382.

96 Kirsch CM, Sigal E, Djokic TD, Graf PD, Nadel JA. An *in vivo* chemotaxis assay in the dog trachea: evidence for chemotactic activity of 8S, 15S-dihydroxyeicosatetraenoic acid. *J. Appl. Physiol.* 1988; In press.

97 Holtzman MJ, Fabbri LM, O'Byrne PH, Gold BD, Aizawa J, Walters EH, Alpert SE, Nadel JA. Importance of airway inflammation for hyperresponsiveness induced by ozone in dogs. *Am. Rev. Respir. Dis.* 1983b; **127**:686–690.

98 Fabbri LM, Aizawa H, Alpert SE, Walters EH, O'Byrne PM, Gold BD, Nadel JA, Holtzman MJ. Airway hyperresponsiveness and changes in cell counts in bronchoalveolar lavage after ozone exposure in dogs. *Am. Rev. Respir. Dis.* 1984; **129**:288–291.

99 Phillips MJ, Gold WM, Goetzl EJ. IgE-dependent and ionophore-induced generation of leukotrienes by dog mastocytoma cells. *J. Immunol.* 1983; **131**:906–910.

11

Nedocromil sodium: a review of its anti-inflammatory properties and clinical activity in the treatment of asthma

R.M. AUTY AND S.T. HOLGATE

The development of nedocromil sodium

Nedocromil sodium (Tilade*), a pyranoquinoline dicarboxylic acid recently introduced for the treatment of asthma, is linked in the minds of many clinicians with the classic antiallergic drug sodium cromoglycate (Intal*). The drugs do indeed share several properties, both *in vitro* and *in vivo*, although in many test systems, such as non-allergic bronchial provocation studies, there are marked potency differences in favour of nedocromil sodium. In addition, the two molecules are structurally quite distinct (Fig. 11.1).

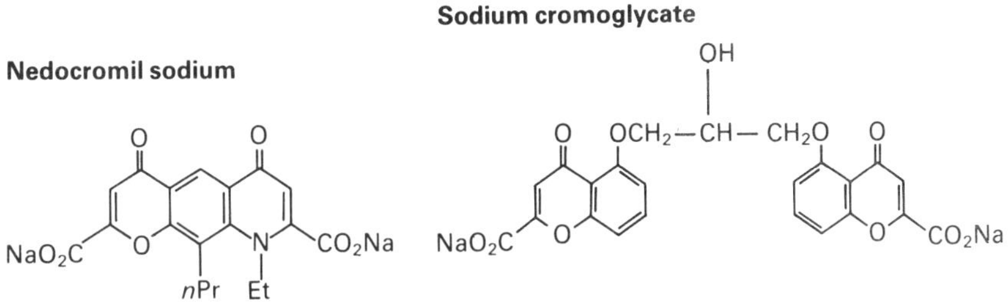

Fig. 11.1 The structures of nedocromil sodium and sodium cromoglycate.

The efficacy of sodium cromoglycate in allergic asthma was first reported over 20 years ago [1]. It was found to inhibit *in vitro* the release of histamine from mast cells [2], and the clinical efficacy of the drug was thus attributed to this property. The pharmaceutical industry consequently embarked on an intensive search for other compounds which inhibited mast-cell mediator release. Many compounds were developed which were very effective mast-cell stabilizers but which failed when put into clinical trials, through either lack of therapeutic activity or problems of toxicity [3–5]. Three conclusions could be drawn:

1 Only part of sodium cromoglycate's clinical efficacy was due to its effect on mast cells.

* Registered trade mark of Fisons p.l.c., Pharmaceutical Division.

2 The classic IgE-mediated screens of antiallergic activity provided only an incomplete model of asthma.
3 The characteristic absorption-limited pharmacokinetics of this drug [6] provided bioavailability in the airways for a longer period than with other 'mast-cell stabilizing' agents.

In the early stages of its development, nedocromil sodium had been shown to be effective in the classical mast-cell screens: the rat passive cutaneous and passive lung anaphylaxis tests [7] and antigen-induced mediator release from rat peritoneal mast cells [8]. The profile of activity which resulted from these models was similar to that of sodium cromoglyçate and gave little stimulus for further investigation of the compound. However, in view of the predictive inadequacies of these models of type I immediate hypersensitivity reactions, the drug development programme at Fisons had been extended to include further *in vitro* and *in vivo* tests which, overall, would comprise a more comprehensive model of clinical asthma. The first of these was a primate model of antigen-induced bronchoconstriction [9–11] which also provided an *in vitro* test for inhibition of mucosal mast-cell mediator release. The macaque monkey (*Macaca arctoides*) was sensitized by repeated infestation with the parasite *Ascaris suum* to produce a pronounced blood eosinophilia, with circulating levels of *A. suum*-specific IgE antibody and basophil sensitivity to *Ascaris* antigen. Bronchoalveolar lavage (BAL) fluid from these animals was found to contain up to 21% of mast cells. Stimulation of the mixed BAL cell population with either *Ascaris* antigen or antihuman IgE, at maximal challenge, resulted in the release of inflammatory mediators in quantities per 10^6 cells of 2–5 μg histamine, 100–300 ng prostaglandin (PG) D_2 and 20–80 ng leukotriene (LT) C_4. The pattern of mediator release and subsequent cell fractionation work suggested that these mediators were mast-cell derived, although an eosinophilic contribution could not be excluded with certainty. Both with antigen and anti-IgE as the mast-cell stimulus, nedocromil sodium proved 200 times more potent than sodium cromoglycate in inhibiting the release of both preformed and newly generated mediators [8]. The *in vivo* screen utilized the bronchoconstriction produced in the infested macaques in response to inhalation of low levels of *Ascaris* antigen. Nedocromil sodium pretreatment (50 breaths of a 2% solution 5 min before antigen challenge) suppressed the changes in lung resistance (85% inhibition) and compliance (55% inhibition), whereas sodium cromoglycate had no significant protective effect on the changes in lung mechanics [12] (Fig. 11.2).

This lack of activity of sodium cromoglycate was surprising but may be partly explained by the type of mast cell involved in this model, which has the histochemical properties of rat intestinal mast cells [13] and is equally unresponsive to compound 48/80, a potent secretagogue of connective tissue mast cells [11]. Mast-cell heterogeneity has been demonstrated also in the

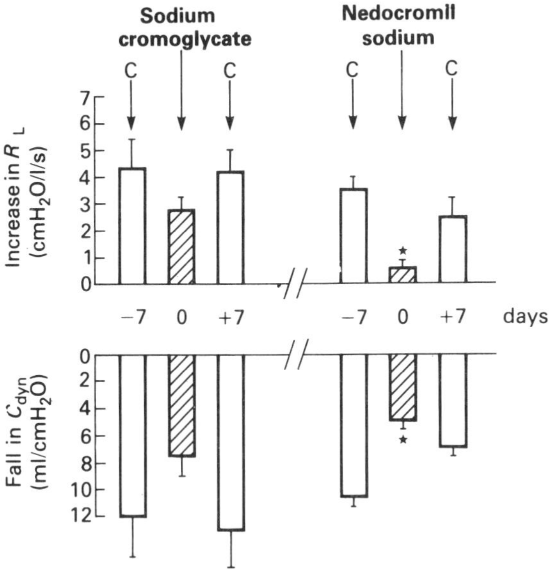

Fig. 11.2 The effect of nedocromil sodium and sodium cromoglycate (hatched bars) on changes in total lung resistance (R_L) and dynamic compliance (C_{dyn}) induced by inhalation of *Ascaris* antigen in sensitized macaques. Open bars represent controls (C) 7 days before and after drug treatment. Results are shown as means ($n = 5$) + standard errors. *$p < 0.05$.

Nippostrongylus-infested rat [14], and sodium cromoglycate was shown to be ineffective in preventing antigen-specific histamine release from the rat intestinal mucosal mast cell, whereas histamine release from the rat peritoneal connective tissue mast cell was inhibited by the drug [15]. Interpretation of these results is complicated, however, since similar findings have been reported for nedocromil sodium [16], and both compounds have been claimed not to have statistically significant inhibitory effects on isolated human intestinal mast cells [16], perhaps indicating the existence of mast-cell subtypes. The lung also contains two histochemically and functionally distinct mast-cell populations [16, 17], those cells from the bronchoalveolar lumen being reported to be more susceptible to the inhibitory effects of both nedocromil sodium and sodium cromoglycate than lung-tissue mast cells [18]. In a comparison of drug activities, using BAL mast cells from patients undergoing diagnostic BAL and mast cells obtained from enzymatic dispersion of lung fragments from sarcoid patients, nedocromil sodium showed significantly more activity than sodium cromoglycate on both BAL and dispersed lung mast cells [18].

The macaque model of antigen-induced bronchoconstriction did not provide any indication of the effect of nedocromil sodium on non-specific bronchial hyperresponsiveness, an important aspect in view of the intended clinical use of the drug. A model was set up in which anaesthetized dogs exposed to the irritant gas sulphur dioxide (400 p.p.m. for 2 h) developed increased bronchial responsiveness to histamine; epithelial damage and an influx of neutrophils into the bronchial lumen were also observed [19]. Exposure to sulphur dioxide caused an immediate increase in the bronchial responsiveness to histamine which returned to pre-exposure baseline in approximately 2 h and was followed by a second, sustained phase of increased responsiveness at 24 h. Nedocromil sodium given pre- and post-sulphur dioxide exposure prevented the immediate increase in lung responsiveness and reduced that occurring at 24 h (Fig. 11.3a). The total number of cells

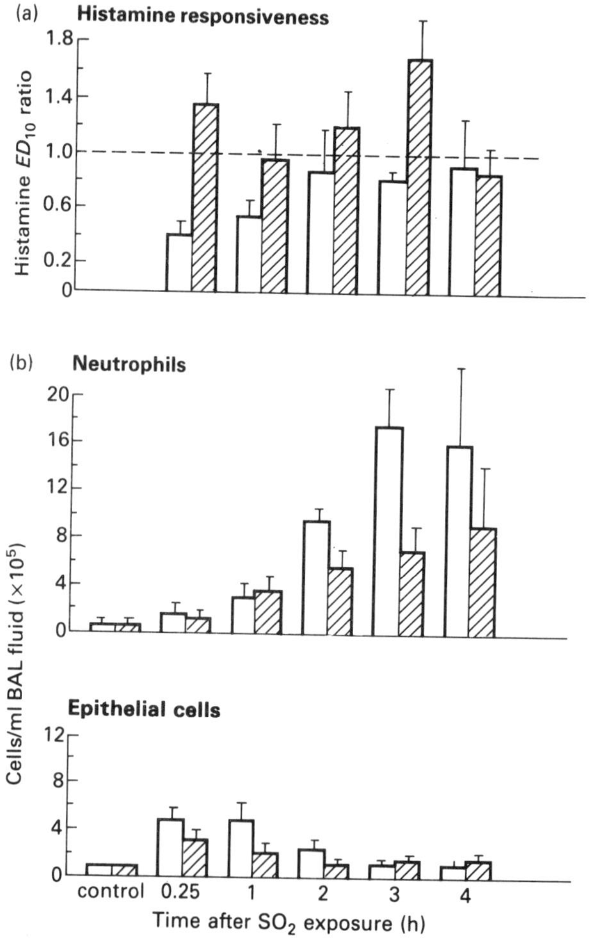

Fig. 11.3 Effect of nedocromil sodium (hatched bars) and saline control (open bars) on (a) histamine responsiveness and (b) BAL cell counts in sulphur dioxide-exposed dogs. Histamine ED_{10} ratio = ratio of log dose of histamine producing a 10 $cmH_2O/l/s$ increase in lung resistance in sulphur dioxide-exposed and control animals (exposed/control). $ED_{10} > 1.0$ = decreased responsiveness; $ED_{10} < 1.0$ = increased responsiveness.

recovered by BAL was increased following exposure, up to 1 h being due to epithelial cells, from 1 to 4 h a profound infiltration with neutrophils and by 3 h neutrophils comprised over 90% of BAL cells. Nedocromil sodium markedly reduced the accumulation of neutrophils for the 4 h following exposure, together with a small reduction in epithelial cells [20] (Fig. 11.3b). Since the increase in responsiveness occurred in the absence of neutrophilia, it may be suggested that the immediate-phase responsiveness resulted from epithelial desquamation, the consequent exposure and stimulation of sensory nerves causing an exaggerated reflex bronchoconstriction. Over several hours this response may become attenuated as the airways become lined with serous exudates and cell infiltrates. The sustained phase of responsiveness may have resulted from the release of inflammatory mediators from the infiltrating cells, particularly the neutrophil, causing sensitization of sensory nerves, facilitation of efferent nerve activity and an increase in bronchial smooth muscle responsiveness. Inhibition of the early increase in responsiveness by nedocromil sodium may involve suppression of sensory nerve activity. The mechanism for the cellular effects is unknown but may relate to the known inhibitory activity of the compound against a variety of inflammatory cell types (see below). In a similar study in dogs in which neutrophilia was induced by inhalation of ozone, neutrophil influx was significantly inhibited by nedocromil sodium pretreatment [21].

Mode(s) of action

Early clinical trials with nedocromil sodium had shown that the drug reduced coughing in asthmatic patients [22], and this observation prompted the development of a conscious dog model in which coughing was induced by inhalation of citric acid [23]. Nedocromil sodium pretreatment significantly prolonged the time to onset of coughing and reduced the total number of coughs; sodium cromoglycate had no significant effect on the response to citric-acid challenge (Fig. 11.4). Since citric acid was thought to activate the cough reflex via an effect upon sensory nerves, further experiments were conducted in anaesthetized dogs to determine which receptor population was being affected by nedocromil sodium. Given intravenously into the aortic arch, nedocromil sodium was found to have a stimulatory effect on bronchial C-fibre endings [Eady, personal communication], causing an increase in the rate of discharge. Stimulation of C-fibre endings in anaesthetized cats by phenyldiguanide has been shown to inhibit the cough reflex [24] and thus it may be suggested that the antitussive effect of nedocromil sodium also relates to C-fibre stimulation. Although a failure to suppress citric acid-induced cough in volunteers and asthmatics has been reported [25, 26], nedocromil sodium did exhibit an antitussive effect in volunteers challenged with

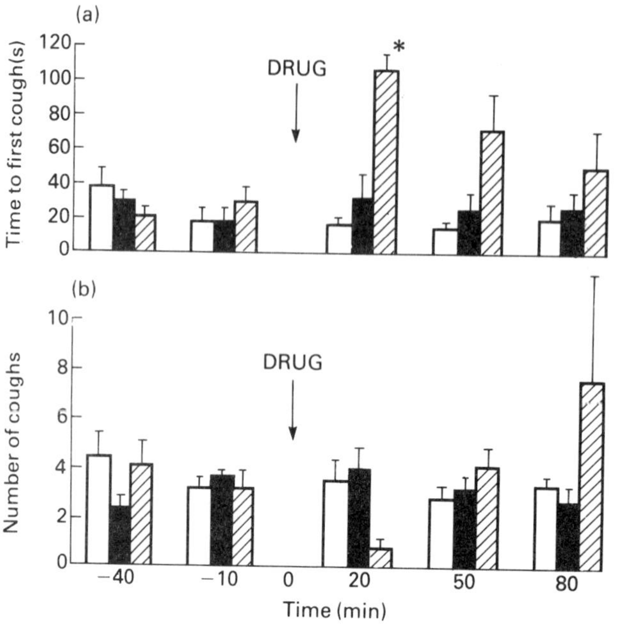

Fig. 11.4 The effect of aerosols of nedocromil sodium 15 mg (hatched bars); sodium cromoglycate 15 mg (solid bars), and saline control (open bars) on (a) the time to first cough and (b) the number of coughs following citric-acid challenge in tracheostomized dogs. Results shown are means ($n=4$) + standard errors. *$p<0.05$.

ultrasonically nebulized distilled water and with the C-fibre stimulant capsaicin [26].

An animal model which has pathophysiological characteristics analogous to human asthma is the sheep naturally sensitized to *Ascaris suum* and displaying early- and late-phase bronchoconstrictor responses to inhalation of *Ascaris* antigen. *In vivo*, both nedocromil sodium and sodium cromoglycate were effective in reducing the early reaction and preventing the late bronchial response [27]. *In vitro*, nedocromil sodium was effective at doses 10-fold lower than sodium cromoglycate in modifying the antigen-induced contractile response of sheep isolated tracheal smooth muscle [27]. In the same model, a second series of experiments investigated the effect of nedocromil sodium on the development of airway responsiveness 24 h after antigen challenge [28]. Nedocromil sodium was given either before or after the early response and in both cases blocked the late response and the antigen-induced increase in airway responsiveness, indicating an inhibitory effect on those mediators eliciting late responses and on the subsequent influx of secondary inflammatory cells.

A useful and recently reported model for investigating the modes of action of antiasthma drugs is the ovalbumin-sensitized guinea pig [29]. This model overcomes some of the drawbacks associated with previous models in that the animal is both sensitized and challenged by inhalation; pretreatment with a histamine H_1-antagonist avoids fatal anaphylaxis whilst allowing the use of sufficiently high doses of ovalbumin to produce a consistent late reaction. Ovalbumin challenge induces a triphasic bronchoconstrictor response consisting of an early-phase response (EPR) at 2 h, a late-phase response (LPR) at 17 h and a late late-phase response (LLPR) at 72 h; the airways are infiltrated with neutrophils at 17 h and with eosinophils at 17 and 72 h. In this system, nedocromil sodium inhaled before challenge protected against the EPR and LPR but not the LLPR; neutrophils at 17 h were reduced but there was no effect on eosinophils. Given between the EPR and LPR, nedocromil sodium inhibited both LPR and LLPR; it was without effect on neutrophil numbers but eosinophils at 72 h were reduced. Given before and after the EPR, the drug inhibited the EPR, LPR and LLPR and the increases in neutrophils at 17 h and eosinophils at 72 h were both reduced [30] (Fig. 11.5).

In vitro anti-inflammatory activity

These results are reflected in the clinical activity of the drug, which prevents both immediate and late asthmatic responses to antigen challenge in asthmatic patients [31–33], and reduces the associated increase in non-specific bronchial responsiveness [34, 35] (Fig. 11.6). The development of increased bronchial responsiveness is thought to be related to the influx into the bronchial lumen of inflammatory cells, particularly eosinophils and neutrophils [36, 37]. Many of these recruited cells show evidence of specific activation, such as enhancement of cell-surface receptors [38], degranulation [36, 39] and increased cytotoxicity [40]. These signs also serve as markers by which the suppressant effect of nedocromil sodium on the activation of secondary inflammatory cells and their mediator release has been investigated *in vitro*.

Assessed as the expression of membrane receptors for complement (C3b) and IgG (Fc), neutrophil activation was markedly inhibited by nedocromil sodium, as was neutrophil killing of schistosomula larvae mediated by release of cytotoxic mediators from neutrophils stimulated by the chemotactic peptide formyl-methionyl-leucyl-phenylalanine (FMLP). Unstimulated receptor expression and schistosome killing were unaffected by the compound [41]. Nedocromil sodium has also been shown to effect up to 80% inhibition of platelet-activating factor (PAF-acether)-induced chemotaxis of human granulocytes [42], which may explain in part the effects of the drug on the *in vivo* neutrophil influx observed in sulphur dioxide-challenged dogs. In addition, the compound inhibits the release of lysosomal enzymes from rabbit neutrophils stimulated by the phorbol ester phorbol dibutyrate (PDB), to a

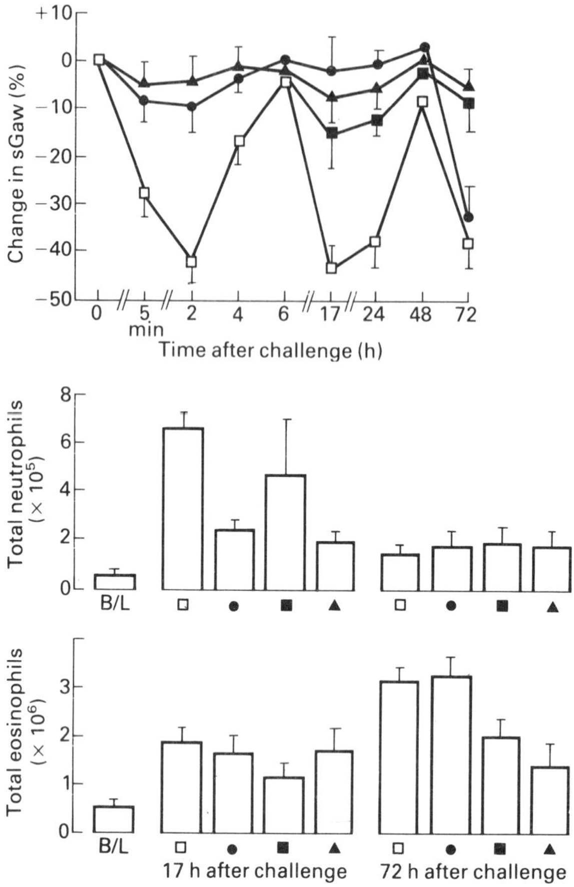

Fig. 11.5 The effect of nedocromil sodium inhalation (10 mg/ml for 2 min) on ovalbumin-induced changes in specific airway conductance (sGaw) and BAL cell counts in sensitized guinea pigs. Nedocromil sodium was given 15 min before challenge (●); 6 h postchallenge (■), or at both times (▲). Saline control (□) was given at the same times. B/L = baseline. Results shown are means + standard errors. For sGaw results $n = 16$ up to 17 h and $n = 8$ thereafter and for the BAL results. (From reference 30 by permission of the publishers.)

greater extent than when FMLP is used as the stimulus [43]. Sodium cromoglycate had no effect on PDB-induced enzyme secretion. Since PDB activates only protein kinase C, it may be suggested that nedocromil sodium's inhibitory effect occurs via an action on protein kinase C.

In the case of eosinophils, as with neutrophils, nedocromil sodium inhibits FMLP-induced receptor expression and schistosomula killing [41]. It inhibits at high concentrations sepharose C3b-stimulated granule protein release [44]; an inhibitory effect on LTC_4 release from human eosinophils stimulated with either opsonized zymosan or the calcium ionophore A-23 187 has also been observed [45].

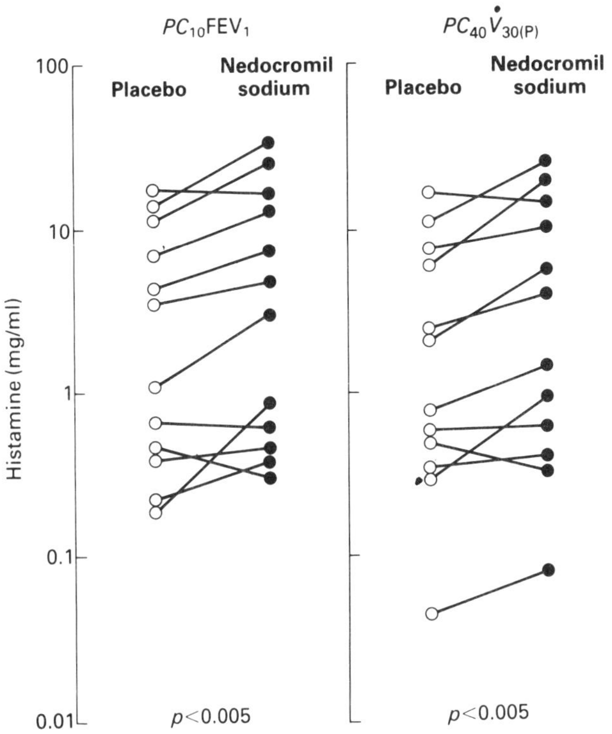

Fig. 11.6 The effect of nedocromil sodium (●) and placebo (○) on histamine airway responsiveness (PC_{10} FEV_1; PC_{40} $\dot{V}_{30(P)}$) in grass pollen-sensitive asthmatics during the pollen season. $n=12$. (From reference 34 by permission of the publishers.)

Nedocromil sodium inhibits FcεRII-mediated activation of rat monocytes or peritoneal macrophages [46] and human monocytes or alveolar macrophages [47]. The compound has been reported to reduce the release of LTB_4 and 5-hydroxyeicosatetraenoic acid (5-HETE) from alveolar macrophages obtained by BAL from asthmatic patients and stimulated by A-23 187 or opsonized zymosan, although results are variable [48].

Platelets also have a low-affinity FcεRII receptor for IgE and nedocromil sodium has been observed to inhibit the anti-IgE or specific antigen-induced release of cytotoxic mediators from platelets obtained from passively sensitized rats or from allergic asthmatics [46, 47]. Interestingly, nedocromil sodium also inhibited *in vitro* the aspirin (or other non-steroidal anti-inflammatory drug) induced activation of platelets from aspirin-sensitive, intrinsic asthmatic patients [49].

Human challenge studies

Clinically, the mechanism of action of nedocromil sodium—and of sodium cromoglycate—is, as yet, unknown. Further insights, however, may be gained from the results of challenge studies with various inhaled, non-allergenic provocants. Nedocromil sodium 4 mg given 30 min before inhalation of ultrasonically nebulized distilled water ('fog') significantly reduced the bronchoconstrictor response in asthmatic patients [50]. It has been proposed that the response to fog inhalation is brought about by hypotonicity of the fluid lining the airways causing mast cells in the surrounding lung tissue to degranulate [51], although in some patients fog-induced bronchospasm is attenuated by inhaled anticholinergic drugs. The bronchoconstrictor response to sulphur dioxide is also likely to depend in part on neural mechanisms, since this substance stimulates afferent nerve activity in animal airways [52], and therefore, not unexpectedly, the response may be decreased by anticholinergic drugs [53]. In asthmatic subjects, nedocromil sodium (2 and 4 mg) reduced the magnitude and duration of sulphur dioxide-induced bronchoconstriction and breathlessness when given 30 min before challenge [54]. Altounyan *et al.* compared nedocromil sodium 4 mg with sodium cromoglycate 10 mg on sulphur dioxide-induced bronchoconstriction in atopic, non-asthmatic subjects, the two drugs being given 0.5, 2 and 4 h before challenge. Nedocromil sodium was significantly more effective than sodium cromoglycate in terms of the maximum reduction in partial expiratory flow rate and duration of effect [55] (Fig. 11.7). The local generation of sulphur dioxide may be involved in the response to inhaled metabisulphite, which induces bronchoconstriction in atopic subjects. This response can be completely inhibited by nedocromil sodium [56, 57]. In the same series of experiments, the muscarinic receptor blocker oxitropium bromide had only a

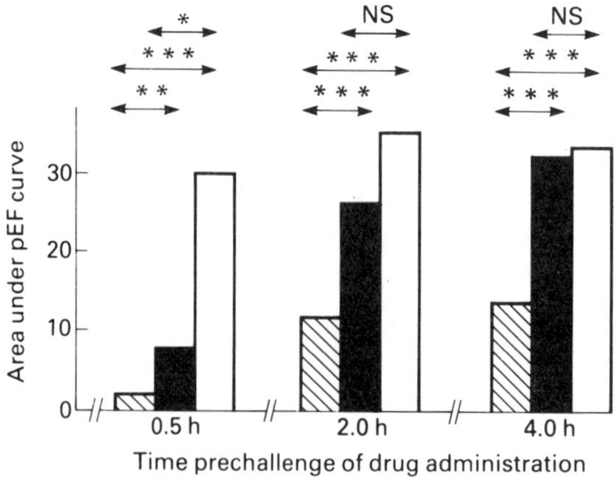

Fig. 11.7 Duration of effect of nedocromil sodium 4 mg (hatched bars), sodium cromoglycate 10 mg (solid bars) and placebo (open bars) on areas under partial expiratory flow (PEF) rate curve after sulphur dioxide challenge in non-asthmatic, atopic subjects. (0.5 h, $n=3$; 2 h, $n=6$; 4 h, $n=5$). *$p<0.05$. **$p<0.01$. ***$p<0.001$. NS = not significant.

slight effect and the antihistamine terfenadine had no significant effect, suggesting that the mechanism of action *in vivo* of nedocromil sodium in this model does not involve mast-cell stabilization [57]. This is quite possible given nedocromil sodium's activity across a range of inflammatory cells.

Adenosine or its endogenous nucleotide, adenosine 5′-monophosphate (AMP), provokes bronchoconstriction in asthmatic patients but not in normal subjects [58]. Adenosine-induced bronchoconstriction is inhibited by nedocromil sodium which, in a comparative study in asthmatic patients, was shown to be more effective at a dose of 4 mg than 2 mg of sodium cromoglycate [59]. These results were confirmed in a study in non-atopic asthmatics, in which nedocromil sodium was found to be statistically significantly more effective than sodium cromoglycate (5.2-fold more potent on a molar basis) in inhibiting AMP-induced bronchoconstriction when the drugs were given by nebulizer prior to challenge [60]. The effect of adenosine is thought to involve potentiation of release of preformed mediators from mast cells [61] and the bronchial response can be prevented by high doses of a selective H_1-antagonist such as terfenadine [62]. Enhancement of vagally mediated neural reflexes has also been proposed as a mechanism of action, but with conflicting results [63, 64].

Bradykinin, a peptide generated from high molecular weight kininogen, is produced in inflammatory reactions. It is a potent bronchoconstrictor of asthmatic airways [65] and has been shown in animals to be a potent and selective stimulator of C-fibre endings, resulting in the release of sensory neuropeptides [66]. In a comparative study, pretreatment with either nedocromil sodium 4 mg or sodium cromoglycate 10 mg was shown to provide effective inhibition of bradykinin-induced bronchoconstriction in asthmatic patients, nedocromil sodium proving approximately twice as effective as sodium cromoglycate at these dosages in terms of forced partial flow volume. It was concluded that the mode of action of the drugs involves mechanisms other than stabilization of mast cells [67].

Neurokinin A is a sensory neuropeptide which has been reported to be more potent than substance P in inducing bronchoconstriction in asthmatics [68]. Nedocromil sodium 4 mg inhaled before challenge protects against neurokinin A-induced bronchoconstriction, lending further support to the hypothesis that part of the beneficial effect of nedocromil sodium is due to inhibition of axon reflex mechanisms in the airways [69].

The many *in vitro* studies showing nedocromil sodium's potent effects on a wide range of inflammatory cells, together with its effect on the late asthmatic reaction and associated increase in bronchial responsiveness, all indicate that it will have a useful anti-inflammatory effect in clinical asthma. The numerous bronchial provocation models also point to a wide spectrum of activity which can be expressed in a range of inflammatory cell types and neural reflexes. The potency differences in favour of nedocromil sodium over sodium cromoglycate in provocation tests where the bronchospastic stimulus is non-immunological may be relevant to the treatment of adult asthma, where the aetiology of the disease is more often non-allergic, and where sodium cromoglycate is perceived to be less effective [70, 71].

Clinical profile

The results of early open trials with nedocromil sodium in adult asthma proved encouraging [72] and the clinical development programme of the new drug progressed to double-blind, placebo-controlled trials of up to 12 weeks duration, in which a large part of the patient population comprised older, more severe and often 'intrinsic' asthmatics, these being the type of patients who would not be expected to derive great benefit from sodium cromoglycate treatment. Efficacy was assessed by a range of variables including diary card monitoring of symptom scores, daily pulmonary function measurement up to three times, reductions in concomitant inhaled and oral bronchodilator therapy, and regular clinic assessments of asthma severity and lung function. The majority of these studies showed convincing superiority of active treatment over placebo [reviewed in 71], with significant improvements typically being observed in several of these efficacy variables. Notably, theophylline-dependent patients were able to discontinue use of sustained-release theophylline while asthma severity continued to improve [73]; similar beneficial effects were seen despite significant reductions in oral [74] and inhaled [75] bronchodilator therapy. Nedocromil sodium was also evaluated as a replacement for maintenance inhaled steroid therapy, with encouraging results [76] (Fig. 11.8). In other studies, the drug was unable to counter completely the acute deterioration in asthma precipitated by abrupt withdrawal of inhaled steroids [e.g. 77] but in several trials patients were able to reduce usage of inhaled steroids gradually under cover of nedocromil sodium treatment [e.g. 22]. An open, 12-month study in patients receiving inhaled steroids showed that it was possible for over half the patients to discontinue inhaled steroid therapy without deterioration of symptoms [78]. Importantly, when nedocromil sodium was added in to existing maintenance therapy with inhaled steroids, significant improvements were seen in the patients' condition, including, in one study, a significant reduction of diurnal variation in peak expiratory flow (PEF) rate (Fig. 11.9) [79]. This ability to confer benefit over and above that achieved with inhaled steroids is a feature which has not been observed with sodium cromoglycate treatment [80].

Long-term studies [74, 78] and post-marketing surveillance [data on file; Fisons p.l.c., Loughborough] have revealed no serious adverse effects. The

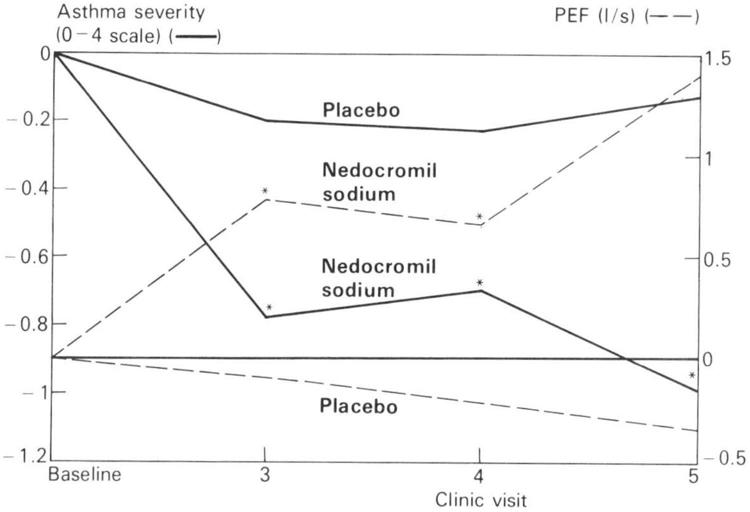

Fig. 11.8 Effect of nedocromil sodium 4 mg twice daily on asthma severity and PEF recorded by physicians at 2-weekly clinic visits, in moderately asthmatic patients (placebo, $n=35$; nedocromil sodium, $n=34$). Inhaled corticosteroids, used by 22 and 24 patients in the active and placebo groups respectively, were discontinued during the baseline period. $^{*}p<0.05$; $^{***}p<0.001$. (From reference 76 by permission of the publishers.)

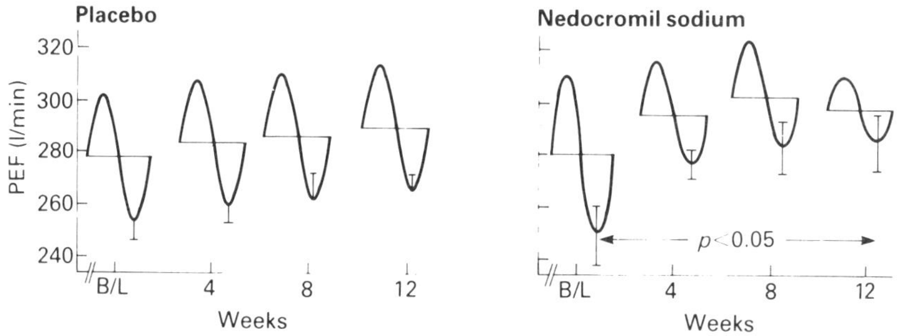

Fig. 11.9 The effect of nedocromil sodium 4 mg four times daily on diurnal variation in peak expiratory flow (PEF) rates in asthmatics receiving inhaled steroid maintenance therapy. Placebo, $n=16$; nedocromil sodium, $n=18$. B/L = baseline. (From reference 79 by permission of the publishers.)

most commonly reported event (in 13.6% of patients) is a bitter taste when the drug is inhaled but this is a perception of taste—many patients taste nothing—and is not a true perversion of the mechanism of taste. It may be a familial characteristic. Headache has been reported in 4.8% of patients, nausea in 4% and vomiting in 1.8% [81].

Concluding remarks

Nedocromil sodium appears to be a clinically effective and safe treatment of adult asthma. Its ability to block both early and late asthmatic responses and to attenuate the development of associated bronchial hyperresponsiveness appear to indicate that nedocromil sodium has a potentially useful place as maintenance therapy in asthma, as an adjunct to inhaled bronchodilator therapy alone and before initiation of inhaled steroid therapy. In the various test systems discussed, nedocromil sodium's increased potency over sodium cromoglycate together with its proven clinical efficacy when administered twice daily provide a non-steroid alternative for mild to moderate adult asthma, irrespective of age and atopic status. It may also occupy a useful position as an add-on treatment to a regimen of inhaled bronchodilator and inhaled steroid, and may prove useful as part of a strategy to reduce regular steroid usage. By virtue of route of administration and its activity at certain points in the inflammatory cascade, nedocromil sodium is not a treatment for acute severe asthma.

References

1 Howell JBL, Altounyan REC. A double blind trial of disodium cromoglycate in the treatment of allergic bronchial asthma. *Lancet* 1967; **ii**:539–542.

2 Cox JSG. Disodium cromoglycate: a specific inhibitor of reaginic antibody: antigen mechanisms. *Nature* 1967; **216**:1328–1329.

3 Church MK. Cromoglycate-like anti-allergic drugs. *Drugs Today* 1978; **14**:281–341.

4 Suschitzky JL, Sheard P. The search for anti-allergic drugs for the treatment of asthma—problems in finding a successor to sodium cromoglycate. *Prog. Med. Chem.* 1984; **21**:1–61.

5 Mann JS, Clement P, Sheridan AQ, Soryal I, Fairfax AJ, Holgate ST. Inhaled lodoxamide in the treatment of bronchial asthma: a double-blind placebo-controlled study. *J. Allergy Clin. Immunol.* 1985; **76**:83–90.

6 Neale MG, Brown K, Hodder RW, Auty RM. The pharmacokinetics of sodium cromoglycate in man after intravenous and inhalation administration. *Br. J. Clin. Pharmacol.* 1986; **22**:373–382.

7 Riley PA, Mather ME, Keogh RW, Eady RP. Activity of nedocromil sodium in mast-cell-dependent reactions in the rat. *Int. Arch. Allergy Appl. Immunol.* 1987; **82**:108–110.

8 Wells E, Jackson CG, Harper ST, Mann J, Eady RP. Characterization of primate bronchoalveolar mast cells. II. Inhibition of histamine, LTC_4, and PGD_2 release from primate bronchoalveolar mast cells and a comparison with rat peritoneal mast cells. *J. Immunol.* 1986; **137**:3941–3945.

9 Pritchard DI, Eady RP, Harper ST, Jackson DM, Orr TSC, Richards IM, Trigg S, Wells E. Laboratory infection of primates with *Ascaris suum* to provide a model of allergic bronchoconstriction. *Clin. Exp. Immunol.* 1983; **54**:469–476.

10 Richards IM, Eady RP, Jackson DM, Orr TSC, Pritchard DI, Vendy K, Wells E. *Ascaris*-induced bronchoconstriction in primates experimentally infected with *Ascaris suum* ova. *Clin. Exp. Immunol.* 1983; **54**:461–468.

11 Wells E, Harper ST, Jackson CG, Mann J, Eady RP. Characterization of primate bronchoalveolar mast cells. I. IgE-dependent release of histamine, leukotrienes, and prostaglandins. *J. Immunol.* 1986; **137**:3933–3940.

12 Eady RP, Greenwood B, Jackson DM, Orr TSC, Wells E. The effect of nedocromil sodium and sodium cromoglycate on antigen-induced bronchoconstriction in the *Ascaris*-sensitive monkey. *Br. J. Pharmacol.* 1985; **85**:323–325.

13 Greenwood B, Orr TSC, Coleman H. The ultrastructure of mucosal mast cells. *Eur. J. Respir. Dis.* 1986; **69** (suppl. 147):210–216.

14 Bienenstock J, Befus AD, Pearce F, Denburg J, Goodacre R. Mast cell heterogeneity: derivation and function, with emphasis on the intestine. *J. Allergy Clin. Immunol.* 1982; **70**:407–412.

15 Shanahan F, Lee TDG, Denburg JA, Bienenstock J, Befus AD. Functional characterization of mast cells generated *in vitro* from the mesenteric lymph node of rats infected with *Nippostrongylus brasiliensis*. *Immunology* 1986; **57**:455–459.

16 Befus AD, Dyck N, Goodacre R, Bienenstock J. Mast cells from the human intestinal lamina propria. Isolation, histochemical subtypes, and functional characterization. *J. Immunol.* 1987; **138**:2604–2610.

17 Leung KBP, Flint KC, Pearce FL, Hudspith B, Brostoff J, Johnson NMcI. A comparison of histamine secretion from human basophils and from human lung mast cells obtained by bronchoalveolar lavage (BAL) and from dispersion of lung fragments. *Clin. Sci.* 1985; **68** (suppl. 11):7P.

18 Leung KBP, Flint KC, Brostoff J, Hudspith BN, Johnson NMcI, Pearce FL. A comparison of nedocromil sodium and sodium cromoglycate on human lung mast cells obtained by bronchoalveolar lavage and by dispersion of lung fragments. *Eur. J. Respir. Dis.* 1986; **69** (suppl. 147):223–226.

19 Jackson DM, Eady RP, Farmer JB. The effect of nedocromil sodium on non-specific bronchial hyperreactivity in the dog. *Eur. J. Respir. Dis.* 1986; **69** (suppl. 147):217–219.

20 Jackson DM, Eady RP. Acute transient SO_2-induced airway hyperreactivity: effects of nedocromil sodium. *J. Appl. Physiol.* 1988; **65**:1119–1124.

21 Chitano P, Finotto S, Bosco V, Mapp C, Fabbri LM. Nedocromil sodium inhibits bronchoalveolar lavage neutrophilia induced by ozone in dogs. In *Abstracts of the 6th Congress of the European Society of Pneumology* (*SEP*) (Eds Stam J, Siebelink J, Vanderschueren R, Wagenaar J), CIP-Gegevens Koninklijke Bibliotheek, Den Haag 1987, p. 121.

22 Chatterjee PC, Fyans PG, Chatterjee SS. A trial comparing nedocromil sodium (Tilade) and placebo in the management of perennial bronchial asthma. *Eur. J. Respir. Dis.* 1986; **69** (suppl. 147):314–316.

23 Jackson DM. The effect of nedocromil sodium, sodium cromoglycate and codeine phosphate on citric acid-induced cough in dogs. *Br. J. Pharmacol.* 1988; **93**:609–612.

24 Tatar M, Webber SE, Widdicombe JG. Lung C-fibre receptor activation and defensive reflexes in anaesthetized cats. *J. Physiol.* 1988; **42**:411–420.

25 Heyrman R, De Backer W, Willemen M, Bogaerts M, Vermeire P. Effect of nedocromil sodium on citric acid induced cough in asthmatic subjects. *Bull. Eur. Physiopathol. Respir.* 1987; **23** (suppl. 12):411s.

26 Lowry RH, Higenbottam TW. Antitussive effect of nedocromil sodium on chemically induced cough. *Thorax* 1988; **43**:256P.

27 Abraham WM, Stevenson JS. Chapman GA, Tallent MW, Jackowski J. The effect of nedocromil sodium and cromolyn sodium on antigen-induced responses in allergic sheep *in vivo* and *in vitro*. *Chest* 1987; **92**:913–917.

28 Abraham WM, Stevenson JS, Eldridge M, Garrido R, Nieves L. Effect of nedocromil sodium on allergen-induced airway hyperresponsiveness in allergic sheep. *J. Appl. Physiol.* 1988; **65**:1062–1068.

29 Hutson PA, Church MK, Clay TP, Miller P, Holgate ST. Early and late-phase bronchoconstriction after allergen challenge of nonanesthetized guinea pigs. I. The association of disordered airway pathology to leukocyte infiltration. *Am. Rev. Respir. Dis.* 1988; **137**:548–557.

30 Hutson PA, Holgate ST, Church MK. Inhibition by nedocromil sodium of early and late phase bronchoconstriction and airway cellular infiltration provoked by ovalbumin inhalation in conscious sensitized guinea-pigs. *Br. J. Pharmacol.* 1988; **94**:6–8.

31 Dahl R, Pedersen B. Influence of nedocromil sodium on the dual asthmatic reaction after allergen challenge: a double-blind, placebo-controlled study. *Eur. J. Respir. Dis.* 1986; **69** (suppl. 147):263–265.

32 Crimi E, Brusasco V, Crimi P. Effect of nedocromil sodium on the dual reaction to bronchial antigen challenge in asthmatic patients. *J. Allergy Clin. Immunol.* 1989; In press.

33 Joubert JR, Davidowitz H. A double-blind crossover trial with nedocromil sodium on the dual asthmatic reaction after antigen challenge. *Bull. Eur. Physiopathol. Respir.* 1986; **22** (suppl. 8):720.

34 Dorward AJ, Roberts JA, Thomson NC. Effect of nedocromil sodium on histamine airway responsiveness in grass-pollen sensitive asthmatics during the pollen season. *Clin. Allergy* 1986; **16**:309–315.

35 Orefice U, Ferrazzano PL, Patalano F, Ruggieri F. Changes in bronchial hyperreactivity to methacoline induced by nedocromil sodium. In *Abstracts of the 6th Congress of the European Society of Pneumology* (Eds Stam J, Siebelink J, Vanderschueren R, Wagenaar J), CIP-Gegevens Koninklijke Bibliotheek, Den Haag 1987, p. 49.

36 De Monchy JGR, Kauffmann HF, Venge P, Koeter GH, Jansen HM, Sluiter HJ, de Vries K. Bronchoalveolar eosinophilia during allergen-induced late asthmatic reactions. *Am. Rev. Respir. Dis.* 1985; **131**:373–376.

37 Diaz P, Gonzalez C, Galleguillos F, Ancic P, Kay AB. Eosinophils and macrophages in bronchial mucus and bronchoalveolar lavage during allergen-induced late-phase asthmatic reactions. *J. Allergy Clin. Immunol.* 1986; **77**:244.

38 Durham SR, Carroll M, Walsh GM, Kay AB. Leucocyte activation in allergen-induced late phase asthmatic reactions. *N. Engl. J. Med.* 1984; **311**:1398–1402.

39 Kay AB. Eosinophils as effector cells in immunity and hypersensitivity disorders. *Clin. Exp. Immunol.* 1985; **62**:1–12.

40 Kay AB, Lee TH, Durham SR, Nagakura T, Cromwell O, Carroll M, Papageorgiou N, Shaw RJ. Mediators of hypersensitivity and inflammatory cells in early and late-phase asthmatic reactions. In *Asthma: Physiology, Immunopharmacology and Treatment* (Eds Kay AB, Austen KF, Lichtenstein LM), Academic Press, London 1983, pp. 211–227.

41 Moqbel R, Walsh GM, Kay AB. Inhibition of human granulocyte activation by nedocromil sodium. *Eur. J. Respir. Dis.* 1986; **69** (suppl. 147):227–229.

42 Bruijnzeel PLB, Griffioen M, Bartels AFM. Inhibition of the platelet activating factor (PAF) induced chemotaxis of polymorphonuclear granulocytes (PMN) by nedocromil sodium. In *Abstracts of the 6th Congress of the European Society of Pneumology* (Eds Stam J *et al.*), CIP-Gegevens Koninklijke Bibliotheek, Den Haag 1987, p. 22.

43 Bradford PG, Rubin RP. The differential effects of nedocromil sodium and sodium cromoglycate on the secretory response of rabbit peritoneal neutrophils. *Eur. J. Respir. Dis.* 1986; **69** (suppl. 147):238–240.

44 Spry CJF, Kumaraswami V, Tai P-C. The effect of nedocromil sodium on secretion from human eosinophils. *Eur. J. Respir. Dis.* 1986; **69** (suppl. 147):241–243.

45 Bruijnzeel PLB, Hamelink ML, Kok PTM. Nedocromil sodium inhibits the A-23 187 and opsonized zymosan induced leukotriene formation by human eosinophils but not by human neutrophils. *Br. J. Pharmacol.* 1989; In press.

46 Thorel T, Joseph M, Vorng H, Capron A. Regulation of IgE-dependent antiparasite functions of rat macrophages and platelets by nedocromil sodium. *Int. Arch. Allergy Appl. Immunol.* 1988; **85**:227–231.

47 Thorel T, Joseph M, Tsicopoulos A, Tonnel AB, Capron A. Inhibition by nedocromil sodium of IgE-mediated activation of human mononuclear phagocytes and platelets in allergy. *Int. Arch. Allergy Appl. Immunol.* 1988; **85**:232–237.

48 Godard P, Chavis C, Daures JP, Crastes de Paulet A, Michel RB, Damon M. Leukotriene B_4 and 5HETE release by alveolar macrophages in asthmatic patients: inhibition by nedocromil sodium. *Am. Rev. Respir. Dis.* 1987; **135**:A318.

49 Thorel T, Ameisen JC, Joseph M, Vorng H, Tonnel AB, Marquette CH, Capron A.

Preventative effect of nedocromil sodium on the abnormal response to aspirin of platelets from aspirin-sensitive asthmatics. *Am. Rev. Respir. Dis.* 1987; **135**:A398.
50 Robuschi M, Vaghi A, Simone P, Bianco S. Prevention of fog-induced bronchospasm by nedocromil sodium. *Clin. Allergy* 1987; **17**:69–74.
51 Schoeffel RE, Anderson SD, Altounyan REC. Bronchial hyperreactivity in response to inhalation of ultrasonically nebulised solutions of distilled water and saline. *Br. Med. J.* 1981; **282**:1285–1287.
52 Boushey HA, Richardson PS, Widdicombe JG, Wise JCM. The response of laryngeal afferent fibres in mechanical and chemical stimuli. *J. Physiol.* 1974; **240**:153–175.
53 Nadel JA, Salem H, Tamplin B, Tokiwa Y. Mechanism of sulphur dioxide induced bronchoconstriction in normal and asthmatic man. *J. Appl. Physiol.* 1965; **20**:164–167.
54 Dixon CMS, Fuller RW, Barnes PJ. Effect of nedocromil sodium on sulphur dioxide induced bronchoconstriction. *Thorax* 1987; **42**:462–465.
55 Altounyan REC, Cole M, Lee TB. Inhibition of sulphur dioxide-induced bronchoconstriction by nedocromil sodium and sodium cromoglycate in non-asthmatic, atopic subjects. *Eur. J. Respir. Dis.* 1986; **69** (suppl. 147):274–276.
56 Chilvers ER, Dixon CMS, Ind PW. Mechanism of metabisulphite-induced bronchoconstriction in atopic non-asthmatic subjects. *Thorax* 1987; **42**:745–746.
57 Dixon CMS, Ind PW. Metabisulphite induced bronchoconstriction does not involve mast cells. *Thorax* 1988; **43**:226–227P.
58 Cushley MJ, Tattersfield AF, Holgate ST. Adenosine-induced bronchoconstriction in asthma. *Am. Rev. Respir. Dis.* 1984; **129**:380–384.
59 Crimi N, Palermo F, Oliveri R, Cacopardo B, Vancheri C, Mistretta A. Adenosine-induced bronchoconstriction: comparison between nedocromil sodium and sodium cromoglycate. *Eur. J. Respir. Dis.* 1986; **69** (suppl. 147):258–262.
60 Scott VL, Phillips GD, Richards R, Holgate ST. Inhibition of AMP induced bronchoconstriction in non-atopic asthma by sodium cromoglycate and nedocromil sodium. *Thorax* 1988; **43**:225–226P.
61 Marquardt DL, Parker CW, Sullivan TJ. Potentiation of mast cell mediator release by adenosine. *J. Immunol.* 1978; **120**:871–878.
62 Rafferty P, Beasley CR, Holgate ST. The inhibitory effect of terfenadine on bronchoconstriction induced by adenosine monophosphate and allergen. *Thorax* 1986; **41**:734.
63 Mann JS, Cushley MJ, Holgate ST. Effect of vagal blockade on adenosine-induced bronchoconstriction in asthma. *Thorax* 1984; **39**:230–231.
64 Okayama M, Ma J-Y, Hataoka I, Kimura K, Ijima H, Inove H, Takishima T. Role of vagal nerve activity on adenosine-induced bronchoconstriction in asthma. *Am. Rev. Respir. Dis.* 1986; **93** (suppl. 133): Abstract.
65 Simonsson BG, Skoogh BE, Bergh NP, Anderson R, Svendmyr N. *In vivo* and *in vitro* effect of bradykinin on bronchial motor tone in normal subjects and in patients with airway obstruction. *Respiration* 1973; **30**:378–388.
66 Kaufman MP, Coleridge HM, Coleridge JCG, Baker DG. Bradykinin stimulates afferent vagal C-fibers in intrapulmonary airways of dogs. *J. Appl. Physiol.* 1980; **48**:511–517.
67 Dixon CMS, Barnes PJ. Bradykinin induced bronchoconstriction: inhibition by nedocromil and cromoglycate. *Thorax* 1988; **43**:225P.
68 Barnes PJ. Asthma as an axon reflex. *Lancet* 1986; **i**:242–245.
69 Joos GS, Pauwels RA, Van Der Straeten ME. The effect of nedocromil sodium on the bronchoconstrictor effect of neurokinin A in asthmatics. *J. Allergy Clin. Immunol.* 1988; **81**:276.
70 Saunders KB, Rudolf M, Brostoff J. Sodium cromoglycate in intrinsic asthma. *Br. Med. J.* 1978; **1**:1184.
71 Holgate ST. Clinical evaluation of nedocromil sodium in asthma. *Eur. J. Respir. Dis.* 1986; **69** (suppl. 147): 149–159.

72 Lal S, Malhotra S, Gribben D, Hodder D. Nedocromil sodium, a new drug for the management of bronchial asthma. *Thorax* 1984; **39**:809–812.

73 Van As A, Chick TW, Bodman SF *et al.* A group comparative study of the safety and efficacy of nedocromil sodium (Tilade) in reversible airways disease: a preliminary report. *Eur. J. Respir. Dis.* 1986; **69** (suppl. 147): 143–148.

74 Carrasco E, Sepulveda R. The acceptability, tolerability and safety of nedocromil sodium in long-term clinical use. *Eur. J. Respir. Dis.* 1986; **69** (suppl. 147):311–313.

75 Fairfax AJ, Allbeson M. A double-blind group comparative trial of nedocromil sodium and placebo in the management of bronchial asthma: a preliminary report. *Eur. J. Respir. Dis.* 1986; **69** (suppl. 147):320–322.

76 Greif J, Fink G, Smorzik Y, Topilsky M, Bruderman I, Spitzer SA. A multicenter, double-blind, parallel-group comparison of nedocromil sodium in the treatment of bronchial asthma. *Chest* 1989; In press.

77 Paananen M, Karakorpi T, Kreus KE. Withdrawal of inhaled corticosteroid under cover of nedocromil sodium. *Eur. J. Respir. Dis.* 1986; **69** (suppl. 147):330–335.

78 Lal S, Malhotra S, Gribben D, Hodder D. An open assessment study of the acceptability, tolerability and safety of nedocromil sodium in long-term clinical use in patients with perennial asthma. *Eur. J. Respir. Dis.* 1986; **69** (suppl. 147):136–142.

79 Williams AJ, Stableforth D. The addition of nedocromil sodium to maintenance therapy in the management of patients with bronchial asthma. *Eur. J. Respir. Dis.* 1986; **69** (suppl. 147):340–343.

80 Toogood JH, Jennings B, Lefcoe NM. A clinical trial of combined cromolyn/beclomethasone treatment for chronic asthma. *J. Allergy Clin. Immunol.* 1981; **67**:317–324.

81 Gonzalez JP, Brogden RN. Nedocromil sodium: a preliminary review of its pharmacodynamic and pharmacokinetic properties, and therapeutic efficacy in the treatment of reversible obstructive airways disease. *Drugs* 1987; **34**:560–577.

Discussion session

BARNES: is there any evidence that nedocromil sodium has any benefit compared with cromoglycate in clinical studies?

HOLGATE: as far as I am aware four clinical trials are in progress to try and answer that question.

KAY: is there anything in the *in vitro* inflammatory cell systems that nedocromil sodium can do which sodium cromoglycate (SCG) cannot?

EADY: the only system where I think there may be a difference is in antigen-induced mediator release from monkey mast cells. We need much higher doses of SCG than nedocromil sodium in this system to get appreciable inhibition.

KAY: but you emphasize the anti-inflammatory effects of nedocromil sodium and so it is on the inflammatory cells that you might expect to see a difference between SCG and nedocromil You have only mentioned differences in the response of the mast cell.

EADY: I should think that one should consider the mast cell to be an inflammatory cell. If you define an inflammatory cell as one that releases inflammatory mediators then the mast cell must be reckoned with. It releases both LTC_4 and PGD_2 at levels higher than that demonstrated for the macrophage, neutrophil or the eosinophil.

12

Cetirizine: a new selective H_1-antagonist with effects on infiltrating inflammatory cells

S.I. WASSERMAN AND C. DE VOS

Introduction

Histamine is a bioactive compound present in many organs and tissues in practically all animal species, including humans. The physiological role of histamine is not clearly defined. In contrast, a very large number of experimental and clinical data indicate that histamine is, at least partly, responsible for clinical aspects of many diseases. In 1910, Dale and Laidlaw [1] showed that the pharmacological effects of histamine were quite comparable to the symptoms of immediate hypersensitivity.

Histamine is stored in mast cells and basophils [2]. These cells are present in the blood (basophils) and the perivascular space of several organs (mast cells) and are found in great quantity in the nasal and ocular mucous membranes, in the bronchial and pulmonary tissue, the gut and in the skin. Degranulation of mast cells and the release of histamine are fundamental mechanisms responsible for some of the clinical findings of allergic conditions [3].

At the respiratory level, histamine is potent as a mediator of bronchospasm. Moreover, asthmatic subjects are almost invariably characterized by hyperreactivity to histamine inhalation, i.e. a marked decrease of the threshold of a bronchospastic response to this amine [4]. At the cutaneous level, a local antigen challenge releases histamine and may cause an increase in histaminaemia [5, 6]. Local appearance of histamine in the skin induces erythema, oedema and pruritus [7].

Clinically, urticaria is extremely frequent and appears at least once in the life of 15–20% of the population [8]. Inhibition of the activity of histamine in these various conditions will produce a definite therapeutic effect.

The first antihistaminic substances were synthesized between 1937 and 1944 [9–11] and they were used clinically by the late 1940s [12]. Their use and clinical efficacy are limited by pharmacological properties other than their antihistaminic activity. These include anticholinergic activity, local anaesthetic activity, and central nervous system (CNS) depression. Additionally, conventional antihistamines do not inhibit some pharmacological effects of histamine. Ash and Schild [13] were the first to put forward the concept of two different receptors for histamine. Black *et al.* [14] confirmed the existence of a second histamine receptor (H_2), responsible for the effect of histamine on gastric secretion. Since then specific anti-H_2 substances have appeared and are used clinically in the treatment of peptic diseases.

A recently described H_3-receptor has been implicated in histamine's regulation of its own synthesis and release from central and peripheral nerve tissue [15, 16].

In recent years, H_1-antihistamines have been developed that have little or no CNS-depressant effect—the non-sedating H_1-antagonists. Their lack of the sedating properties so characteristic of earlier H_1-blockers has been ascribed to the very limited ability of these agents to cross the blood–brain barrier.

Cetirizine, a new representative of this class of drugs, is characterized by two additional features: no affinity has been found for any other agonist receptor other than the H_1-receptor, and it is active (and almost completely excreted) as the unmetabolized compound. The former feature may make cetirizine into an exquisitely precise tool for the study of the role of histamine in physiological and pathological processes, while the latter implies that what may be observed *in vitro* has validity *in vivo*. Furthermore, cetirizine possesses a novel property, the inhibition of allergic/inflammatory cell accumulation.

Animal pharmacology [17–19]

The pharmacological profile of cetirizine is essentially characterized by its potency and its selectivity for histamine H_1-receptors.

In vivo, oral cetirizine potently protects against the systemic effect of histamine. Histamine-induced guinea pig lethality is decreased. The ED_{100} of cetirizine on this model is 0.1 mg/kg. Cetirizine strongly and selectively inhibits the pulmonary effects of histamine in the guinea pig (ID_{50}=0.09 mg/kg). The drug is 40 times less potent on serotonin-induced bronchospasm and is inactive versus acetylcholine-induced bronchospasm. *In vitro*, cetirizine strongly inhibits the histamine-induced contraction of the isolated guinea pig trachea with an IC_{50} of 0.9 μg/ml. This activity is quite selective: carbachol-induced contraction is minimally inhibited at cetirizine concentrations as high as 100 μg/ml. Histamine-induced contraction of isolated human bronchi is inhibited with an IC_{50} of 0.4 μg/ml. The cutaneous effects of histamine are also potently inhibited by oral cetirizine in the four animal species studied.

In rats, the dose that reduced the reaction to histamine by 50% was 2 mg/kg and between 0.05 and 0.1 mg/kg for the other species (mice, guinea pigs and dogs). Cetirizine also inhibited the immediate hypersensitivity reaction induced by *Ascaris* extract in dogs. In this model, cetirizine, by the oral route, had a more potent and longer-lasting activity than mepyramine, clemastine, terfenadine, astemizole and hydroxyzine. After oral cetirizine, the cutaneous reactivity of monkeys to histamine, compound 48/80, substance P and vasoactive intestinal peptide (VIP) was significantly decreased [20]. Using computer-assisted active anterior rhinomanometry in monkeys, intranasal histamine causes increased nasal resistance. Cetirizine given as a single dose markedly inhibited histamine-provoked increase in nasal resistance [21].

Cetirizine *in vivo* and *in vitro* is devoid of α-blocking activities, calcium antagonism and positive or negative ionotropic or chronotropic effects in rats and dogs. Cetirizine does not exert a relevant effect on the cardiovascular system up to the dose of 10 mg/kg i.v. which is 100 times higher than the antihistaminic dose.

Neuropharmacological studies fail to show any activity of cetirizine on the CNS up to a dose of 15 mg/kg i.v., i.p. or p.o. It is devoid of sedative properties and does not potentiate the central effects of pentobarbital (mouse) or ethanol (rat).

The radioactive distribution of ^{14}C-cetirizine in the rat gives no indication of significant penetration into the CNS. This is confirmed by the absence of a relevant occupancy of CNS H_1-receptors *in vivo* in mice with a dose 20 times higher than the skin ID_{50} [22]. In contrast, at 10 mg/kg, D-chlorpheniramine and hydroxyzine occupy a majority of rat central histamine H_1-receptors while neither cetirizine nor terfenadine at this dose occupy significant numbers of CNS histamine H_1-receptors. At 30 mg/kg, terfenadine occupies 70% of brain H_1 sites, while cetirizine occupies about half as many [23].

Cetirizine's reduced incidence of sedative side effects may stem partly from its relative exclusion from the CNS but also partly from its exquisite selectivity for H_1-receptors. Indeed, the sedating effects of classic antihistamines may well be due to actions at receptors other than those involved in their antihistaminic effects. For example, the anticholinergic effects of antihistamines may cause sedation [24]. Blockade of serotonin receptors also could influence the state of consciousness, as could effects on α-adrenergic and calcium-channel receptors [25, 26]. Thus, an ideal antihistamine would not affect any brain receptor other than histamine H_1 sites. To investigate such receptor selectivity, Snyder and Snowman [23] compared the influence of cetirizine, hydroxyzine and terfenadine on calcium antagonist, α_1-adrenergic dopamine D_2, serotonin $5HT_2$, muscarinic–cholinergic and histamine H_1-receptors (Table 12.1). The three agents differed greatly in their effects on various receptor sites. Cetirizine was similar in potency to terfenadine at H_1-receptors, and both were somewhat less potent than hydroxyzine.

Table 12.1 Differential effects of antihistamines on neuronal receptors.* (From reference 23 by permission of the publisher)

	IC_{50} μmol/l		
Radiolabelled ligand (receptor site)	Cetirizine	Hydroxyzine	Terfenadine
^{3}H-desmethoxy verapamil (Ca^{2+} channel)	> 10	3.4	0.19
^{3}H-nitrendipine (Ca^{2+} channel)	> 10	> 10	0.44
^{3}H-prazosin (α_1-adrenergic)	> 10	0.46	6.0
^{3}H-spiperone (dopamine D_2)	> 10	0.56	1.6
^{3}H-spiperone (serotonin $5HT_2$)	> 10	0.17	0.44
^{3}H-quinuclidinyl benzilate (muscarinic–cholinergic)	> 10	> 10	> 10
^{3}H-mepyramine (histamine H_1)	0.65	0.10	0.42

* Concentration–response relationships were based on assays in triplicate at six to 10 concentrations of each drug. ^{3}H-ligands were employed at concentrations several-fold lower than their K_D values, so that IC_{50} values approximate the K_i.

At concentrations as high as 10 μmol/l (4.6 μg/ml), cetirizine failed to affect any of the other receptor sites examined. In contrast, hydroxyzine and terfenadine did influence other receptor sites, Hydroxyzine was as potent a $5HT_2$-receptor antagonist as it was an antihistamine. It exhibited an ability to block dopamine D_2 and α_1-adrenergic receptors that was only several times less than its ability to block histamine H_1-receptors. Hydroxyzine was about one-sixth as potent a calcium antagonist at the verapamil-type sites than it was a dopamine D_2-receptor binder. Terfenadine differed from the other two agents in that it was equally potent, or more potent at three other receptor sites as it was an antihistamine. Thus, terfenadine was about as active in blocking both serotonin and dihydropyridine calcium-antagonist receptors as it was in blocking histamine H_1-receptors. It was about twice as potent a calcium antagonist of the verapamil type as it was an antihistamine. The only receptor site for which all three drugs were weaker than 10 μmol/l was the muscarinic–cholinergic receptor site [23].

These animal pharmacological results thus establish cetirizine as a potent and highly specific antihistamine.

Human pharmacology

Experimental models have proven useful for evaluating antihistamine compounds in humans. In these model systems, a test drug may be employed to block one or more of the known effects of exogenously administered histamine or a histamine agonist. Alternatively, the release of endogenous histamine may be brought about in a controlled fashion by agents such as allergens, opiates or compound 48/80, and the drug's effects on this process may then be measured. The dose response and duration of action of orally administered antihistamines can be determined in a simple skin model by their blocking of the wheal and flare resulting from an intradermal challenge with histamine [27]. With such a model, cetirizine has been compared with terfenadine, a minimally sedating antihistamine [28].

The histamine cutaneous reactivity of healthy volunteers was measured after a single oral intake of placebo, terfenadine 60 mg and 180 mg, and cetirizine 10 mg. Central side effects were evaluated by visual analogue scales (VAS) for drowsiness and movement coordination. The anti-H_1 cutaneous effect of cetirizine proved to be significantly more rapid, more pronounced, and longer lasting than that of terfenadine 60 mg. Cetirizine and terfenadine 180 mg were equipotent. The VAS for CNS effects did not show any difference between cetirizine, terfenadine and placebo [29].

Using the same model, Rihoux [personal communication] confirmed the increased potency of 10 mg cetirizine, not only as compared with terfenadine 60 mg but also with dexchlorpheniramine 6 mg, loratadine 10 mg, astemizole 10 mg, clemastine 1 mg and mequitazine 5 mg.

In another investigation conducted by Juhlin *et al.* [30], a single oral dose of cetirizine 10 mg was administered to eight healthy subjects. It markedly inhibited the wheal and flare induced 4 h later by intracutaneously injected histamine and compound 48/80. Dermographism was produced by different pressures (100–500 g/15 mm^2) in 10 patients with factitial urticaria. Four hours after 10 mg cetirizine, the whealing was absent in eight patients and markedly reduced in the other two. In 12 patients with cold urticaria, wheals were induced by application of an ice cube for 30 s to 12 min. Four hours after 10 mg cetirizine, the urticarial reaction had disappeared in five patients and was decreased in the other seven. No itching was experienced by any of the patients after cetirizine, but the tested areas showed an erythema that lasted from 20 to 60 min. An interesting characteristic of cetirizine is the regularity (i.e. interindividual reproducibility) of its pharmacoclinical effects in comparison with other antihistamines.

With 10 mg cetirizine, the histamine-induced skin reaction is inhibited within a range of 50 to 100% (mean inhibition 84%). This is not the case with terfenadine 60 mg whose effects ranged between 90% aggravation and 100% inhibition (mean inhibition 42%), nor with astemizole 10 mg whose effects ranged between 80% aggravation and 70% inhibition (mean inhibition 21%) [Rihoux, personal communication].

This regularity of effect of cetirizine is probably due to the minimal metabolism of this compound in humans and to the reproducibility of its pharmacokinetic characteristics. Single oral doses of ^{14}C-cetirizine (10 mg) were administered to six healthy male volunteers. The drug was rapidly absorbed: the mean peak concentrations of radioactivity (359 ng equivalents/ml) and of unchanged drug (341 ng/ml) were achieved within 1 h. Mean concentrations of cetirizine declined biexponentially and showed a mean elimination half-life of 7.9 ± 1.4 h. The drug was excreted quite rapidly, with 60% of the dose recovered in the 24-h urine. An additional 10% was excreted in urine over the next 4 days. Approximately 10% of the dose was excreted in faeces over the 5-day study period. The dose was excreted mainly as the unchanged drug. Examination of the radioactive compounds present in the plasma, and excreted in the urine and faeces indicates that there is little metabolism of cetirizine. One minor metabolite, formed by oxidative O-dealkylation of the cetirizine side chain, was detected in plasma and faeces [31]. These characteristics confer to cetirizine the benefit of avoiding the problem of low and high metabolizers in the clinic.

The most common adverse effect of antihistamines is impairment of mental performance. This may range from simple drowsiness to a variety of behavioural disturbances. The latter include cognitive deficits, altered attention span, abnormal stimulation and changes in such vegetative functions as sleep and hunger. Most classic H_1-antihistamines do cross the blood–brain barrier and interact with CNS receptors [32, 33]. Compounds that are most active at the H_1-receptors tend to be lipophilic and so readily pass the blood–brain barrier.

Recent data with newer agents now challenge the long-held concept that sedation is an inseparable component of classic antihistaminic effects [34, 35]. These newer agents tend to have a smaller volume of distribution in the body, and some suggest that lower CNS penetration is responsible for the lower incidence of CNS effects. Simons and Simons [36] showed that classic antihistamines have apparent volumes of distribution ranging from 3.4 to 18.5 l/kg. Cetirizine, with a volume of distribution of only 0.56 l/kg, is an exception [37].

Thus, pharmacokinetic characteristics of cetirizine, its property of poorly crossing the blood–brain barrier, and its selective affinity for the H_1-receptor, augured well for a good clinical toleration. This assumption was confirmed by several pharmacoclinical investigations using different CNS models. Gengo *et al.* [38] performed a sophisticated study comparing the relative antihistaminic and psychomotor effects of hydroxyzine and cetirizine. Like other 'classic H_1-antihistamines', the CNS effects of hydroxyzine are well appreciated and can at times make the daytime treatment of allergic conditions difficult [33]. Approximately 45–60% of orally administered hydroxyzine undergoes oxidative metabolism to a carboxylic acid metabolite which, in fact, is cetirizine.

Twelve atopic, but otherwise healthy subjects received single doses of hydroxyzine 25 mg, its metabolite cetirizine 10 and 20 mg, and placebo, in a four-way, cross-over study randomized by latin-square design. Skin wheal response to histamine, psychomotor effects, and serum concentrations of each drug were measured for 36 h after each dose. Central nervous system effects were measured with critical flicker frequency, Stroop word testing, and VAS. All three active treatments (cetirizine 10 mg, cetirizine 20 mg, and hydroxyzine 25 mg) produced an equivalent suppression of skin wheal response to histamine that was significantly greater than placebo ($p<0.01$). Hydroxyzine produced a significant change compared with placebo in all three CNS parameters. Neither 10 mg nor 20 mg of cetirizine produced any significant change in CNS function. Both the intensity and time course of CNS effects were related significantly ($p<0.05$) to hydroxyzine plasma concentrations. The CNS changes measured after oral hydroxyzine are the result of the parent drug, whereas its metabolite cetirizine, when administered alone, produced significant antihistaminic effects without CNS changes.

The same authors [39] also compared the effects of cetirizine and diphenhydramine using driving simulators to assess CNS impairment. This was a placebo-controlled, double-blind, five-way, cross-over study in 15 healthy subjects, comparing cetirizine 5, 10 and 20 mg with diphenhydramine 50 mg. Simulated driving reaction time (SDRT), serum drug levels and suppression of skin wheals produced by intradermal histamine were assessed over time. Diphenhydramine 50 mg caused significantly longer prolongation of SDRT than all three cetirizine doses studied. The maximum effect of diphenhydramine on SDRT was measured at 2 h and corresponded to the peak of diphenhydramine serum levels. No cetirizine dose produced changes in SDRT different from placebo. Diphenhydramine produced more drowsiness than any of the cetirizine doses. Cetirizine 20 mg produced some drowsiness; following cetirizine 5 and 10 mg, drowsiness was not different from placebo. The conclusion was that doses of cetirizine that produce antihistaminic effects cause little or no effects on the CNS function.

Seidel *et al.* [40] studied cetirizine effects on objective measures of daytime sleepiness and performance. Parallel groups of healthy subjects ($n=60$) received either 5, 10 or 20 mg cetirizine, or 25 mg hydroxyzine or placebo. Multiple sleep latency tests (MSLT = 20-min opportunities to fall asleep in bed while EEG and eye movements are recorded) were given at 2-h intervals during daytime after oral intake of the drug, and a 30-min vigilance performance test was given at 10.00 a.m. and 4.00 p.m. Subjects receiving cetirizine in doses of 5–20 mg did not differ from placebo controls in any objective or subjective measure of daytime alertness. Subjects receiving hydroxyzine were significantly more sedated and showed slower reaction times than the placebo control group for at least 4 h after treatment.

Finally, Pechadre *et al.* [41] compared central and peripheral effects of terfenadine 60 mg and cetirizine 10 mg. This placebo-controlled, double-blind, cross-over study was carried out with nine healthy volunteers. The central effects were evaluated by quantified EEG (qEEG) and VAS, and the peripheral activities were evaluated by the objective measurement of the skin reactivity to intradermal injection of histamine.

Two hours after dosing, there were no significant central effects for either terfenadine or cetirizine. However, the skin reaction was significantly inhibited by cetirizine and not by terfenadine. Six hours after dosing, there were no significant modifications in the VAS under terfenadine or cetirizine. However, the qEEG was significantly modified by terfenadine (increased δ- and θ-frequency waves, decreased α_1-frequency waves) and not by cetirizine. At the same time, the histamine-induced skin reaction was reduced by 59.7% with terfenadine, and 70% after cetirizine.

Thus, the pharmacoclinical properties of cetirizine with regard to specificity, potency, duration of action and absence of CNS side effects offered a good rationale for its clinical testing in the classic indications for antihistamines: chronic urticaria and allergic rhinitis.

Clinical results in urticaria and allergic rhinitis

Numerous clinical investigations in chronic urticaria, seasonal and perennial allergic rhinitis were performed in Europe and the USA, involving over 5000 patients. A few are summarized here.

In a double-blind, cross-over, randomized study including 29 patients suffering from daily attacks of chronic urticaria, cetirizine 10 mg o.d., terfenadine 60 mg b.i.d., and placebo were compared. Each treatment was given during 2 consecutive weeks. Symptoms, side effects and a global evaluation of the patients' condition (on a 100-mm VAS) were assessed at the outset of the study and after each treatment period. At the end of the trial, patients were asked to state which treatment period(s) they preferred. Cetirizine was shown to be significantly active in the treatment of urticaria. The trend in favour of cetirizine over terfenadine has been confirmed in other studies. Sedation incidence was low on all three treatments [42].

In seasonal allergic rhinitis Broide *et al.* [43] presented data obtained from 88 patients included in a 2-week, double-blind, parallel, placebo-controlled trial and who were given 5 mg cetirizine b.i.d., or 10 mg cetirizine in one single intake either in the morning or in the evening. Both treatments significantly relieved sneezing, nasal itch, conjunctival itch and improved the overall symptom severity score of patients. The adverse experiences reported were generally mild to moderate in severity, of short duration, and subsided spontaneously. The incidence of sedation with cetirizine was not statistically different from the placebo group.

In perennial allergic rhinitis, a double blind, placebo-controlled multicenter study including 220 patients was reported by Berman *et al.* [44].

Cetirizine 10 or 20 mg was given in the morning for a 4-week period. Both the patients' and the investigators' scores showed that cetirizine significantly improved postnasal discharge and sneezing as well as significantly alleviated overall rhinitis symptoms. The incidence of headache, somnolence and dry mouth was comparable in all treatment groups.

In another multicenter study, cetirizine 10 mg o.d., terfenadine 60 mg b.i.d. and placebo were compared in 199 patients suffering from perennial allergic rhinitis [45, 46]. The protocol followed a double-blind, cross-over design with a duration of 14 days for the different treatments. The patients and the investigators were asked to indicate which 14-day period they preferred and what kind of side effects were noticed. Patients' and investigators' opinions were both significantly in favour of cetirizine. The incidence of side effects was lower during the cetirizine period than during the terfenadine period. In conclusion of this representative clinical trial, cetirizine seemed to be safe and effective in relieving the symptoms of chronic urticaria or seasonal and perennial allergic rhinitis.

Novel antiallergic properties of cetirizine

Inhibition of eosinophil infiltration

The IgE-mediated mechanisms leading to mast-cell degranulation and rapid release of mediators probably account almost entirely for the clinical manifestations of immediate hypersensitivity: rhinorrhea, nasal blockade, sneezing, wheeze, pruritus, abdominal discomfort and diarrhoea. With a single exposure to an allergen, these symptoms are self-limiting and usually subside spontaneously. However, in most models of allergenic challenge, the immediate reaction is followed by a late-phase reaction. The late reaction is associated with the migration of inflammatory cells to the site of the immediate reaction [47–49], the principal component of this infiltrate being the eosinophilic polymorphonuclear leucocyte. This becomes the predominant feature in the case of repeated or continued exposure to allergen. Several mechanisms may contribute to eosinophilic recruitment.

During the immediate response, some of the mediators released are in fact chemotactic and can thus attract these leucocytes [47–49]. Allergen, through IgE fixed to low-affinity FcεRII, can also directly activate eosinophils, monocytes, and platelets, providing additional chemotactic mediators [50].

The pathophysiological concept of allergy has developed greatly over the past few years. Not long ago, mast-cell degranulation, with the associated immediate reaction, was considered to be the key event in an allergic response.

Today the allergic reaction, whether it affects the ear, nose and throat (ENT) system, the skin or the lungs, may be compared with a chronic disease, progressing from an acute state to a chronic condition. The study of the late-phase reaction, the association between inflammation and the allergic response, and the changes that these events can cause, serve to explain the complex situation in the allergic patient.

The cutaneous domain obviously is easier to investigate, and the early and late cutaneous reactions in normal and allergic subjects have been assessed by different techniques: one, an intradermal allergenic injection followed by a skin abrasion at the cutaneous reaction sites [51, 52] permitting analysis of the subsequent cellular infiltration by the use of the Rebuck window technique [53], and a second, using the suction-induced blisters, followed by deroofing and the placement of skin chambers [54]. The latter technique permits the measurement in the chamber fluid of the concentration of those mediators which are released once the allergic reaction has been triggered, as well as the investigation of cellular infiltration which follows employing the Rebuck window technique [53]. In Fadel's study [51], the cellular migration seen 24 h after an intradermal injection of normal saline was predominantly monocytic, whereas after diffusion in a chamber, Michel *et al.* [55] identified polymorphonuclear neutrophils.

As neutrophils generally precede monocytes in inflammatory reaction [56] it is very probable that the physiological events measured 24 h after stimulation by diffusion correspond to those generally observed 6–8 h after intradermal injection.

Fadel *et al.* [51] (Fig. 12.1a) have shown that, in healthy subjects ($n=10$), the cellular migration induced by histamine is no different to that provoked by the control medium, and no eosinophils are found. Similarly, in allergic subjects ($n=10$) stimulated by either control medium or histamine, there is no migration of eosinophils (Fig. 12.1b).

Henocq and Vargaftig [57] have also shown that an intradermal injection of platelet-activating factor (PAF-acether) will only cause a migration of eosinophils in sensitized individuals and not in normal subjects. Bryant and Kay showed in 1977 [58] that, in atopic and non-atopic individuals with similar blood eosinophils, topical application of eosinophilic chemotactic factor of anaphylaxis (ECF-A) to the abraded skin would provoke, 24 h later, the selective migration of eosinophils only in the atopic patients, and that this migration was always less than that obtained by application of allergen. The two determinant factors which appear to be required to provoke the migration of eosinophils are the triggering of a reaction which will release at least one factor chemotactic for eosinophils, and the presence of atopy. Allergen is a far more potent triggering agent for eosinophil infiltration than isolated chemotactic factors or mast-cell degranulating agents.

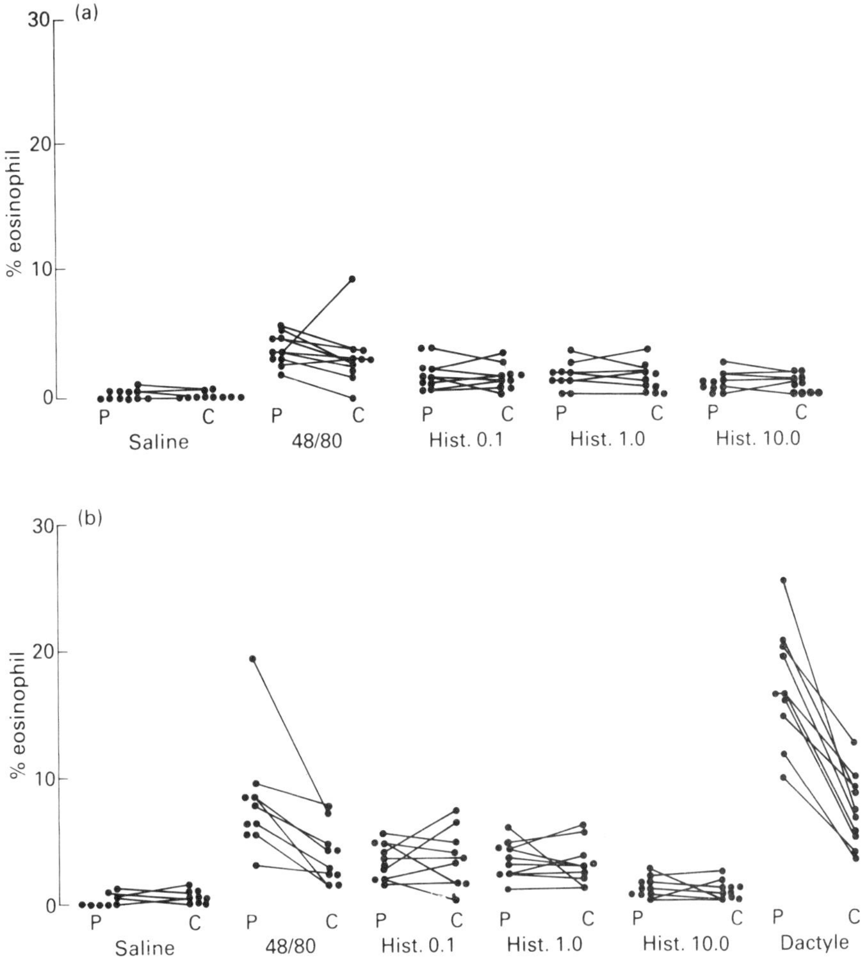

Fig. 12.1 Double-blind, cross-over study with cetirizine 10 mg/day (C) and placebo (P) in 10 healthy volunteers (a) and 10 grass-pollen-sensitive patients (b). On day 4, intradermal injections of saline (10 μl), 48/80 (10 μg/10 μl per site), histamine (0.1, 1.0 and 10 μg/10 μl per site) and dactyle pollen (1/10 000 solution, 10 μl per site) were performed, followed by a skin window measurement of eosinophil migration.

In a series of experiments carried out in several laboratories, oral cetirizine has been shown to be an effective inhibitor of eosinophil accumulation in allergic reaction sites.

Fadel *et al.* [51] (Fig. 12.1b) first showed that a single oral dose of 10 mg of cetirizine significantly inhibited eosinophil migration at the abraded skin site 24 h after an intradermal injection of pollen (60% inhibition, $n=8$) in hypersensitive subjects.

Similar inhibition was demonstrated after repeated oral daily doses of 10 mg cetirizine (69% inhibition at 4 h and 62% inhibition at 24 h after an intradermal injection of pollen, $n=10$).

Michel *et al.* [55] have confirmed this property of cetirizine by using a model where the Rebuck window is placed not on the abraded skin but onto the area denuded by a suction-induced blister. In this study, it was shown that a 10 mg/day dose would significantly inhibit the migration of eosinophils (65% inhibition at 24+1 h, $n=6$) (Fig. 12.2).

Finally, Charlesworth *et al.* [59], using a similar suction-induced blister technique together with a cutaneous chamber, have also shown that 20 mg/day of cetirizine significantly inhibited an eosinophilic cellular infiltrate produced by an allergenic stimulus. In this case, cellular recruitment was measured directly by counting the leucocytes in the chamber fluid 6–8 h after the onset of the reaction. A 65–83% inhibition of eosinophil accumulation was noted ($n=10$).

Thus, it appears that administration of a single 10 mg dose of cetirizine was sufficient to provide considerable inhibition of tissue eosinophil infiltration.

Inhibition of eosinophil recruitment is a property not previously observed with anti-H_1 agents. In studies of cellular migration using the Rebuck technique [53], H_1-blockade by promethazine (50 mg q.i.d.) did not inhibit the influx of eosinophils [60]. Ting *et al.* [61] have also shown the migration of cutaneous eosinophils in allergic subjects after stimulation by allergen, a phenomenon not inhibited by the injection of sodium cromoglycate at the site [62]. Corticosteroids can inhibit eosinophil influx due to antigen but do not possess antihistaminic activity and do not prevent allergen-induced wheal-and-flare reactions [63–65].

No other antihistamine has yet been shown to possess this antieosinophil property, and it is not yet possible to attribute this to an anti-H_1 action of the molecule or to some other activity of cetirizine.

It is also not yet known whether cetirizine affects the eosinophils directly or rather the process of their recruitment. In this regard it has been shown that cetirizine (0.1 μg/ml) inhibits the migration of purified human eosinophils to PAF-acether or formyl-methionyl-leucyl-phenylalanine (FMLP) [66, 67], while polaramine is less active (Fig. 12.3) [68]. Cetirizine is only minimally metabolized in humans [69], which lends support to these *in vitro* data. The ginkgolide BN-52 021, a specific PAF-acether antagonist, is very active in the model employed to demonstrate PAF-acether-induced chemotaxis of human eosinophils. Sodium cromoglycate, nedocromil sodium, salbutamol and dexamethasone had no effect in this particular model [70]. Despite preliminary evidence for a direct effect of cetirizine on eosinophils, one can, at this stage, not exclude that the eosinophil inhibition produced by cetirizine might

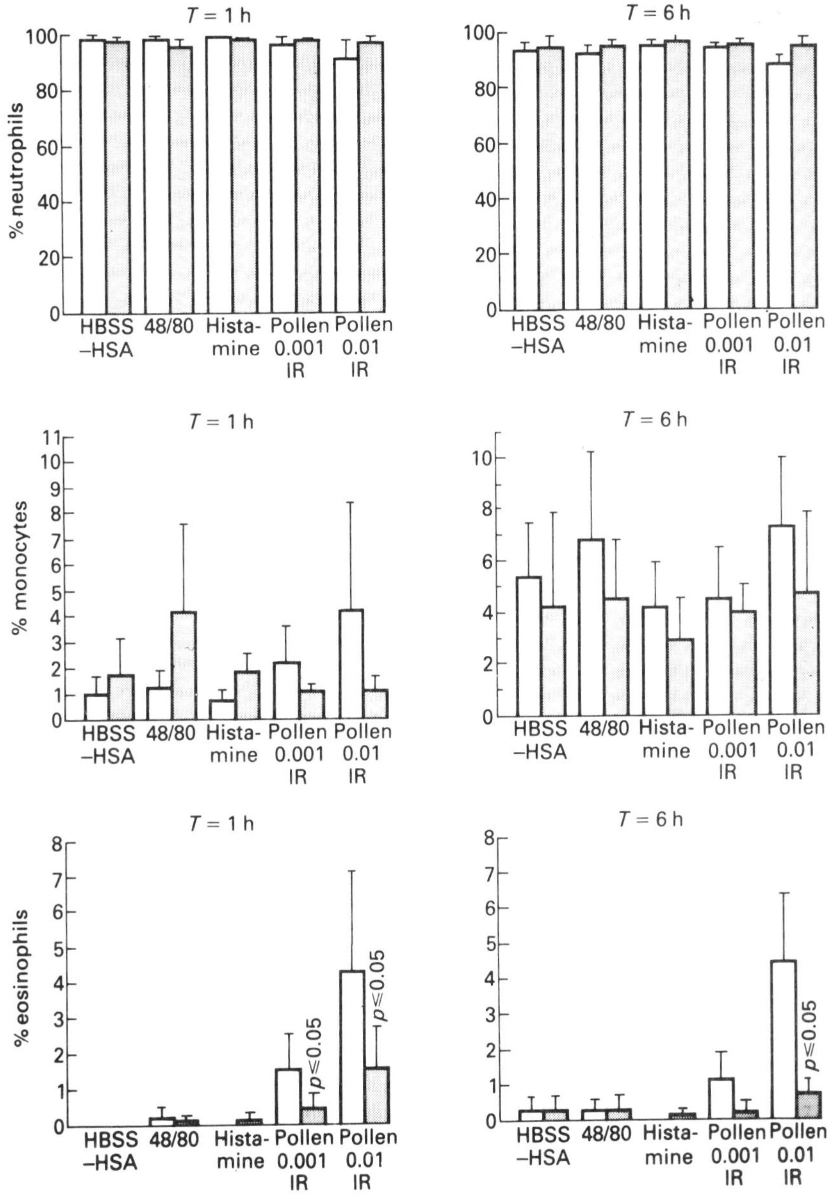

Fig. 12.2 Percentage of recruited inflammatory cells (neutrophils, monocytes, eosinophils) on the superficial dermis 24 h after challenge with HBSS-HSA, compound 48/80 (1 μg), histamine (1 μg), pollen 0.001 IR and pollen 0.01 IR in pollen-sensitive patients (mean % ± SD, $n = 6$) under placebo (white bars) and cetirizine 2HCl 10 mg/day (shaded bars) oral administration. Unless p values are indicated, there is no significant inhibition of recruited cell percentage under cetirizine 2HCl treatment as compared with the placebo sequence according to Wilcoxon's statistical test. (From reference 55 by permission of the publisher.)

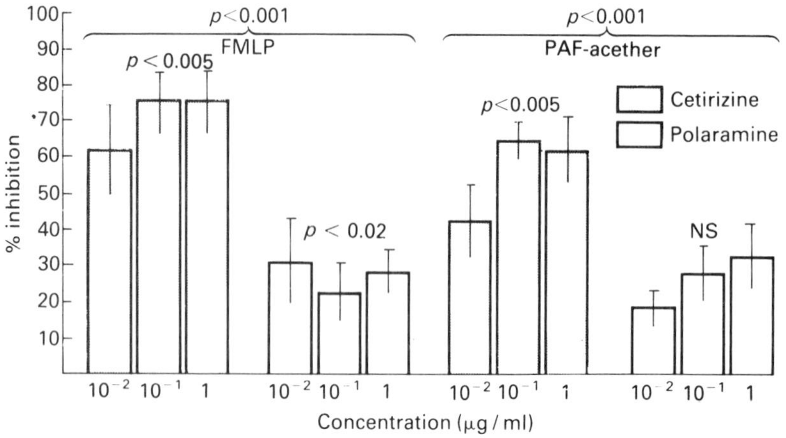

Fig. 12.3 Inhibition of eosinophil chemotactic response to FMLP 10^{-6} M and to PAF-acether 10^{-6} M in the presence of various concentrations of cetirizine or polaramine. Each histogram represents the mean ± standard error of 14 experiments. (From reference 66 by permission of the publisher.)

also be the result of its potent anti-H_1 characteristics. A complete blockage by cetirizine of the vascular effects of histamine could lead to mechanical inhibition of cellular diapedesis by reducing the vascular permeability or by inhibiting histamine-induced endothelial cell synthesis of PAF-acether. This hypothesis is not likely in view of the fact that the antihistamine promethazine failed to mimic cetirizine [60].

Cetirizine and the pharmacology of mediator release

In addition to eosinophil-directed and H_1-receptor-directed anti-inflammatory actions, cetirizine may also act upon mast cells to inhibit release of preformed or generated mediators.

Michel *et al.* [55] showed that cetirizine 10 mg/day moderately inhibited histamine penetration into skin chambers after pollen introduction. Similarly Charlesworth *et al.* [59] obtained a moderate inhibition of prostaglandin (PG) D_2 release 3–5 h after pollen instillation.

More experiments are necessary to confirm a mediator-release inhibitory activity of cetirizine.

Cetirizine and the late cutaneous allergic response

In the skin, late reactions to allergen or anti-IgE injections exist [71–73] and are inhibited by beclomethasone but not by terbutaline or the H_1-antagonist mepyramine [74], or by cromolyn [75] or indomethacin [76].

Dorsh [personal communication] demonstrated that cetirizine, when given orally at a dose of 10 mg b.i.d. over 4 days, would significantly inhibit not only the immediate cutaneous response to allergen but also the induration associated with the late-phase reaction, an effect which has not yet been described for any other anti-H_1 agent. Smith *et al.* [77] have reported that chlorpheniramine, when given orally, inhibited the cutaneous reaction to histamine, as well as the immediate cutaneous response to allergen, but not the late-phase cutaneous response to allergen.

It can be speculated that there is a relationship between inhibition of the cellular migration of eosinophils to the site of the allergic reaction by cetirizine, and inhibition of the appearance of some mediators. These different properties could thus explain the inhibition that cetirizine exerts on the late cutaneous reaction. Present knowledge regarding the physiological role of histamine, and the pharmacology of the antihistamines, does not allow us to state whether these unexpected properties of cetirizine are novel features of the molecule, or are the result of its selective antihistamine activity [23].

Bronchoalveolar eosinophilic infiltration and the late-phase asthmatic response

Upon bronchial challenge with allergen, besides an immediate bronchoconstriction, many asthmatic patients experience late-onset (6–8 h) bronchospasm. The hallmark of the late asthmatic reaction is the presence of bronchoalveolar eosinophilia.

Recent work by Kirby *et al.* [78] confirms that there is an inflammatory cellular infiltrate in the allergic asthmatic but not in the healthy subject. This infiltrate consists mainly of metachromatic cells (mast cells and basophils) and eosinophils, and even persists once the asthma has been controlled. In asthma, the eosinophil is pro-inflammatory due to constituents released from its granules which are capable of destroying bronchial epithelium. These lesions probably contribute to bronchial hyperreactivity [79, 80]. The degree of bronchial hyperreactivity, blood eosinophilia and bronchoconstriction noted in the late phase after allergen provocation are correlated [81].

After provocation with a specific allergen, the late-phase bronchoconstriction is accompanied by migration of eosinophils to the airway. This phenomenon is not observed in those patients who only present with the immediate reaction, nor is it found in the healthy controls [82].

Potential place of cetirizine in the pharmacological treatment of asthma

Research models and clinical studies in atopic asthma are in progress with cetirizine. Cetirizine provides a significant dose-dependent protection against histamine-induced bronchoconstriction which is superior to that produced by hydroxyzine in subjects with mild asthma [83]. In a double-blind, randomized, cross-over study of protection against bronchial challenge with house-dust mite allergen in 20 patients with perennial asthma, cetirizine was compared with placebo. Therapy with 20 mg of cetirizine for 14 days significantly increased the PD_{20} FEV_1 [84].

Cetirizine 15 mg/day was reported to protect the patients suffering from pollen asthma [85]. More recently [86], a 6-week clinical trial involving 39 patients in the Netherlands compared the prophylactic effects of 10 mg cetirizine b.i.d. with 60 mg terfenadine b.i.d. to prevent the appearance of pollen asthma during the hay-fever season of 1987. The conclusions of this study demonstrated a protective effect of cetirizine with a significant therapeutic superiority over terfenadine ($p = 0.007$).

While these data are intriguing, their magnitude and relevance to clinical asthma remain uncertain. Moreover, the mechanisms (anti-H_1, antieosinophil or other) by which such effects occur are unknown.

Concluding remarks

As an antihistamine, cetirizine possesses several valuable properties. As a research tool it could become the choice compound to experimentally unravel the role of histamine in physiological and pathological processes because its minimal metabolism ensures that *in vitro* effects are relevant *in vivo*, and because its exquisitely selective affinity for the H_1-receptor provides for clear interpretation of experimental results. As a drug in the classical indications for antihistamines, its potency, its regularity of clinical effects and its poor CNS penetration, are major assets. Clinical studies of cetirizine have established an excellent therapeutic index, and a satisfactory duration of action allowing once a day oral intake.

Moreover, as a novel feature, cetirizine inhibits the recruitment of infiltrating inflammatory cells during the cutaneous allergenic response. This property is at least in part due to a direct effect on eosinophils and may extend important antiallergic properties to cetirizine. All these characteristics may well lead to its use in ameliorating late-phase allergic inflammatory changes intimately linked to the release of mediators from cells recruited to the immunologically stimulated sites, and which are felt responsible for the chronic nature of severe allergic problems.

References

1 Dale HH, Laidlaw PP. The physiological action of β-iminazolylethylamine. *J. Physiol.* 1910; **41**:318–344.

2 Riley JF, West GB. The presence of histamine in tissue mast cells. *J. Physiol.* 1953; **120**:528–537.
3 Piper PJ. Anaphylaxis and the release of active substances in the lungs. *Pharmacol. Ther.* (Part B) 1977; **3**:75–98.
4 Boushey HA, Holtzman MJ, Sheller JR, Nadel JA. Bronchial hyperreactivity. *Am. Rev. Respir. Dis.* 1980; **121**:389–413.
5 Winkelmann RK, Wilhelmj CM, Horner FA. Experimental studies on dermographism. *Arch. Dermatol.* 1965; **92**:436–442.
6 Greaves MW, Sondergaard J. Urticaria pigmentosa and factitious urticaria. Direct evidence for release of histamine and other smooth muscle-contracting agents in dermographic skin. *Arch. Dermatol.* 1970; **101**:418–425.
7 Marks R, Greaves MW. Vascular reactions to histamine and compound 48/80 in human skin: suppression by a histamine H_2-receptor blocking agent. *Br. J. Clin. Pharmacol.* 1977; **4**:367–369.
8 Burland WL, Mills JG. The pathophysiological role of histamine and potential therapeutic uses of H_1 and H_2 antihistamines. In *Pharmacology of Histamine Receptors* (Eds Ganellin CR, Parsons ME), J. Wright & Sons, London 1982, pp. 436–481.
9 Bovet D, Staub A. Action protectrice des éthers phénoliques au cours de l'intoxication histaminique. *C.R. Séances Soc. Biol. Fil.* 1937: **124**:547–549.
10 Halpern BN. Les antihistaminiques de synthèse. Essais de chimiothérapie des états allergiques. *Arch. Int. Pharmacodyn. Thér.* 1942; **68**:339–408.
11 Bovet D, Horclois R, Walthert F. Propriétés antihistaminiques de la *N*-*p*-méthoxybenzyl-*N*-diméthylaminoéthyl α amino-pyridine. *C.R. Séances Soc. Biol. Fil.* 1944; **138**:99–100.
12 Loew ER, MacMillan R, Kaiser ME. The antihistamine properties of Benadryl, β-dimethylaminoethyl benzhydryl ether hydrochloride. *J. Pharmacol. Exp. Ther.* 1946; **86**:229–238.
13 Ash ASF, Schild HO. Receptors mediating some actions of histamine. *Br. J. Pharmacol. Chemother.* 1966; **27**:427–439.
14 Black JW, Duncan WAM, Durant CJ, Ganellin CR, Parsons ME, Definition and antagonism of histamine H_2-receptors. *Nature* 1972; **236**:385–390.
15 Ishikawa S, Sperelakis N. A novel class (H_3) of histamine receptor on perivascular nerve terminals. *Nature* 1987; **327**:158–160.
16 Arrang JM, Garbarg M, Lancelot JC, Lecomte JM, Pollard H, Robba M, Schunack W, Schwartz JC. Highly potent and selective ligands for histamine H_3-receptors. *Nature* 1987; **327**:117–122.
17 De Vos C, Maleux MR, Gobert J. *In vitro* pharmacologic profile of cetirizine 2HCl, a new non sedating antiallergic drug. *Alergol. Immunol. Clin.* 1987; **2**:82.
18 De Vos C, Maleux MR, Gobert J. Antiallergic properties of cetirizine 2HCl in various models of anaphylaxis in animals. *Alergol. Immunol. Clin.* 1987; **2**:81.
19 De Vos C, Maleux MR, Baltes E, Gobert J. Inhibition of histamine and allergen skin wheal by cetirizine in four animal species. *Ann. Allergy* 1987; **59**:278–281.
20 Van Cauwenberge PB, Doyle WJ, Rihoux JP, Skoner DP, Tanner E. Effect of cetirizine 2HCl and cimetidine on the cutaneous reactivity of the rhesus monkey to histamine, 48/80, substance P and vasoactive intestinal polypeptide (VIP). *Clin. Exp. Dermatol.* 1987; **12**:149.
21 Fireman P, Skoner D, Tanner E, Doyle W. A primate model for the evaluation of antihistamines. *Ann. Allergy* 1987; **59** (Part II):9–12.
22 De Vos C, Rihoux JP, Körner M, Redouane K. Comparison between cutaneous anti-H_1 action and occupation of cerebral H_1 receptors with cetirizine 2HCl in mice. *XIIIth Congress of the European Academy of Allergology and Clinical Immunology*, Budapest 1986, p. 312 (Abstract Book).
23 Snyder SH, Snowman AM. Receptor effects of cetirizine. *Ann. Allergy* 1987; **59** (Part II):4–8.
24. Diffley D, Tran VT, Snyder SH. Histamine H_1 receptors labelled *in vivo*: antidepressant and antihistamine interactions. *Eur. J. Pharmacol.* 1980; **64**:177–181.
25 Peroutka SJ, U'Prichard DC, Greenberg D, Snyder SH. Neuroleptic drug interactions with norepinephrine alpha receptor binding sites in rat brain. *Neuropharmacology* 1977; **16**:549–556.

26 U'Prichard DC, Greenberg D, Sheehan PP, Snyder SH. Tricyclic antidepressants: therapeutic properties and affinity for alpha-noradrenergic receptor binding sites in the brain. *Science* 1978; **199**:197–198.

27 Harvey RP, Schocket AL. The effect of H_1 and H_2 blockade on cutaneous histamine response in man. *J. Allergy Clin. Immunol.* 1980; **65**:136–139.

28 Brandon ML, Murray-Weiner MD. Clinical investigation of terfenadine, a non sedating antihistamine. *Ann. Allergy* 1980; **44**:71–75.

29 Rihoux JP, Dupont P. Comparative study of the peripheral and central effects of terfenadine and cetirizine 2HCl. *Ann. Allergy* 1987; **59**:235–238.

30 Juhlin L, De Vos C, Rihoux JP. Inhibiting effect of cetirizine on histamine induced and 48/80 induced wheals and flares, experimental dermographism, and cold-induced urticaria. *J. Allergy Clin. Immunol.* 1987; **80**:599–602.

31 Wood SG, John BA, Chasseaud LF, Yeh J, Chung M. The metabolism and pharmacokinetics of ^{14}C-certirizine in humans. *Ann. Allergy* 1987; **59** (Part II):31–34.

32 Clarke CH, Nicholson AN. Performance studies with antihistamines. *Br. J. Clin. Pharmacol.* 1978; **6**:31–35.

33 Pishkin V, Sengel RA, Lovallo WR, Shurley JT. Cognitive and psychomotor evaluation of clemastine, diphenhydramine HCl and hydroxyzine in a double blind study. *Curr. Ther. Res.* 1983; **33**:230–247.

34 Lefur G, Margoulis C, Uzan A. Effect of a non sedating antihistamine on brain H_1 receptors. *Life Sci.* 1981; **29**:547–552.

35 Cohen AF, Hamilton M, Philipson R, Peck AW. The acute effects of acrivastine (BW 825C), a new antihistamine, compared with triprolidine on measures of central nervous system performance and subjective effects. *Clin. Pharmacol. Ther.* 1985; **38**:381–386.

36 Simons FER, Simons KJ. H_1 receptor antagonists: clinical pharmacology and use in allergic disease. *Ped. Clin. North Am.* 1983; **30**:899–914.

37 Simons FER, Simons KJ, Chung H, Yeh J. The comparative pharmacokinetics of H_1 receptor antagonists. *Ann. Allergy* 1987; **59** (Part II):20–24.

38 Gengo F, Dabronzo J. Yurchak A, Love S, Miller JK. The relative antihistaminic and psychomotor effects of hydroxyzine and cetirizine. *Clin. Pharmacol. Ther.* 1987; **42**:267–272.

39 Gengo F, Gabos C, Mechtler L. The effects of diphenhydramine and increasing doses of cetirizine on simulated automobile driving. *J. Allergy Clin. Immunol.* 1988; **81**:211.

40 Seidel WF, Cohen S, Bliwise NG, Dement WC. Cetirizine effects on objective measures of daytime sleepiness and performance. *Ann. Allergy* 1987; **59** (Part II):58–62.

41 Pechadre JC, Trolese JF, Bloom M, Dupont P, Rihoux JP. Compared central and peripheral effects of terfenadine and cetirizine 2HCl. *Alergol. Immunol. Clin.* 1987; **2**:83.

42 Dockx P, Lambert J, Arendt C, Darte D. Comparison of cetirizine and terfenadine in idiopathic urticaria. *Alergol. Immunol. Clin.* 1987; **2**:113.

43 Broide DH, Love S, Altman R, Wasserman SI. Evaluation of cetirizine in the treatment of patients with seasonal allergic rhinitis. *J. Allergy Clin. Immunol.* 1988; **81**:176.

44 Berman B, Buchman E, Dockhorn R, Leese P, Mansmann H, Middleton E. Cetirizine therapy of perennial allergic rhinitis. *J. Allergy Clin. Immunol.* 1988; **81**:177.

45 Sabbah A, Bonneau JC, Arendt C, Dupont P. Comparison of cetirizine and terfenadine in perennial allergic rhinitis. *Alergol. Immunol. Clin.* 1987; **2**:322.

46 Sabbah A, Boniver R, Gastpar H, Haguenauer JP, Bruttmann G, Schmeisser KJ, Paquelin F, Arendt C, Darte D, Rihoux JP. Comparison between cetirizine 2HCl and terfenadine in the treatment of perennial allergic rhinitis. Results of a multicenter clinical trial. *XIIth Congress of the European Rhinologic Society*, Amsterdam, June 1988.

47 Fowler JW III, Lowell FC. The accumulation of eosinophils as an allergic response to allergen applied to the denuded skin surface. *J. Allergy Clin. Immunol.* 1966; **37**:19–28.

48 Durham SR, Carroll M, Walsh GM, Kay AB. Leukocyte activation in allergen induced late phase asthmatic reactions. *N. Engl. J. Med.* 1984; **311**:1398–1402.

49 Wei Y, Heghinian K, Bell RL, Jakschik BA. Contribution of macrophages to immediate hypersensitivity reaction. *J. Immunol.* 1986; **137**:1993–2000.

50 Capron A, Dessaint JP, Capron M, Joseph M, Ameisen JC, Tonnel AB. From parasities to allergy: a second receptor for IgE. *Immunol. Today* 1986; **7**:15–18.

51 Fadel R, Herpin-Richard N, Rihoux JP, Henocq E. Inhibitory effect of cetirizine 2HCl on eosinophil migration *in vivo*. *Clin. Allergy* 1987; **7**:373–379.

52 Fadel R, Herpin-Richard N, Henocq E, Rihoux JP. Effect of cetirizine on *in vivo* cutaneous leukocytes migration in allergic rhinitis. *J. Allergy Clin. Immunol.* 1988; **81**:178.

53 Rebuck JW, Crowley JH. A method of studying leukocytic functions *in vivo*. *Ann. NY Acad. Sci.* 1955; **59**:757–805.

54 Dubertret L, Lebreton C, Touraine R. Neutrophil studies in psoriatics: *in vivo* migration, phagocytosis, and bacterial killing. *J. Invest. Dermatol.* 1982; **79**:74–78.

55 Michel L, De Vos C, Rihoux JP, Burtin C, Benveniste J, Dubertret L. Inhibitory effect of oral cetirizine on *in vivo* antigen induced histamine and PAF-acether release and eosinophil recruitment in human skin. *J. Allergy Clin. Immunol.* 1988; **82**:101–109.

56 Issekutz AC. Quantitation of acute inflammation in the skin: recent methodological advances and their application to the study of inflammatory reactions. *Surv. Synth. Pathol. Res.* 1983; **1**:89–110.

57 Henocq E, Vargaftig BB. Accumulation of eosinophils in response to intracutaneous PAF-acether and allergens in man. *Lancet* 1986; **1**:1378–1379 (Letter).

58 Bryant DH, Kay AB. Cutaneous eosinophil accumulation in atopic and non atopic individuals: the effect of an ECF-A tetrapeptide and histamine. *Clin. Allergy* 1977; **7**:211–217.

59 Charlesworth EN, Kagey-Sobotka A, Norman PS, Lichtenstein LM. Effects of cetirizine on mast cell mediator release and cellular traffic during the cutaneous late phase response. *J. Allergy Clin. Immunol.* 1988; **81**:212.

60 Eidinger D, Wilkinson R, Rose B. A study of cellular responses in immune reactions utilizing the skin window technique. *J. Allergy* 1964; **35**:77–85.

61 Ting S, Dunsky EH, Lavker RM, Zweiman B. Patterns of mast cell alterations and *in vivo* mediator release in human allergic skin reactions. *J. Allergy Clin. Immunol.* 1980; **66**:417–423.

62 Ting S, Zweiman B, Lavker RM. Cromolyn does not modulate human allergic skin reactions *in vivo*. *J. Allergy Clin. Immunol.* 1983; **71**:12–17.

63 Slott RI, Zweiman B. Histological studies of human skin test responses to ragweed and compound 48/80. II. Effects of corticosteroid therapy. *J. Allergy Clin. Immunol.* 1975; **55**:232–240.

64 Zweiman B, Slott RI, Atkins PC. Histologic studies of human skin test responses to ragweed and compound 48/80. III. Effects of alternate day steroid therapy. *J. Allergy Clin. Immunol.* 1976; **58**:657–663.

65 Dunsky EH, Atkins PC, Zweiman B. Histologic responses in human skin test reactions to ragweed. IV. Effects of a single intravenous injection of steroids. *J. Allergy Clin. Immunol.* 1977; **59**:142–146.

66 Leprevost C, Capron M, De Vos C, Tomassini M, Capron A. Inhibition of eosinophil chemotaxis by a new antiallergic compound (cetirizine). *Int. Arch. Allergy Appl. Immunol.* 1988; In press.

67 De Vos C, Joseph M, Leprevost C, Vorng H, Tomassini M, Capron M, Capron A. Inhibition of human eosinophil chemotaxis and of the IgE dependent stimulation of human blood platelets by cetirizine. *Int. Arch. Allergy Appl. Immunol.* 1988; In press.

68 Huang SA, Athanikar NK, Sridhar K, Huang YC, Chiou WL. Pharmacokinetics of chlorpheniramine after intravenous and oral administration in normal adults. *Eur. J. Pharmacol.* 1982; **22**:359–365.

69 Simons FE, Simons KJ, Chung M, Yeh J. The comparative pharmacokinetics of H1 receptor antagonists. *Ann. Allergy* 1987; **59**:20–24.

70 Kurihara K, Wardlaw AJ, Moqbel R, Kay AB. The ginkgolide (BN 52021) inhibits PAF induced chemotaxis and specific binding to human eosinophils and neutrophils. *J. Allergy Clin. Immunol.* 1988; **81**:208.

71 Dolovich J, Hargreave FE, Chalmers R, Shier KJ, Gauldie J, Bienenstock J. Late cutaneous allergic responses in isolated IgE-dependent reactions. *J. Allergy Clin. Immunol.* 1973; **52**:38–46.
72 Solley GO, Gleich GJ, Jordon RE, Schroeter AL. The late phase of the immediate wheal and flare skin reaction. *J. Clin. Invest.* 1976; **58**:408–420.
73 Zetterström O. Dual skin reactions and serum antibodies to subtilisin and *Aspergillus fumigatus* extracts. *Clin. Allergy* 1978; **8**:77–91.
74 Grönneberg R, Sirandberg K, Stålenheim G, Zetterström O. Effect in man of antiallergic drugs on the immediate and late phase cutaneous allergic reactions induced by anti IgE. *Allergy* 1981; **36**:201–208.
75 Grönneberg R, Zetterström O. Inhibition of the late phase response to anti IgE in humans by indomethacin. *Allergy* 1985; **40**:36–41.
76 Grönneberg R, Zetterström O. Effect of disodium cromoglycate on anti IgE induced early and late skin response in humans. *Clin. Allergy* 1985; **15**:167–171.
77 Smith JA, Mansfield LE, de Shazo RD. An evaluation of the pharmacologic inhibition of the immediate and late cutaneous reaction to allergen. *J. Allergy Clin. Immunol.* 1980; **65**:118–121.
78 Kirby JG, Hargreave FE, Gleich GJ, O'Byrne PM. Bronchoalveolar cell profiles of asthmatic and non asthmatic subjects. *Am. Rev. Respir. Dis.* 1987; **136**:379–383.
79 Gleich GJ. The role of the eosinophilic leukocyte in bronchial asthma. *Bull. Eur. Physiopathol. Respir.* 1986; **22**:62–69.
80 Frigas E, Gleich GJ. The eosinophil and the pathophysiology of asthma. *J. Allergy Clin. Immunol.* 1986; **77**:527–537.
81 Durham SR, Kay AB. Eosinophils, bronchial hyperreactivity and late phase asthmatic reactions. *Clin. Allergy* 1985; **15**:411–418.
82 De Monchy JGR, Kauffman HF, Venge P, Koëter GH, Jansen HM, Sluiter HJ, De Vries K. Bronchoalveolar eosinophilia during allergen induced late asthmatic reactions. *Am. Rev. Respir. Dis.* 1985; **131**:373–376.
83 Brik A, Tashkin DP, Gong H, Dauphinee B, Lee E. Effect of cetirizine, a new histamine H_1 antagonist, on airway dynamics and responsiveness to inhaled histamine in mild asthma. *J. Allergy Clin. Immunol.* 1987; **80**:51–56.
84 Thomas KE, Ferguson H, Herdman MJ, Davies RJ. The effect of multiple dose therapy with cetirizine on the immediate asthmatic response to allergen provocation testing. Accepted for presentation at the *XIIIth International Congress of Allergology and Clinical Immunology*, October 1988.
85 Bruttmann G, Pedrali P, Arendt C, Rihoux JP. Therapeutic effect of cetirizine 2HCl on pollen asthma. *XIIth World Congress of Asthmology*, Barcelona 1987, pp. 287–288.
86 Van Ganse E, Nierop G, Molkenboer JF, Vanderschueren R, Hekking PRM, Dijkman JH. Comparison between cetirizine and terfenadine in the prophylactic treatment of pollinic asthma. Accepted for presentation at the *Annual Meeting of the European Academy of Allergology and Clinical Immunology*, June 1988.

Discussion session

BARNES: are you saying that prevention of eosinophil infiltration and PAF-acether generation is a property which will be shared by all other H_1-antagonists?

WASSERMAN: yes, perhaps in pharmacological concentrations.

BARNES: that certainly questions the whole idea of eosinophils and asthma.

WASSERMAN: I am not sure it does.

BARNES: because H_1-antagonists are not that effective in clinical asthma or in blocking hyperresponsiveness or preventing the late response.

WASSERMAN:, we have never given adequate concentrations in these circumstances I believe.

BARNES: I disagree, Professor Holgate has shown that with terfenadine you can get extremely high doses.

WASSERMAN: yes, but has he ever done BAL and looked at eosinophil activation?

HOLGATE: there have only been two studies to my knowledge which have looked at the highly selective H_1-antagonists on the late reaction. Where astemizole was administered when the late reaction was evolving it was shown to attenuate the late-phase reaction. A similar study has been undertaken with terfenadine, but I am unsure of the outcome of this trial.

13

Lipocortin

R.J. FLOWER

Introduction

The glucocorticosteroids are dramatically effective in the treatment of asthma and allergic airway diseases but their precise mechanism of action remains obscure.

Based upon experience with other types of inflammation it would probably be true to say that the steroids act to control asthma at practically every level: the signs and symptoms are quenched, the production of mucus decreased, the migration of leucocytes diminished, the biosynthesis of pro-inflammatory and bronchoconstrictor metabolites inhibited and the immune response suppressed.

Clearly no single mechanistic explanation could suffice to account for all the biological effects that are observed. In this chapter the author will explore a line of research which has led to the discovery of a family of proteins whose control by steroids could account for some of the antiasthma properties of these drugs. These proteins have recently been sequenced and cloned and have turned out, surprisingly, to be related to (and in some cases identical with) a group of membrane proteins hitherto thought to be ultrastructural components and substrates for oncogene kinases. The history of their discovery goes back to the mid 1970s.

Asthma, inflammation and lipid mediator production

Asthma has a strong inflammatory component and most research workers would argue that the inflammatory response itself is largely regulated by the release of chemicals—mediators—from dying or activated cells and that these mediators are responsible for initiating the familiar signs and symptoms of inflammation together with many of the underlying pathological changes and systemic effects which are also seen. Histamine, serotonin and bradykinin, are amongst the local hormones which have been implicated in the development of inflammation, but it is the lipid mediators such as the prostaglandins, leukotrienes and platelet-activating factor (PAF-acether) which have dominated the imagination of those interested in the pathogenesis of asthma.

It is not my intention to review the compelling evidence that suggests a role for these mediators in asthma or other types of inflammatory response [cf. 1–3 for a detailed guide to the bibliography covering this section], but an outline of their biosynthesis will have to be presented here so that the arguments made later in this paper will appear clear.

Both the prostaglandins and the leukotrienes (collectively known as the eicosanoids) are biosynthesized from polyunsaturated fatty acids, pre-eminent amongst which in land-dwelling mammals, is arachidonic acid. This fatty acid (as well as some other related compounds) can be transformed by the fatty acid cyclooxygenase into the cyclic endoperoxides and thence into prostaglandins, thromboxanes and prostacyclin. Other enzymes convert this fatty acid into leukotrienes. The first of these transformations may occur in practically any cell in the body (with the exception of the erythrocyte) whilst the latter series of reactions occur mainly in leucocytes [cf. 4–7].

Platelet-activating factor is probably biosynthesized in a two-stage process in which a specific phosphatide precursor is catalytically hydrolysed to lyso-PAF by phospholipase A_2 and then acetylated by a specific acetyl transferase [8]. It has been observed that a high proportion of this ether phosphatide precursor also contains arachidonic acid esterified to the 2′ position such that cleavage by phospholipase liberates both the precursors for the eicosanoids, and that for PAF simultaneously [9].

Biologically active lipids are, of course, not the only chemicals involved in the inflammation of airways or other tissues. For example, complement plays a role in cell migration and oedema formation, and fever—a common sequel to inflammation—is regulated by the release from macrophages of yet another type of chemical, the polypeptide interleukin (IL) 1. Other, less well-characterized factors regulating immune and other functions are also generated by cells such as monocytes. There is evidence for substantial interactions between inflammatory mediators [10].

Glucocorticosteroids and the inhibition of lipid mediator synthesis

When the inhibitory action of aspirin on the prostaglandin-forming cyclooxygenase was discovered, many other types of drugs were tested as putative inhibitors [11–14]. Amongst inactive compounds in this test were the narcotic analgesics and the anti-inflammatory glucocorticoids. The negative effects observed were puzzling, because although they were many times more active in experimental models of inflammation, they were all inactive against the cyclooxygenase enzyme, and thus, by implication on prostaglandin synthesis.

These findings suggested that the glucocorticoids had no effect on the prostaglandin system, but a number of observations soon began to appear in the literature which apparently contradicted this conclusion. In 1974,

Herbaczynska-Cedro and Staszewska-Barczak demonstrated that the release of a prostaglandin-like substance into the venous blood of exercising dogs was blocked by hydrocortisone [15]. Shortly after this came a report by Lewis and Piper in which it was observed that steroids could antagonize the release from isolated fat pads of prostaglandin (PG) E_2 which accompanies adrenocorticotrophic hormone (ACTH)-induced vasodilation [16]. The suggestion made at that time to explain these findings was that the glucocorticoids were preventing the release of prostaglandins from the adipocytes, rather than their biosynthesis, but some other experiments reported by different groups suggested another interpretation. In 1975, Gryglewski and his colleagues found that two glucocorticoids, hydrocortisone and dexamethasone, prevented the noradrenaline-induced release of prostaglandins from rabbit perfused mesenteric vascular bed as well as the immunologically induced release of prostaglandins from guinea pig lungs [17]. The direct conversion of arachidonic acid into prostaglandins in these preparations was not blocked by the glucocorticoids indicating that they had no inhibitory activity on the cyclooxygenase itself.

The rate-limiting step in prostaglandin synthesis seemed to be the release of arachidonic acid from some intracellular lipid pool—probably the phosphatide pool. It was suggested that the glucocorticoids were interfering with substrate release from membrane phospholipids. This conclusion was supported by some elegant experiments from Levine's group published in 1976. Using cultured fibroblasts labelled with tritiated arachidonic acid, this group was able to demonstrate that the anti-inflammatory glucocorticoids prevented phospholipid deacylation and arachidonic acid release which accompanied stimulation by various agents [18]. Levine's group proposed that the mechanism of action of these drugs was to prevent substrate release and they suggested that this might underlie their inflammatory action. Several other groups also reported that steroids could prevent synthesis of prostaglandins by intact cells and in each case the mechanism was consistent with an action on the supply of substrate.

Early studies on glucocorticoid action in the lung

The author and his colleagues, then at the Wellcome Foundation, were also studying the effect of steroids upon prostaglandin synthesis in the light of some early work implicating the enzyme phospholipase A_2 in this event [19]. The system used was the guinea pig perfused isolated lung, a simple technique which has proved extremely useful to workers in the prostaglandin field [20].

The perfused lungs can be induced to release thromboxane and other prostaglandin endoperoxide derivatives in two ways: firstly, by the injection or infusion into the lungs of the substrate arachidonic acid, and secondly by the injection of 'releasing factors' such as histamine, bradykinin, slow-reacting substance of anaphylaxis (as it used to be known) or antigen. All these substances apparently release eicosanoids (as detected by superfusion bioassay) by liberating arachidonic acid within the lung. Under investigation at that time was another substance, RCS–RF (rabbit aorta contracting substance–releasing factor), a low molecular weight peptide found in immunologically shocked lung effluent [20]. RCS–RF had no direct effect upon the assay tissues but when injected into the lung it caused a release of biologically active substances identified as a mixture of thromboxane and prostaglandin endoperoxides. Arachidonic acid injections induced a similar effect.

When a glucocorticoid such as dexamethasone was infused into this preparation, it was observed that the generation of biologically active substances elicited by the releasing factor was blocked whereas the conversion of arachidonic acid was unimpaired [21]. There was always a time delay with the first effects of steroid being observed after about 30 min of infusion. To produce greater than 50% inhibition it was often necessary to continue the infusion for 45–60 min or even longer. Of course, when indomethacin was administered, the generation of products in response to both arachidonic acid and RCS–RF was blocked.

Dexamethasone was not the only steroid able to produce this effect: all the common glucocorticoids shared this activity.

Mineralocorticoids and most sex steroids were inactive in this test, although in some experiments oestradiol seemed to have a similar action. The releasing activity of some other agents such as 5-hydroxytryptamine (5-HT) and histamine was also blocked by the steroids, although curiously, the action of bradykinin was not. The steroid-induced block of eicosanoid release was not irreversible. If the infusion of the drug was stopped, the inhibition gradually faded with (in most cases) a complete return to control values within 60–90 min.

When the arachidonic acid content of the perfusate was measured by gas chromatography, it was observed that releasing factors induced a transient release of arachidonic acid from the lung, and that during steroid infusion, this release was inhibited or reduced. This inhibition was easily reversed when the steroid infusion was terminated [21].

The most widely accepted idea at that time was that the major store of polyunsaturated fatty acid precursors in the cell was the phospholipids, and that it was the liberation of arachidonic acid from this pool under the influence of the enzyme phospholipase A_2 which was the first step in the generation of prostaglandins. Virtually all mammalian cells contain phospholipases A_2 and there are several types which differ in their pH optima, calcium requirement and subcellular location [22], but it seemed highly likely that the enzyme most relevant to the generation of arachidonic acid was the plasma membrane phospholipase A_2. This protein is a constituent of most plasma membranes, it has a requirement for calcium, and a neutral or alkaline pH optimum. Presumably the enzyme must be present as some functionally inactive complex which can be 'switched on' in some way by stimuli known to release arachidonic acid.

To test if this concept obtained in the lung, a specifically labelled phospholipid substrate of phospholipase A_2 was synthesized with a radioactive fatty acid in the 2′ position. The author and his colleagues reasoned that if a cell membrane phospholipase was attached to the surface of some cell-type in the lung (as it is with several other cell types) then it should be revealed by examining the hydrolysis of the radioactive substrate upon passage through the lung.

When aliquots of the labelled isotope were injected into the pulmonary circulation of the lung we found that there was a 'background' hydrolysis of the labelled phospholipid and that this was strongly inhibited by infusions of glucocorticoids such as dexamethasone [23]. Again, the onset of the steroid effect occurred after a time delay and was reversible upon the termination of the infusion. Experiments of a similar nature also determined that phospholipase A_2-like activity was increased by the injection of releasing factors such as RCS–RF, histamine and slow-reacting substance of anaphylaxis (SRS-A) and that this elevated hydrolytic activity was also blocked by glucocorticoids.

Did the glucocorticoids cause inhibition of phospholipase activity by the 'classical pathway' of steroid action? We found that there was indeed a high affinity glucocorticoid binding protein in the cell-free supernatants of the guinea pig lung [24] and that the antiphospholipase effect of steroids in the lungs could be reversed by inhibition of RNA or protein synthesis. Danon and Assouline had also observed that the glucocorticoid effects on prostaglandin production by renal interstitial cells depended upon unimpaired RNA and protein synthesis [25] and going even further, Russo-Marie and her colleagues demonstrated unequivocally that the inhibitory action of corticosteroids on prostaglandin synthesis by rat renomedullary cells is mediated through receptor occupancy and requires ongoing RNA and protein synthesis [26].

All these experiments strongly suggested that steroids inhibited phospholipase A_2 activity by a mechanism depending upon glucocorticoid receptor occupancy, followed by *de novo* RNA and protein synthesis and begged the question of whether the glucocorticoids were inducing the synthesis or release of an inhibitor of phospholipase A_2.

Detection of a steroid-induced 'second messenger'

To test this hypothesis we devised a bioassay experiment in which two guinea pig isolated lungs were perfused in series [24]. The design of the experiment was such that the outflow from one lung (generator lung) was pumped into a second lung (target lung) in which phospholipase activity was assessed by the radiochemical assay or bioassay (i.e. release of prostaglandins). The second lung was rendered insensitive to steroids by the continuous infusion of cycloheximide. The design of this experiment meant that when a steroid was infused directly into the second lung, there was little or no effect on phospholipase activity. However, when the drug was infused into the first lung, we observed that the phospholipase activity of the second test lung declined after a short lag period. The most obvious interpretation of these experiments was that a soluble inhibitor of phospholipase was being secreted from the first lung under steroid stimulation and was being transported in the perfusate into the second lung.

The action of glucocorticoids was investigated in some other systems and found to be more complex than was originally envisaged. Bray and Gordon had found in 1976 that corticosteroids blocked prostaglandin synthesis by guinea pig macrophages [27], and in 1979 Di Rosa and his colleagues at the University of Naples demonstrated that rat peritoneal lavage cells (about 80% macrophages) were also highly sensitive to the prostaglandin inhibitory effect of steroids and that this effect was also dependent on *de novo* RNA and protein synthesis [28]. They obtained evidence that this effect was also caused by the release from the cells of a phospholipase inhibitory protein [29]. Further experiments revealed that the release of this inhibitor was dependent upon the concentration of the inducing steroid and that the amount of the protein present depended upon the cell number as well as the drug concentration.

When the biological profile of the lung and macrophage-derived inhibitors were compared, it was determined that they were both proteins and had similar molecular weights (about 15 kD) as judged by a gel exclusion chromatography. Both inhibitors had similar biological actions, i.e. the macrophage-derived material was active in the guinea pig lung system and vice versa and both proteins shared a similar resistance to heating being stable for up to 10 min at 70 °C but being destroyed at higher temperatures.

These and other results encouraged the author and his colleagues to believe that they were dealing with a single protein which was named 'macrocortin' [30]: 'cortin' to indicate that the protein was induced (or released) specifically by the glucocorticoids and mimicked their action in the two systems, while the prefix 'macro' indicated the most likely cellular origin (macrophages) and also that their molecular weight (15 kD) was considerably in excess of the steroids themselves.

Other researchers had independently arrived at a similar conclusion concerning the mechanism of action of steroids on phospholipase. At the National Institute of Health (NIH) in Bethesda, Hirata and his colleagues had been investigating the mechanism of neutrophil chemotaxis and observed that when neutrophils were stimulated with chemoattractants such as formyl-methionyl-leucyl-phenylalanine (FMLP) there were rapid changes in membrane biochemistry which included phospholipid methylation and arachidonic acid release [31]. These authors demonstrated that both events were preceded by an action of phospholipase A_2 in the cell membrane, and that drugs which inhibited this enzyme such as the antimalarial mepacrine blocked both the phospholipase A_2 activation and the chemotactic response.

When steroids were tested as inhibitors in the chemotaxis assay, both cell movement and phospholipase A_2 were blocked and only glucocorticoids were effective in this respect. As in the guinea pig lung experiments, steroid receptors were demonstrated in the neutrophils and displacement of labelled dexamethasone from the receptor was observed in the presence of inducing steroids. The inhibitory activity of the steroids was also abrogated by inhibitors of protein and RNA synthesis.

Particulate fractions of the steroid-treated neutrophils were solubilized with the non-ionic detergent NP40 and were found to contain a phospholipase inhibitory protein which chromatographed on Sephadex with an apparent molecular weight of 40 kD [31]. If neutrophils were incubated with labelled lysine, the Sephadex fractions containing the 40 kD protein were found to have a greater incorporation of the isotope when prepared from steroid-treated cells than did similar fractions derived from untreated cells. The conclusion from these experiments was that glucocorticoids blocked chemotaxis by inhibiting phospholipase A_2 and this was the result of an induction (or at least an increase) of protein inhibitor of this enzyme. This protein was subsequently named 'lipomodulin' by the NIH group.

Russo-Marie and her colleagues at the Necker Hospital in Paris [26, 32] observed that renomedullary interstitial cells in culture produce large amounts of prostaglandins, and both non-steroidal and steroidal drugs have inhibitory effects on the generation of these lipids. In a series of beautiful experiments they were able to demonstrate that glucocorticoids inhibited prostaglandin synthesis and that the concentration required to do this generally correlated well with that required to occupy glucocorticoid binding sites. Once again, only steroids with the glucocorticoid properties prevented prostaglandin secretion, this effect was not caused by a depression of cell growth or cyclooxygenase activity and was blocked by inhibitors of macromolecular synthesis.

In later experiments this group discovered that the inhibition of prostaglandin synthesis was secondary to phospholipase inhibition and this in turn was caused by the generation of a heat-labile non-dialysable factor subsequently dubbed renocortin. Analysis of renocortin demonstrated that it was a mixture of two proteins, one having a molecular weight of 15 kD and the other a molecular weight of 30 kD [33].

Gupta and his colleagues demonstrated that dexamethasone induces the formation of phospholipase inhibitory proteins (PLIP) in cultures of embryonic palate cells and thymocytes. Once again, the molecular weight species observed (55 kD, 40 kD, 28 kD and 15 kD) are very similar to those observed by the other groups [34, 35].

Longenecker and his colleagues have also isolated from human peritoneal dialysis fluid, a 40 kD phospholipase inhibitor [36]. This protein displays many properties common with the other inhibitors referred to above in that it inhibits eicosanoid release from cells. No information is available on the steroid inducibility and the available evidence tends to suggest that it is not identical to the protein now named lipocortin 1 (see below) although it may well be related in some way.

It was originally thought that macrocortin, lipomodulin and renocortin were three different proteins induced by the glucocorticoids, but a comparison of the conditions of their generation and a more rigorous examination of the distribution of molecular weights, their antienzyme and immunological properties led the Wellcome group, the NIH group and the Necker Hospital group to the conclusion that all these proteins were functionally identical active fragments of the same precursor. A unified nomenclature was agreed and the name 'lipocortin' proposed [37]. It was suggested that the disparity in molecular weights was a consequence of proteolysis.

Distribution of lipocortin

The peritoneal macrophage was used as a source of lipocortin in many of the early experiments and this remains one of the richest known cellular sources of the protein, although there is also a great deal in placenta [38]. It is possible that the lipocortin generated by the guinea pig perfused lung is actually derived from alveolar macrophages and indeed in some unpublished experiments Parente and Flower observed that alveolar macrophages (obtained by bronchial lavage) from several species release lipocortin in response to glucocorticoids [39]. That other types of cell can also generate this protein was illustrated by studies such as Hirata's early work on lipocortin in the neutrophil, and the work of Russo-Marie and her colleagues with kidney interstitial cells. Endothelial and mesothelial cells also generate lipocortin-like substances [40–43] as do skin fibroblasts [44] and human endometrium [45] in response to glucocorticoids.

Coote *et al.* [unpublished data cited in 46] studied the distribution of lipocortin in the rat using a labelled antibody and a solid-phase assay system and noted that it was widely distributed and could be detected in every organ tested. Thymus, spleen, lung and brain had especially high amounts of the protein under 'resting' conditions (i.e. with no exogenous steroids). When treated with dexamethasone the immunoreactivity of all organ extracts (except some gastrointestinal tissues) was increased within 2 h in some cases by four- to five-fold.

Purification of lipocortin

Several groups, using different strategies, tried to purify and sequence lipocortin. Whilst the highly purified protein was soon available for biological work, it was some time before the structural work on the molecule was completed.

The author's group prepared partially purified antiphospholipase proteins from rat lavage fluid using DEAE ion-exchange chromatography and gel filtration. A complex pattern of active species was seen eluting from DEAE columns and several of these proteins appeared, on subsequent analysis, to have the same molecular weight. Initially, Sephadex G-75 was used as a gel filtration medium for molecular weight analysis but later gel-exclusion chromatography was performed by high-performance liquid chromatography (HPLC) using a TDK 200 SW column. Using these methods, the most prominent species detected was the 40 kD protein although a 15 kD species was also often found. In other experiments, Sephacryl SF300 was employed and led to the detection of more antiphospholipase proteins with much higher molecular weights (principal species 125 kD [47]). These proteins also reduced prostaglandin production by cells and seemed to behave in every way as the lower molecular weight species.

Although considerable purification can be achieved by combination of ion exchange and gel filtration chromatography this was not sufficient to give homogeneous protein bands on SDS gel electrophoresis when rat peritoneal lavage fluid was used as a starting material. The neutralizing monoclonal antilipocortin antibody RM 23 proved difficult to use when coupled to an affinity matrix, but phospholipase A_2-affinity chromatography substantially improved the procedure. Typically, ex DEAE fractions containing a mixture of proteins were passed through the column (phospholipase A_2 linked to agarose beads) whereupon 80–90% of the biological activity was retained by the bound ligand. This could be subsequently eluted with 2 M KCl and a considerable purification now achieved.

Isolation, cloning and sequencing of lipocortin 1

At the time of writing, the most complete analysis of the phospholipase inhibitory activity in rat peritoneal lavage fluid has been furnished by Pepinsky and his colleagues at the Biogen Research Corporation [48]. These workers purified the predominant phospholipase inhibitory protein from lavage fluid and characterized it as a 37 kD protein which is also present in a series of lower molecular weight fragments including a 30 kD, 24 kD and 15 kD form. This inhibitor, which probably corresponded to 'polypeptide 1' according to the author's purification scheme, was purified to homogeneity using a combination of ion-exchange and gel-exclusion chromatography and submitted to partial sequence analysis.

The sequences of several tryptic fragments of the rat peritoneal lavage protein were determined and this information was used to chemically synthesize pools of oligonucleotide probes. Using these probes it was possible to locate, in a U937 cDNA library, the human gene. From this it was possible to deduce the complete amino acid sequence, and thus the primary structure of the protein [49]. Human lipocortin is very closely related to the rat protein isolated by Pepinsky and his colleagues suggesting that it is highly conserved throughout evolution. It is a very polar molecule, with approximately one third of its amino acids being charged. These are distributed throughout the molecule separated by short stretches of hydrophobic residues. There are four cysteines close to the C-terminal (263, 270, 324 and 343) which presumably can form disulphide bridges. The rat protein has two additional cysteines.

Recombinant human lipocortin does not apparently contain a leader peptide, and although it is clearly recovered as an extracellular protein, the mechanism whereby its release from cells is affected is unclear at the moment. The protein contains a single potential glycosylation site (Asn-43-Ser-45) and also contains the consensus sequences for both tyrosine and threonine phosphorylation sites (Tyr-21 and Thr-212). The latter point is of special significance since it had already been demonstrated that the naturally occurring molecule could be phosphorylated, and that the phosphorylated form was inactive as a phospholipase A_2 inhibitor [50].

A full-length coding sequence of human lipocortin was expressed in *Escherichia coli* and the action of crude extracts of these *E. coli* organisms, or highly purified fractions containing the recombinant protein were found to be highly inhibitory in a popular conventional cell-free phospholipase A_2 enzyme assay thus confirming the identity of the protein.

Using Northern blot analysis, it was possible to demonstrate that the lipocortin gene was indeed glucocorticoid-sensitive: mRNA levels in peritoneal lavage cells were increased by approximately six-fold 2 h after the injection of an inducing steroid.

Huang *et al.* observed that human placental extracts contained two types of lipocortin [38]. One species (subsequently named lipocortin 1) was identical to the recombinant protein previously cloned but the other (called lipocortin 2) was a different protein. Like lipocortin 1, lipocortin 2 does not contain a signal sequence: both proteins are very polar molecules with approximately one-third of their total amino acids being charged. Overall there was approximately 50% sequence homology suggesting that the genes for the two proteins arose by gene duplication. It was within the central region of lipocortin 2 that the greatest homology with lipocortin 1 was observed. Interestingly, the sequences near the N-terminus of the two proteins were substantially different although both contained a sequence which can be phosphorylated by tyrosine kinases. Both proteins have very comparable molecular weights and similar pI values (7.9), both are inhibitors of phospholipase activity and require the presence of calcium before association with membranes can occur.

We do not know anything about the biological activity of lipocortin 2, other than the fact that it has approximately the same antiphospholipase activity *in vitro* as lipocortin 1. However, the protein has now been cloned and it will presumably only be a matter of time before such data is available.

Relationship of lipocortin to other membrane-associated proteins

Kretsinger and Creutz observed the presence of three repeating consensus sequences in lipocortin 1 [51], and Munn and Mues [quoted in 38] have gone further and suggested that, apart from the first 43 amino acids, the primary structure of lipocortin 1 is built from four repeats of a single unit. The same basic structure has been observed in lipocortin 2 and similar sequences are also seen in some other calcium- and lipid-binding proteins such as calpactin. In fact, some very recent work has demonstrated that the membrane-associated protein previously referred to as calpactin 2 is in all probability identical to lipocortin 1 [52].

It is likely that lipocortin 1 and lipocortin 2 are related, or identical to, some other hitherto unsequenced proteins as well and, indeed, may belong to a superfamily of membrane-associated proteins. For example, a very important discovery made by the Biogen group [38, 53] was that lipocortin 1 is identical to the PP60 src kinase substrate. These results are important, for whilst the molecular weight of the major cellular substrates for these kinases has been known for a long time, their precise function was a mystery. The fact that lipocortin 1 and 2 are inhibitors of phospholipase activity may provide a valuable clue as to the role of these proteins in the regulation of cell growth and division, and promises to be a fruitful and rewarding area of future research.

Biological properties of lipocortin

Antienzyme activity in vitro

The original property of the lipocortin family which evoked such interest was their ability to inhibit phospholipase A_2 activity. In many of the original experiments, the evidence that phospholipase was directly inhibited was compelling, but ultimately only circumstantial, the site of action of the protein being deduced from its behaviour in complex cellular systems.

One problem was that the most likely target for these proteins was the membrane-bound phospholipase, an enzyme notoriously difficult to assay. A partial solution was achieved by using the more readily available soluble pancreatic enzyme as a model 'target'. This was done in the full knowledge that there were obviously differences between the membrane-bound enzyme and the pancreatic protein. There are considerable problems in assaying even this latter species since its behaviour does not comply with the tenets of classical Michaelis–Menton kinetics and exhibits a number of anomalies [cf. 54].

The inhibition of phospholipase A_2 activity in a cell-free system was first demonstrated by Hirata [50]. Lipocortin was found to exert a concentration-dependent inhibition of pancreatic phospholipase as assessed in a micellar assay using labelled phosphatidylcholine as a substrate. The pancreatic phospholipase A_2 from the snake (*Naja naja*) and bee venom was also inhibited to comparable degrees. Phospholipase C from two bacterial sources was also blocked, but the inhibition was not so pronounced. Phospholipase D (from cabbage) was also inhibited but again it was not so susceptible to inhibition as the A_2 enzymes. Equilibrium dialysis experiments suggested that protein might bind Ca^{2+} and Hirata suggested that this action underlined its ability to inhibit phospholipase. Recently, the Ca^{2+} properties of lipocortin have been confirmed. According to Schlaepfer and Haigler, the molecule contains four Ca^{2+} binding sites although it has little affinity for the cation in the absence of phosphatidylserine [55]. The ability of lipocortin to form a ternary complex with Ca^{2+} phospholipids is extremely interesting and may be important from the mechanistic viewpoint.

Hirata demonstrated that lipocortin reduced the V_{max} of the enzyme-catalysed reaction but not the K_m value for the substrate, and it was suggested that a stoichiometric complex was formed between the protein and the inhibitor [56]. This complex was disrupted by some detergents suggesting that hydrophobic binding was important for the inhibitory action.

The author's group studied the release of antiphospholipase proteins into peritoneal lavage fluid of the rat also using the pancreatic enzyme as a target and micellar phosphatidylcholine as a substrate and used the assay as a basis for estimating the amount of lipocortin present [57]. The lavage fluid from 'normal' rats was found to contain abundant antiphospholipase activity as estimated in the *in vitro* assay, but this was greatly reduced in the lavage fluid from adrenalectomized animals. Administration of steroids to both groups caused an increase within 1 h. In addition to these studies, there are several other reports of proteins present in plasma or cell (especially leucocyte) extracts which have antiphospholipase activity in *in vitro* assay systems [58, 59].

The results presented this far suggest that lipocortin binds to phospholipase itself to bring about inhibition of enzyme activity, and indeed this property was sometimes utilized as a means of purifying the protein (i.e. in affinity chromatography). Recently, however, the mechanism of action of these proteins in bringing about an inhibition of the enzyme has been disputed.

Davidson and her colleagues studied the inhibition of pancreatic and snake venom phospholipases by preparations of calpactin II (probably identical to lipocortin 1) and were unable to demonstrate inhibition of the enzyme at saturating concentrations of substrate [60]. Inhibition was seen only when the substrate was present in rate-limiting amounts and the inhibition could be overcome by adding substrate, but not more enzyme. The authors suggested that lipocortin produced inhibition of phospholipase activity by sequestering substrate and that it did so by binding to it in a Ca^{2+}-dependent manner.

Published work from some other groups also supports their contention: Aarsman *et al.* observed that the inhibition by a human monocyte lipocortin of phosphatidylethanolamine hydrolysis by the pancreatic enzyme or by partially purified intracellular phospholipases was substrate dependent [61] and the same group found that this inhibitor blocked hydrolysis of a negatively charged, but not zwitterionic phospholipid substrate [62]. Haigler *et al.* also failed to find an interaction between lipocortin and phospholipase A_2 and suggested that the inhibitory action of the protein was caused by an effect on the substrate [63].

Because of its structure, lipocortin is indeed able to bind to certain phosphatides such as the anionic substrates used by many of these groups, but since it does not apparently bind to phosphatidylcholine vesicles [55]—the substrate used in all the early studies—it is difficult to see how this notion explains all the antiphospholipase results and in any case could not explain the phospholipase affinity column data.

It is certainly true, however, that the special Ca^{2+} and lipid-binding properties of this molecule could give rise to spurious results when the substrate concentration is low as it is in the widely used *E. coli* assay. It is evident that we still do not fully understand the behaviour of this protein in the apparently straightforward assay systems used routinely by most authors and perhaps the simplest explanation which would accommodate both sets of data is that lipocortin can have effects both on the enzyme and its substrate.

Suppression of mediator production by cells in vitro

It was the steroid-induced inhibition of prostaglandin production by cells which originally initiated research into this field. The antiphospholipase proteins mimic the action of steroids in many systems quite closely, although little latency of action is observed and their effect is resistant to the action of protein/RNA synthesis inhibitors. The inhibition of cellular prostaglandin synthesis by these proteins is dose-dependent, leukotriene production is also blocked as one would expect [64], and the inhibitory effect of these proteins—like that of the steroids themselves—is readily reversed by the addition of exogenous arachidonic acid. The inhibitory effects of the aspirin-like drugs are, of course, not reversed by this manoeuvre indicating two distinct sites of action.

Another putative inflammatory mediator, important to those interested in asthma, which is also formed by the action of phospholipase A_2 is PAF-acether. Parente and Flower demonstrated that affinity-purified lipocortin blocked the release of lyso-PAF (the precursor of PAF-acether) from rat resident macrophages stimulated with zymosan, and that this action was prevented by the neutralizing monoclonal RM 23 when this was present in the mixture [39]. Interestingly, the latter monoclonal can also inhibit the blocking action of glucocorticoids themselves providing additional evidence that the inhibitory action of glucocorticoids is mediated by lipocortin.

Many experiments support the idea that lipocortin inactivates membrane phospholipase by interacting with the enzyme from the *outside* of the cell. In the guinea pig perfused lung system for example, phospholipase activity is inhibited almost as soon as lipocortin infusions are begun, and declines shortly after the infusions are terminated [57]. It seems inherently improbable that a protein of a minimum size 15 kD could so rapidly gain access to the interior of cells as it would have to do if the target enzyme was cytosolic. Again, Hirata *et al.* observed that the steroid inhibition of neutrophil chemotaxis (which was associated with an inhibition of phospholipase A_2) was completely reversed if the neutrophils were exposed to the proteolytic enzyme pronase [65]. This experiment suggests that at least part of the inhibitory protein is accessible to the (extracellular) hydrolytic activity of the enzyme. Again, the experiment with the monoclonal RM 23 cited above also supports this idea.

Cirino and Flower observed that lipocortin-like activity was easily eliminated from cells when they were washed in low Ca^{2+} media and that the *ex vivo* inhibition of eicosanoid generation by steroids could be abrogated if cells were subjected to this procedure [66]. Authentic recombinant human lipocortin 1 also required the presence of Ca^{2+} before it produced an inhibition of eicosanoid synthesis in macrophages. Even repetitive washing of cells by vigorous centrifugation in Ca^{2+}-containing solutions is often sufficient to remove lipocortin [67].

An important conclusion to be derived from all this work is that the *ex vivo* actions of steroids may be missed if the protocol involves a step where cells are exposed to low calcium environments or to multiple washing procedures. This could explain many discrepancies in the literature.

Cloning of the human lipocortin 1 gene was a major breakthrough because it enabled large amounts of the protein to be produced for biological testing. Because there were several apparently distinct antiphospholipase components in peritoneal lavage fluid, an obvious question was whether lipocortin 1 could account for the properties observed in the early experiments with macrocortin-containing extracts.

Cirino and Flower demonstrated that recombinant human lipocortin 1 (ex *E. coli*) strongly inhibited the release of prostacyclin by human umbilical arterial endothelial cells, with complete suppression occurring in concentrations above 0.1 μM [68]. Cirino and his colleagues also tested recombinant lipocortin 1 in the guinea pig perfused lung preparation [69]. Like macrocortin, and the glucocorticoids themselves, lipocortin was able to block the release of thromboxane A_2 from the perfused lung when this was elicited by leukotriene (LT) C_4 or formyl-methionyl-leucyl-phenylalanine (FMLP) but not when elicited by bradykinin or arachidonic acid. Unlike steroids, but like macrocortin itself, the action of lipocortin was very rapid and was easily reversed when the infusion was stopped. Similar to the steroids and macrocortin, the effect was cumulative with time and the actual amount of inhibition was dependent upon the concentration of the agonist as well as the concentration of lipocortin itself.

Anti-inflammatory activity in vivo

What evidence is there that these induced proteins actually produce an anti-inflammatory effect *in vivo*?

The author's group used crude protein extracts from peritoneal lavage fluid from dexamethasone or saline-treated animals [57]. The proteins were concentrated, dialysed to remove steroid and injected together with carrageenin into the pleural cavity of anaesthetized rats. Four hours later the animals were killed and the extent of the inflammation assessed by measuring the fluid exudation and leucocyte infiltration. Protein extracts from rats treated with saline alone did not alter the pleurisy when compared to that produced by carrageenin alone, but a substantial inhibition occurred when proteins from dexamethasone-treated rats were used. This was caused by the presence of an induced protein in the exudate because boiling destroyed the activity, and treatment of the rats in which the protein was raised with actinomycin D prevented the anti-inflammatory principle from appearing. Fractionation of the active extracts on DEAE ion-exchange cellulose indicated that the peaks of anti-inflammatory activity coincided with antiphospholipase activity.

In later studies, partially purified (ex DEAE and Sephadex G-75) anti-phospholipase proteins were used as anti-inflammatory agents in the carrageenin paw oedema model [64]. Again this test showed that lipocortin-containing fractions suppressed the oedema with a duration of action of about 3 h, and that the effect was reversed by a supra-injection of arachidonic acid.

Human recombinant lipocortin 1 also possesses anti-inflammatory properties. Cirino *et al.* studied the action of lipocortin 1 given locally in the rat paw oedema assay [70], a very well-characterized model of inflammation sensitive both to steroidal and non-steroidal drugs. The swelling of the paw was substantially inhibited by lipocortin, when the inflammogenic agent was carrageenin which acts principally by releasing eicosanoids, but not when PAF-acether or dextran were used. The latter two irritants probably act by releasing mast cell amines and although they are sensitive to steroid they are not sensitive to inhibition by the non-steroidal, aspirin-like drugs. Another agent which also works by liberating mast cell amines is the enzyme phospholipase A_2 itself. In the particular case, the purified phospholipase A_2 from the venom of *Naja naja mocambique* was employed. The pro-inflammatory action was greatly reduced when this agent was premixed with lipocortin 1 before injection—presumably because the protein was binding to, and inactivating the inhibitor.

These experiments suggest that lipocortin 1 duplicates those actions of steroids which are mediated by the release of eicosanoids or by the direct action of phospholipase itself but not those caused by the direct action of substances such as PAF-acether, histamine or 5-HT. Since the glucocorticoids can inhibit *all* these types of oedema it would strongly suggest that another protein (or mechanism) is involved. Di Rosa and his colleagues have indeed suggested that such a protein exists and have named it 'vasocortin'.

Concluding remarks

It is far too early to say whether or not lipocortin 1—or indeed any of the other related proteins—are important in the control of asthma or allergic airway diseases. Indeed, we have only just begun to scratch the surface of lipocortin biology and there are many important questions which need to be settled. How is the protein released from cells? At what levels do the steroid hormones regulate its synthesis and/or release? What biological functions do the other lipocortins have and are they under steroid control? At present, the evidence suggests the existence of a system which can diminish the production of inflammatory mediators in the lung and which is under the control of steroid hormones and it would be difficult to accept that such a mechanism is without relevance to the understanding of glucocorticoid therapy in asthma and related diseases.

References

1 Ferreira SH, Vane JR. The mode of action of anti-inflammatory agents which are prostaglandin synthetase inhibitors. In *Handbook of Experimental Pharmacology, 50(II) Anti-inflammatory Drugs* (Eds Vane JR, Ferreira SH), Springer-Verlag, Berlin 1979, pp. 348–398.
2 Piper PJ. Pharmacology of leukotrienes. *Br. Med. Bull.* 1983; **39**:255–259.
3 Bray MA. Pharmacology and pathophysiology of leukotriene B_4. *Br. Med. Bull.* 1983; **39**:249–254.
4 Hamberg M, Svensson J, Samuelsson B. Prostaglandin endoperoxides. A new concept concerning the mode of action and release of prostaglandins. *Proc. Natl. Acad. Sci. USA* 1974; **71**:3824–3828.
5 Hamberg M, Svensson J, Samuelsson B. Thromboxanes: a new group of biologically active compounds derived from prostaglandin endoperoxides. *Proc. Natl. Acad. Sci. USA* 1975; **72**:2994–2998.
6 Murphy RC, Hammarstrom S, Samuelsson B. Leukotriene C. A slow-reacting substance from murine mastocytoma cells. *Proc. Natl. Acad. Sci. USA* 1979; **76**:4275–4279.
7 Moncada S. Biological importance of prostacyclin. VIII Gaddum Lecture. *Br. J. Pharmacol.* 1982; **76**:3–31.
8 Albert DH, Snyder F. Biosynthesis of 1-alkyl-2-acetyl-*sn*-glycero-3-phosphocholine (platelet-activating factor) from 1-alkyl-2-acyl-3n-glycerophosphocholine by rat alveolar macrophages. Phospholipase A_2 and acetyltransferase activities during phagocytosis and ionophore stimulation. *J. Biol. Chem.* 1983; **258**:97–102.
9 Albert DH, Snyder F. Release of arachidonic acid from 1-alkyl-2-acyl-*sn*-glycero-3-phosphocholine, a precursor of platelet-activating factor, in rat alveolar macrophages. *Biochim. Biophys. Acta.* 1984; **796**:92–101.
10 Williams TJ. Interactions between prostaglandins, leukotrienes and other mediators of inflammation. *Br. Med. Bull.* 1983; **39**:239–242.
11 Vane JR. Inhibition of prostaglandin synthesis as a mechanism of action of aspirin-like drugs. *Nature* 1971; **231**:232–235.
12 Ferreira SH, Moncada S, Vane JR. Indomethacin and aspirin abolish prostaglandin release from the spleen. *Nature* 1971; **231**:237–239.
13 Smith JB, Willis AL. Aspirin selectivity inhibits prostaglandin production in human platelets. *Nature* 1971; **231**:235–237.
14 Flower RJ, Gryglewski R, Herbaczynska-Cedro K, Vane JR. The effects of anti-inflammatory drugs on prostaglandin biosynthesis. *Nature* 1972; **238**:104–106.
15 Herbaczynska-Cedro K, Staszewska-Barczak J. Adrenocortical hormones and the release of prostaglandin-like substances (PLS). *Abstracts of II Congress of Hungarian Pharmacological Society*, Budapest 1974, p. 19.
16 Lewis GP, Piper PJ. Inhibition of release of prostaglandins as an explanation of some of the actions of anti-inflammatory corticosteroids. *Nature* 1975; **254**:308–311.
17 Gryglewski RJ, Panczenko B, Korbut R, Grodzinska L, Ocetkiewicz A. Corticosteroids inhibit prostaglandin release from perfused mesenteric blood vessels of rabbit and from perfused lungs of sensitized guinea-pigs. *Prostaglandins* 1975; **10**:343–355.
18 Hong SC, Levine L. Inhibition of arachidonic acid release from cells as the biochemical action of anti-inflammatory steroids. *Proc. Natl. Acad. Sci. USA* 1976; **73**:1730–1734.
19 Flower RJ, Blackwell GJ. The importance of phospholipase A_2 in prostaglandin biosynthesis. *Biochem. Pharmacol.* 1976; **25**:285–291.
20 Piper PJ, Vane JR. Release of additional factors in anaphylaxis and its antagonism by anti-inflammatory drugs. *Nature* 1969; **223**:29–35.

21 Nijkamp FP, Flower RJ, Moncada S, Vane JR. Partial purification of RCS–RF (rabbit aorta contracting substance-releasing factor) and inhibition of its activity by anti-inflammatory steroids. *Nature* 1976; **263**:479–482.

22 Van Den Bosch H. Intracellular phospholipases A. *Biochem. Biophys. Res. Commun.* 1980; **604**:191–246.

23 Blackwell GJ, Flower RJ, Nijkamp FP, Vane JR. Phospholipase A_2 activity of guinea-pig isolated perfused lungs: stimulation and inhibition by anti-inflammatory steroids. *Br. J. Pharmacol.* 1978; **62**:79–89.

24 Flower RJ, Blackwell GJ. Anti-inflammatory steroids induce biosynthesis of a phospholipase A_2 inhibitor which prevents prostaglandin generation. *Nature* 1979; **278**:456–459.

25 Danon A, Assouline G. Inhibition of prostaglandin biosynthesis by corticosteroids requires RNA and protein synthesis. *Nature* 1978; **273**:552–554.

26 Russo-Marie F, Duval D. Dexamethasone-induced inhibition of prostaglandin production does not result from a direct action on phospholipase activities but is mediated through a steroid-inducible factor. *Biochim. Biophys. Acta.* 1982; **712**:177–185.

27 Bray MA, Gordon D. Effects of anti-inflammatory drugs on macrophage prostaglandin biosynthesis. *Br. J. Pharmacol. Chemother.* 1976; **57**:466P.

28 Di Rosa M, Persico P. Mechanism of inhibition of prostaglandin biosynthesis by hydrocortisone in rat leucocytes. *Br. J. Pharmacol. Chemother.* 1979; **66**:161–163.

29 Carnuccio R, Di Rosa M, Persico P. Hydrocortisone induced inhibitor of prostaglandin biosynthesis in rat leucocytes. *Br. J. Pharmacol. Chemother.* 1980; **68**:14–16.

30 Blackwell GJ, Carnuccio R, Di Rosa M, Flower RJ, Parente L, Persico P. Macrocortin: a polypeptide causing the anti-phospholipase effect of glucocorticoids. *Nature* 1980; **287**:147–149.

31 Hirata F, Corcoran BA, Venkatasubramanian K, Schiffmann E, Axelrod J. Chemoattractants stimulate degradation of methylated phospholipids and release of arachidonic acid in rabbit leukocytes. *Proc. Natl. Acad. Sci. USA* 1979; **76**:2640–2643.

32 Russo-Marie F, Paing M, Duval D. Involvement of glucocorticoid receptors in steroid-induced inhibition of prostaglandin secretion. *J. Biol. Chem.* 1979; **254**:8498–8504.

33 Cloix JF, Colard O, Rothhut B, Russo-Marie F. Characterisation and partial purification of 'renocortins': two polypeptides formed in renal cells causing the anti-phospholipase-like action of glucocorticoids. *Br. J. Pharmacol.* 1983; **79**:313–321.

34 Gupta C, Katsumata M, Goldman AS, Piddington R, Herold R. Glucocorticoid-induced phospholipase A_2 inhibitory proteins mediate glucocorticoid teratogenicity in vitro. *Proc. Natl. Acad. Sci. USA* 1984; **81**:1140–1143.

35 Gupta C, Goldman AS. Dexamethasone-induced phospholipase A_2 inhibitory proteins (PLIP) influenced by the H-2 histocompatibility region (41980). *Proc. Soc. Exp. Biol. Med.* 1985; **178**:29–35.

36 Longenecker JP, Rose JW, Giffin K, Shepard D, Johnson LK. Isolation of a human, endogenous, phospholipase A_2 inhibitory, anti-inflammatory protein. *Adv. Prostaglandin Thromboxane Leukotriene Res.* 1987; **17**:581–586.

37 Di Rosa M, Flower FJ, Hirata F, Parente L, Russo-Marie F. Nomenclature announcement. Anti-phospholipase proteins. *Prostaglandins* 1984; **28**:441–442.

38 Huang K-S, Wallner BP, Mattaliano RJ, Tizard R, Burne C, Frey A, Hession C, McGray P, Sinclair LK, Pingchong Chow E, Browning JL, Ramachandran KL, Tang J, Smart JE, Pepinsky RB. Two human 35 kD inhibitors of phospholipase A_2 are related to substrates of pp60 v-SRC and of the epidermal growth factor receptor/kinase. *Cell* 1986; **46**:191–199.

39 Parente L, Flower RJ. Hydrocortisone and macrocortin inhibit the zymosan-induced release of lyso-PAF from rat peritoneal leucocytes. *Life Sci.* 1985; **36**:1225–1231.

40 De Caterina R, Weksler BB. Modulation of arachidonic acid metabolism in human endothelial cells by glucocorticoids. *Thromb. Haemost.* 1986; **55**:369–374.

41 Rosenbaum RM, Cheli CD, Gerritsen ME. Dexamethasone inhibits prostaglandin release from rabbit coronary microvessel endothelium. *Am. J. Physiol.* 1986; **250**:C970–C977.

42 Van de Velde VJ, Bult H, Herman AG. Dexamethasone and prostacyclin biosynthesis by serosal membranes of the rabbit peritoneal cavity. *Agents Actions* 1986; **17**:308–309.
43 Van de Velde VJ, Herman AG, Bult H. Effects of dexamethasone on prostacyclin biosynthesis in rabbit mesothelial cells. *Prostaglandins* 1986; **32**:169–178.
44 Errasfa M, Rothhut B, Fradin A, Billardon C, Junien JL, Bure J, Russo-Marie F. The presence of lipocortin in human embryonic skin fibroblasts and its regulation by anti-inflammatory steroids. *Biochim. Biophys. Acta* 1985; **847**:247–254.
45 Gurpide E, Markiewicz L, Schatz F, Hirata F. Lipocortin output by human endometrium *in vitro*. *J. Clin. Endocrinol. Metab.* 1986; **63**:162–166.
46 Flower RJ. Macrocortin and the anti-phospholipase proteins. In *Advances in Inflammation Research*, vol. 8 (Ed. Weissmann G), Raven Press, New York 1984, pp. 1–33.
47 Coote PR, Di Rosa M, Flower RJ, Parente L, Merrett M, Wood JN. Detection and isolation of a steroid-induced antiphospholipase protein of high molecular weight. *Br. J. Pharmacol.* 1983; **80**:597P.
48 Pepinsky RB, Sinclair LK. Epidermal growth factor-dependent phosphorylation of lipocortin. *Nature* 1986; **321**:81–84.
49 Wallner BP, Mattaliano RJ, Hession C, Cate RL, Tizard R, Sinclair LK, Foeller C, Pingchong Chow E, Browning JL, Ramachandran KL, Pepinsky RB. Cloning and expression of human lipocortin, a phospholipase A_2 inhibitor with potential anti-inflammatory activity. *Nature* 1986; **320**:77–81.
50 Hirata F. The regulation of lipomodulin, a phospholipase inhibitory protein, in rabbit neutrophils by phosphorylation. *J. Biol. Chem.* 1981; **256**:7730–7733.
51 Kretsinger RH, Creutz CE. Consensus in exocytosis. *Nature* 1986; **320**:573.
52 Glenney JR, Tack B, Powell MA. Calpactins: two distinct Ca^{2+}-regulated phospholipid- and actin-binding proteins isolated from lung and placenta. *J. Cell. Biol.* 1987; **104**:503–511.
53 Pepinsky RB, Sinclair LK, Browning JL, Mattaliano RJ, Smart JE, Chow EP, Falbel T, Ribolini A, Garwin J, Wallner BP. Purification and partial sequence analysis of a 37 kDa protein that inhibits phospholipase A_2 activity from rat peritoneal exudates. *J. Biol. Chem.* 1986; **261**:4239–4246.
54 Chang J, Musser JH, McGregor H. Phospholipase A_2: function and pharmacological regulation. *Biochem. Pharmacol.* 1987; **36**:2429–2436.
55 Schlaepfer DD, Haigler HT. Characterisation of Ca^{2+}-dependent phospholipid binding and phosphorylation of lipocortin 1. *J. Biol. Chem.* 1987; **262**:6931–6937.
56 Hirata F. Lipomodulin: a possible mediator of the action of glucocorticoids. *Adv. Prostaglandin Thromboxane Leukotriene Res.* 1983; **11**:73–78.
57 Blackwell GJ, Carnuccio R, Di Rosa M, Flower RJ, Langham CSJ, Parente L, Persico P, Russell-Smith NC, Stone D. Glucocorticoids induce the formation and release of anti-inflammatory and anti-phospholipase proteins into the peritoneal cavity of the rat. *Br. J. Pharmacol.* 1982; **76**:185–194.
58 Authi KS, Solanky A, Traynor JR. Inhibition of an inflammatory exudate phospholipase A_2 by an endogenous inhibitor of polymorphonuclear leucocytes. *Pharm. Res. Commun.* 1982; **14**:401–407.
59 Bartolf M, Franson RC. Modulation by cytosol and commercial proteins of acid-active phospholipase A_2 from adrenal medulla. *Biochim. Biophys. Acta* 1987; **917**:308–317.
60 Davidson FF, Dennis EA, Powell M, Glenney JR. Inhibition of phospholipase A_2 by lipocortins and calpactins — an effect of binding to substrate phospholipids. *J. Biol. Chem.* 1987; **262**:1698–1705.
61 Aarsman AJ, Mynbeek G, Van Den Bosch H, Rothhut B, Prieur B, Comera C, Jordon L, Russo-Marie F. Lipocortin inhibition of extracellular and intracellular phospholipases A_2 is substrate concentration dependent. *FEBS Lett.* 1987; **219**:176–180.
62 Rothhut B, Camera C, Prieur B, Errasfa M, Minassian G, Russo-Marie F. Purification and characterisation of a 32 kDa phospholipase A_2 inhibitory protein (lipocortin) from human peripheral blood mononuclear cells. *FEBS Lett.* 1987; **219**:169–175.

63 Haigler HT, Schlaepfer DD, Burgess WH. Characterisation of lipocortin 1 and an immunologically unrelated 33 kDa protein as epidermal growth factor receptor/kinase substrates and phospholipase A_2 inhibitors. *J. Biol. Chem.* 1987; **262**:6921–6930.
64 Parente L, Di Rosa M, Flower RJ, Ghiara P, Meli R, Persico P, Salmon JA, Wood JN. Relationship between the anti-phospholipase and anti-inflammatory effects of glucocorticoid-induced proteins. *Eur. J. Pharmacol.* 1984; **99**:233–239.
65 Hirata F, Schiffmann E, Venkatasubramanian K, Salomon D, Axelrod J. A phospholipase A_2 inhibitory protein in rabbit neutrophils induced by glucocorticoids. *Proc. Natl. Acad. Sci. USA* 1980; **77**:2533–2536.
66 Cirino G, Flower RJ. The inhibitory effect of lipocortin on eicosanoid synthesis is dependent upon Ca^{2+} ions. *Br. J. Pharmacol.* 1987; **92**:521P.
67 Carnuccio R, Di Rosa M, Flower RJ, Pinto A. The inhibition by hydrocortisone of prostaglandin biosynthesis in rat peritoneal leucocytes is correlated with intracellular macrocortin levels. *Br. J. Pharmacol.* 1981; **74**:322–324.
68 Cirino G, Flower RJ. Human recombinant lipocortin 1 inhibits prostacyclin production by human umbilical artery *in vitro. Prostaglandins* 1987; **34**:59–62.
69 Cirino C, Flower RJ, Browning JL, Sinclair LK, Pepinsky RB. Recombinant human lipocortin inhibits thromboxane release from guinea-pig isolated perfused lung. *Nature* 1987; **328**:270–272.
70 Cirino G, Peers SH, Flower RJ, Browning J, Sinclair LK, Pepinsky RB. Anti-inflammatory action of human recombinant lipocortin 1 in the rat paw oedema test. *Proc. Natl. Acad. Sci. USA* 1989; In press.

Discussion session

QUESTIONER (unidentified): what are the plasma levels of lipocortin?

FLOWER: I do not know. That is a difficult one to answer because we now know that lipocortin is affected by citrate and the absence of calcium. We realize now that this undermines all our previous estimates.

KATZ: do the autoantibodies react with glucocorticoids themselves?

FLOWER: no.

KATZ: can you inhibit the binding of your monoclonal antibody to lipocortin with these autoantibodies?

FLOWER: we have not done that but what we have done with some of the sera is to show that it can prevent dexamethasone activity in a simple model system. For example, if you take monocytes and macrophages and then add dexamethasone, after an hour or so PGE_2 synthesis begins to go down. If you put suitably purified serum from these patients into the medium, it can prevent that.

14

Azelastine—a novel oral antiasthma compound with several modes of action

J.L. PERHACH, N. CHAND, W. DIAMANTIS, R.D. SOFIA AND A. ROSENBERG

Summary

Azelastine hydrochloride, a phthalazinone derivative—1(2H)-phthalazinone; 4-([4-chlorophenyl]methyl)-2-(hexahydro-1-methyl-1H-azepin-4-yl) monohydrochloride—has been shown to be a potent, orally effective antiallergic/antiasthma drug in several experimental models and clinical trials.

It possesses a broad spectrum of pharmacological activities including:

1 Inhibition of histamine secretion from rabbit and human basophils and rat mast cells with IC_{50}s of 1.9–4.8 μM.

2 Blockade of antigen- and calcium ionophore A-23 187-induced formation of leukotriene (LT) C_4 in guinea pig chopped lung and rat peritoneal mixed cells with IC_{50} of 14 and 22.85 μM respectively.

3 Antagonism of LTC_4, LTD_4 and rat slow-reacting substance of anaphylaxis (SRS-A) in isolated guinea pig ileum ($IC_{50}=7.9$–8.4 μM, 2 min).

4 Relaxation of LTC_4-induced sustained contractions in guinea pig ileum ($IC_{50}=3.25$ μM).

5 Inhibition of antihistamine-resistant, leukotriene-mediated acute allergic bronchoconstriction in guinea pigs (oral $ID_{50}=0.063$ mg/kg, 2 h and 0.120 mg/kg, 24 h).

6 Inhibition of acute lung anaphylactic responses following oral or aerosol administration.

7 Its long-lasting antiallergic activity is dissociated from its histamine H_1-receptor antagonism.

8 Azelastine has also been shown to interfere with some inflammatory processes in several *in vitro/in vivo* model systems.

Based on experimental data, azelastine seems to exert its action by inhibition of secretion of histamine and synthesis of leukotriene and superoxide free radicals ($\cdot O_2^-$), antagonism of chemical mediators such as leukotrienes and histamine (and perhaps other mediators) and Ca^{2+} in target organs.

Clinical studies have shown that these pharmacological properties of azelastine result in significant bronchodilation of long duration in moderate to severe asthmatics. Long-term administration of azelastine to asthmatics results in an overall improvement in airway function while reducing the requirement for backup medications (aerosol β-agonists and theophylline). Furthermore, azelastine provides protection against exercise and allergen provocation. Azelastine's effectiveness has also been demonstrated in seasonal and perennial allergic rhinitis. This phthalazinone derivative is capable of interfering with the synthesis/antagonism of a wide variety of mediators of immediate hypersensitivity and provides significant protection and bronchodilation in allergic hay fever and allergic asthma respectively.

Introduction

The therapeutic failure of traditional antihistamines (histamine H_1-receptor blockers) in the treatment of asthma has led to the discovery of other mediators and how they contribute to allergic airway diseases. Several specific factors (platelet-activating factor [PAF]-acether, leukotrienes, thromboxanes, superoxide free radicals, lipid radicals, chemotactic factor, etc.) have been suggested to contribute to airway inflammation and hyperresponsiveness. Although some of these factors have been shown to be bronchoconstrictor agents, the specific antagonists of PAF-acether, leukotrienes and thromboxane have not been shown to date to provide the anticipated protection in asthma. It is now generally accepted that multiple chemical mediators act and possibly interact to produce airway bronchoconstriction, inflammation and hyperresponsiveness [1–7].

Allergic diseases, including asthma, are multifactorial inflammatory processes characterized by early- and late-phase responses. The allergic responses are initiated by the synthesis/secretion of chemical mediators from mast cells and basophils. These mediators cause bronchoconstriction, increased mucus production/secretion and increased capillary permeability (oedema). The continued synthesis and secretion of mediators (5*S*-hydroperoxy-6,8,11,14-eicosatetraenoic acid [5-HPETE], 5-hydroxy-6,8,11,14-eicosatetraenoic acid [5-HETE], LTB_4, PAF-acether, free oxygen radicals and lipid radicals, chemotactic factors, etc.) promote the recruitment and/or activation of polymorphonuclear cells, eosinophils, alveolar macrophages, platelets and perhaps other cell types which produce cellular infiltration and hyperresponsiveness both in the upper and lower airways. The drug which is capable of inhibiting inflammatory mediator synthesis/secretion in the mast cells, leucocytes, macrophages and airways and can provide beneficial effects in both the early and late phase of allergic responses may be a suitable treatment for asthma and rhinitis [8].

Chemical structure

Azelastine, 4-(*p*-chlorobenzyl)-2-(hexahydro-1-methyl-1H-azepine-4-yl)-1(2H)-phthalazinone hydrochloride, shown in the centre of Fig. 14.1 is a chemically distinct novel structure for an antiallergic/antiasthma drug. Azelastine is structurally different from the older H_1-receptor antagonists (chlorphenira-mine) as well as newer derivatives (astemizole and terfenadine) shown to have clinical antiallergic activity, but possess no sustained clinical effect on lower airways even at multiples of the clinical dose. Azelastine has no structural similarity to agents used to treat lower airway dysfunction represented by the oral medications theophylline and ketotifen and by the aerosol agents, disodium cromoglycate and the β-agonists (albuterol is used as a representative of this chemical class of drugs with oral activity as well). Although azelastine has been shown to interfere with certain inflammatory processes in both animals and humans, azelastine does not resemble molecules with the characteristic steroid structure which possess significant anti-inflammatory properties. Azelastine represents a new chemical class of drugs to treat airway disorders and possesses a distinct pharmacological profile.

Pharmacological profile

Pharmacological studies in laboratory animals and *in vitro* model systems indicate that azelastine HCl is a chemically novel, orally active, long-acting agent with multiple pharmacological actions. These actions include anti-allergic properties, ability to modulate airway smooth-muscle response and interference with inflammatory processes [8–68].

Antiallergic properties

Azelastine has been shown to inhibit allergic and non-allergic histamine secretion (Table 14.1) [9–26, 28, 29] as well as act synergistically with β-adrenergics to inhibit rat peritoneal mast-cell histamine release (Table 14.2) [12, 13]. Inhibition of histamine release from human basophils and chopped lung tissue has also been demonstrated [30–33]. Allergen-induced broncho-spasm in the rat [34] and guinea pig (Fig. 14.2) [35–37] and IgE-dependent passive cutaneous anaphylaxis in the rat [25, 42–46] have been blocked by azelastine. Furthermore, azelastine has been shown to interfere with the synthesis and release of leukotrienes from rat mixed peritoneal cells, guinea pig chopped lung, rabbit and human polymorphonuclear leucocytes [27, 47–51] and in passive peritoneal anaphylaxis in rats [52].

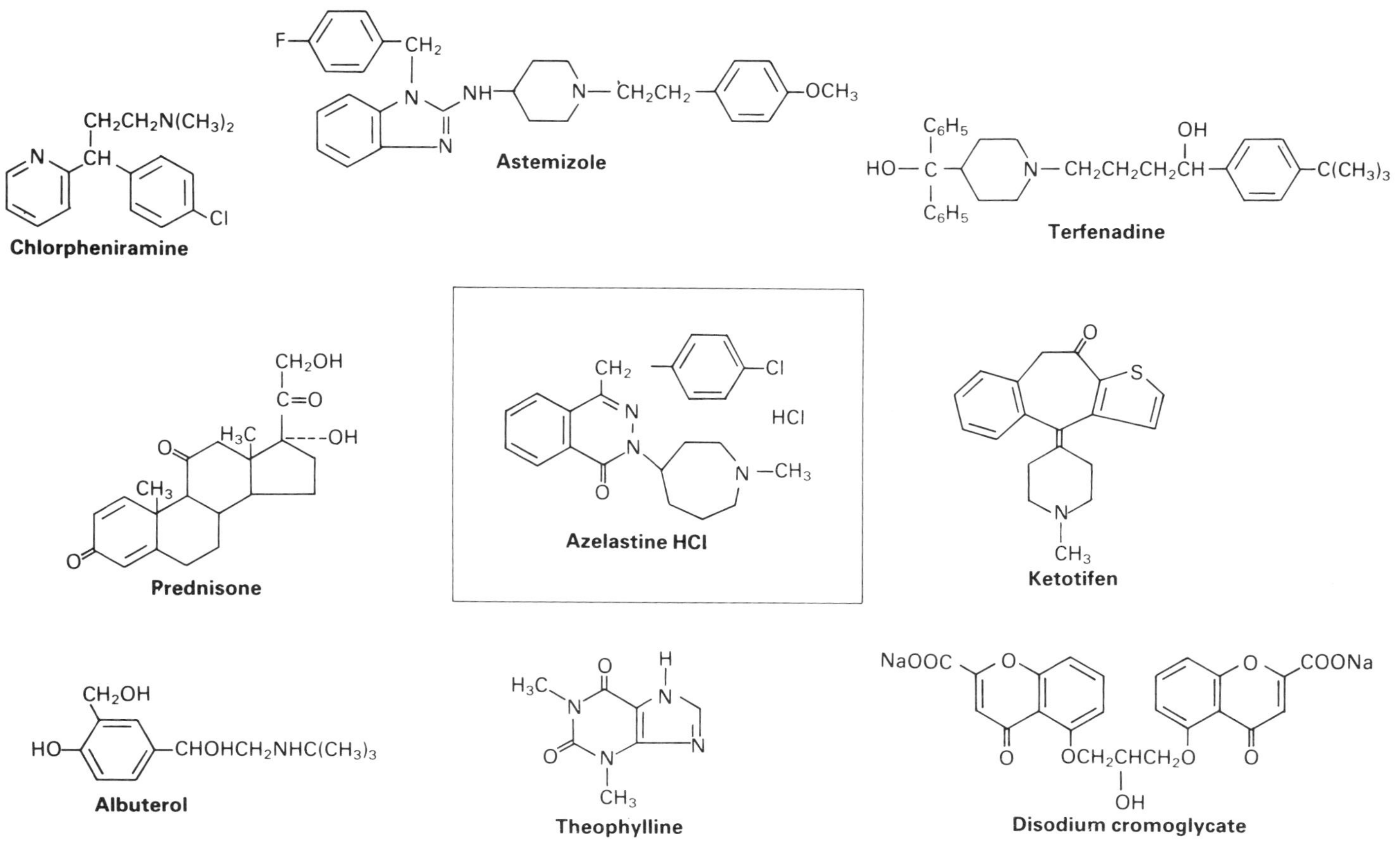

Fig. 14.1 Chemical structures of azelastine and other drugs used in treatment of upper and lower airway disorders.

Table 14.1 IC_{50} values of azelastine on allergic and non-allergic histamine release

Cell source	Secretagogue	IC_{50} (μM) Duration of preincubation 0 min	10 min
Human basophils	Ragweed antigen extract (10 μg/ml)	—	3.3
Rabbit basophils	Ragweed antigen extract (10 μg/ml)	2.4	4.5 1.9 (replicate)
Rat peritoneal mast cells	Ovalbumin (10 μg/ml) + phosphatidyl serine (10 μg/ml)	7.6	4.8
	Concanavalin A (10 μg/ml) + phosphatidyl serine (10 μg/ml)	7	2
	Calcium ionophore A-23 187 (0.1 μg/ml)	7.7	5 9.1
	Compound 48/80 (0.1 μg/ml)	49	42

Table 14.2 Synergistic effect of azelastine and albuterol on allergic histamine release from rat peritoneal mast cells *in vitro*

Albuterol* conc. (μM)	IC_{30} azelastine† (μM)
—	0.51
0.01	0.03**
0.1	0.09**
1.0	0.12**

* Preincubation time 10 min.
† Preincubation time 0 min.
** $p \leqslant 0.05$ compared with azelastine alone.

Modulation of airway smooth-muscle response

Azelastine has been shown to effectively antagonize a number of mediators known to influence airway hyperreactivity (Table 14.3) [49–58] as well as calcium and to inhibit leukotriene-mediated (H_1-antihistamine resistant) bronchospasm in the guinea pig [35–37]. Azelastine can also dissociate its antiallergic properties from its H_1 and 5-HT antagonist activities in rat skin

Table 14.3 *In vitro* antagonistic activity of azelastine

Tissue	Spasmogen	IC_{50} (μM)
Guinea pig ileum	Histamine (0.1–0.2 μg/ml)	0.00065
	Acetylcholine (5–20 ng/ml)	3.1
	Calcium (1 mM)*	6.05
	LTC_4 (2–10 nM)	8.27
	LTD_4 (10–200 nM)	8.39
	Bradykinin (50–100 ng/ml)	10.3
Guinea pig tracheal segments	Calcium (1 mM)*	9.84
	5-HT (2 nM)	0.56
Guinea pig vas deferens	Norepinephrine (2 μg/ml)	0.53
Rat fundus	5-HT (2–5 ng/ml)	0.43

* Ca^{2+}-depleted, KCl-depolarized smooth-muscle preparations.

Table 14.4 Dissociation of antiallergic (anti-PCA) activities of azelastine from its H_1 and 5-HT antagonist activities in rat skin

	ID_{50} (mg/kg)—post-treatment time					
	2 h			24 h		
Drug	PCA	Histamine	Serotonin	PCA	Histamine	Serotonin
Azelastine HCl	2.6	3.1	>10	3.7	>10	>10

(Table 14.4) [42]. In addition, azelastine effectively interferes with antigen- and phospholipase A_2-induced airway hyperresponsiveness to cooling in rat tracheal smooth muscle [59–61] and antigen-induced contractible responses in guinea pig [62, 63] and rat [59, 61] tracheal smooth muscle preparations (Table 14.5).

Anti-inflammatory

Chemotactic responsiveness in guinea pig macrophages has been shown to be inhibited by azelastine [64]. Azelastine inhibits superoxide radical ($\cdot O_2^-$) generation in guinea pig and rabbit polymorphonuclear leucocytes and in human neutrophils and eosinophils (Table 14.6) [50, 65, 66].

It interferes with inflammatory processes [25, 42, 46, 67] by inhibiting the Ca^{2+}-dependent steps in the synthesis/secretion of inflammatory chemical mediators from mast cells, leucocytes and other cell types in experimental

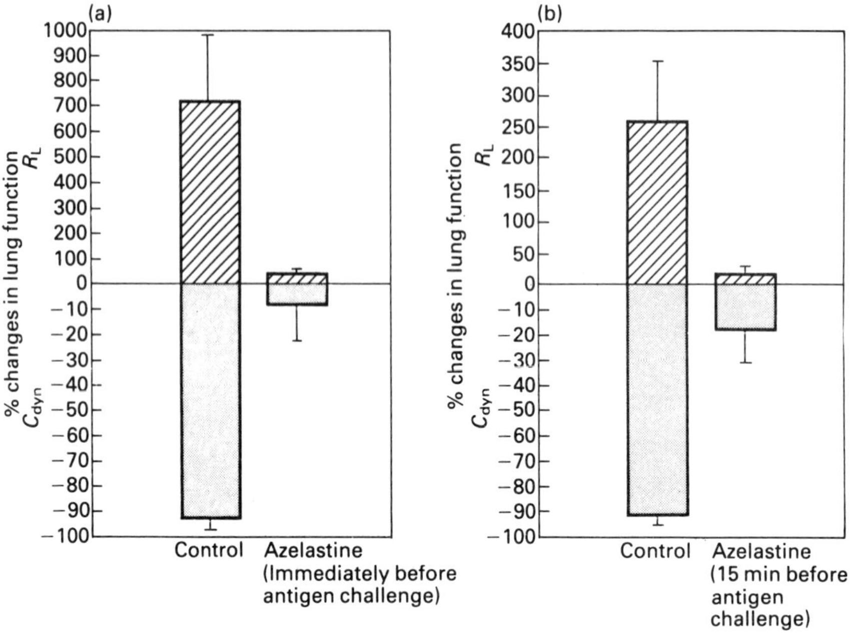

Fig. 14.2 Inhibition of acute lung anaphylaxis by aerosolized azelastine immediately (a) or 15 min (b) before antigen challenge (ovalbumin = 0.6 mg/kg i.v.) in aerosol-sensitized guinea pigs. Values are means ± SEM. * $p < 0.05$ as compared with per cent increase in pulmonary resistance (% R_L) and per cent decline in dynamic lung compliance (% C_{dyn}) in the control 'untreated' guinea pigs.

Table 14.5 Effect of azelastine on phospholipase A_2 (PLA_2)-induced airway hyperreactivity to cold provocation in isolated rat tracheal segments

Azelastine* conc. (μM)	Mean % inhibition ± SE of cold provocation (10 °C) responses
1.0	27.2 ± 10.7
5.0	66.5 ± 10.4†
10.0	66.0 ± 11.5†

* Two-hour preincubation.
† $p \leq 0.05$.

animals and humans [9–33, 47–52]. Calcium and calmodulin play an essential and important role in the synthesis/secretion of chemical mediators [68–70]. Azelastine is capable of antagonizing Ca^{2+} and calmodulin (Table 14.7) [10, 55, 71]. It appears to interfere with the influx/utilization of calcium in the inflammatory cells and target organs [10, 50, 55].

Table 14.6 Inhibition of generation of superoxide from human neutrophils and eosinophils by azelastine

	IC_{50} (μM)	
Activator	Neutrophils	Eosinophils
PMA	0.9	1.8
FMLP	2.1	3.0
Calcium ionophore	1.6	1.7

Table 14.7 Effect of azelastine as a calmodulin inhibitor (NPN–calmodulin fluorescence assay)

	IC_{50} μM
Azelastine	4.0
W7	2.0

Other activities that may relate directly or indirectly to the activity of azelastine include:

1 Moderate inhibition of α-adrenergic receptors *in vitro.*

2 Relatively weak relaxant activity on tracheal smooth muscle *in vitro* compared with isoproterenol.

3 Activation of adenylate cyclase and inhibition of phosphodiesterase *in vitro* which occurs in relatively high concentrations (100 μM and above).

4 No direct effect on the antigen–antibody interaction *in vitro.*

Tissue distribution studies in rat and guinea pigs using radiolabelled azelastine clearly demonstrate that the lungs, the site of action of the drug, preferentially take up the drug. Moreover, the concentration of radioactivity in the lung is greater than the concentrations required *in vitro* to inhibit mediator synthesis, release or activity [9–33, 47–66, 97].

The pharmacological profile of azelastine HCl does not permit it to be classified into one of the existing drug armamentariums for treatment of upper and lower airway disorders, i.e. β-sympathomimetic, steroids, theophylline or disodium cromoglycate-like drugs. It is a new chemical entity with a novel mechanism(s) of overall action with tissue selectivity for the airways.

Clinical activity

Airway challenges

The hyperreactivity of the bronchial tree, which is a common feature of lower airway disorders such as asthma, can be assessed by various provocation tests.

Motojima *et al.* [72] carried out an exercise test using a treadmill. Protective effects of azelastine have been demonstrated with a single oral dose of 3 mg. The maximum percentage fall in FEV_1 (forced expiratory volume in one second) for placebo was 38.9% and only 11.3% after a single oral dose of 3 mg azelastine. This difference was statistically significant.

Several clinical trials have shown protection by azelastine against allergen provocation [72–74]. In a trial performed in Japan, a single dose of 4 mg provided protection against the immediate, as well as the late reaction compared with placebo when evaluated in asymptomatic asthma patients. The maximum percentage fall in FEV_1 immediately after inhalation of allergen was 37.2% for placebo treated and 17.3% for those receiving azelastine. During the late phase, the fall in FEV_1 was 36.0% for placebo and 10.0% for the azelastine group [72]. In another allergen provocation trial, azelastine 4 mg, but not 2 mg, administered twice daily for 21 days was shown to significantly increase the amount of inhaled allergen required to produce a 35% fall in specific airway conductance [74].

Azelastine has also been shown to provide significant protection against histamine-induced provocation after single oral doses of 4 mg in asthmatics [75–77]. Protection against LTC_4 was not shown [78] which may be due to azelastine's more pronounced effect on the synthesis than on antagonism.

Clinical activity in asthma

During a multicentre clinical trial, adolescent and adult subjects with moderate to moderately severe asthma were randomized into a double-blind, placebo-controlled, parallel-group study. The treatment groups were single oral doses of placebo or azelastine (2.0, 4.0, 8.0, 12.0 or 16.0 mg). Spirometric pulmonary function was measured up to 8 h postdosing. Statistical significance was determined from comparisons with placebo therapy. All azelastine doses produced statistically and clinically significant improvement in FEV_1 within 3 h postdosing. The 4.0 mg dose produced sustained mean improvement in FEV_1 for at least 8 h (Fig. 14.3). Higher doses of azelastine were as effective as the 4.0 mg dose [79–82].

In a short-term, multicentre study, adult and adolescent subjects with moderate to moderately severe bronchial asthma were randomized into a two-week, double-blind, placebo- and positive-controlled, parallel-group trial [83]. The oral treatments were placebo, azelastine 4.0 mg b.i.d. and albuterol tablets 4.0 mg t.i.d. Spirometric pulmonary function was measured for 6 h after the first dose and weekly thereafter. The frequency of asthma attacks and the use of backup antiasthma medications (immediate-release theophylline and aerosolized albuterol) were recorded. Both the azelastine- and albuterol-treated groups achieved statistically and clinically significant mean maximum

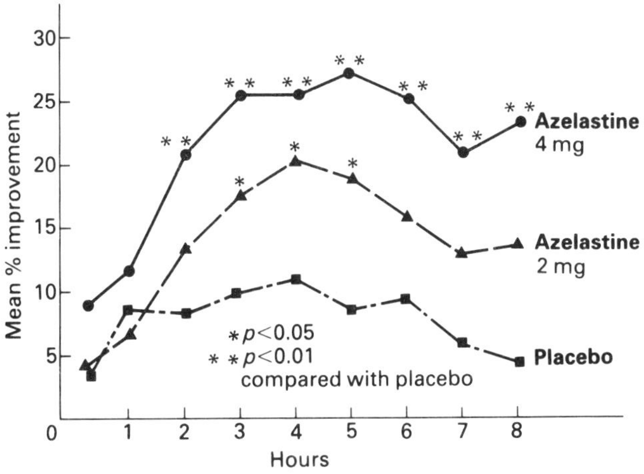

Fig. 14.3 Mean per cent improvement from baseline in FEV_1 for placebo and 2 mg and 4 mg of azelastine.

improvement in FEV_1 following the first dose when compared with placebo therapy. Pulmonary function tests were repeated after 1 and 2 weeks of therapy; neither the azelastine- nor the albuterol-treatment group achieved statistically significant mean improvement in FEV_1, but responses for each treatment exceeded the clinical standard of 15% improvement. Both active-treatment groups used less theophylline than did the placebo-treated group. The difference from placebo therapy was statistically significant for the azelastine-treated group, but not for the albuterol-treated group. Both active-treatment groups also used less aerosolized albuterol than did the placebo-treated group. The greatest decreases occurred in the azelastine-treated subjects, but these differences did not reach statistical significance. Only the azelastine-treated group experienced a statistically significant reduction in the mean daily number of asthma attacks after 2 weeks of therapy.

During a multicentre, long-term trial, subjects aged 12–60 years and presenting with mild to severe asthma were randomized into a double-blind, placebo-controlled, parallel-group study [84]. The oral treatment groups were placebo or azelastine (2.0, 4.0 or 6.0 mg b.i.d.) for 12 weeks. Spirometric pulmonary function was measured up to 8 h postdosing after the first dose of double-blind medication and after 1, 4, 8 and 12 weeks of therapy. Statistical significance was determined from comparisons with placebo therapy.

Beginning 2–3 h after the first dose of azelastine 4.0 mg, the subjects experienced clinically and statistically significant mean improvement in FEV_1 and forced expiratory flow $(FEF)_{25-75\%}$ which persisted throughout 8 h of observation. Improvement in pulmonary function was clinically detectable throughout the course of the study, and at the end of the study the improvement in baseline (zero hour) FEV_1 had statistically significantly improved for the 6 mg azelastine group and also approached statistical significance for the 4 mg group.

Analyses of backup aerosolized albuterol and immediate-release theophylline use were done separately and by a standardized equivalence scoring system defined in *Guidelines for Clinical Investigation of Non-Bronchodilator Anti-Asthmatic Drugs* [98]. The azelastine 4.0 and 6.0 mg dosage regimens resulted in clinically and statistically significant reductions in the need for backup albuterol and theophylline compared with that used by the placebo-treated group (Fig. 14.4). The latter group was maintained through increased use of these antiasthmatic agents.

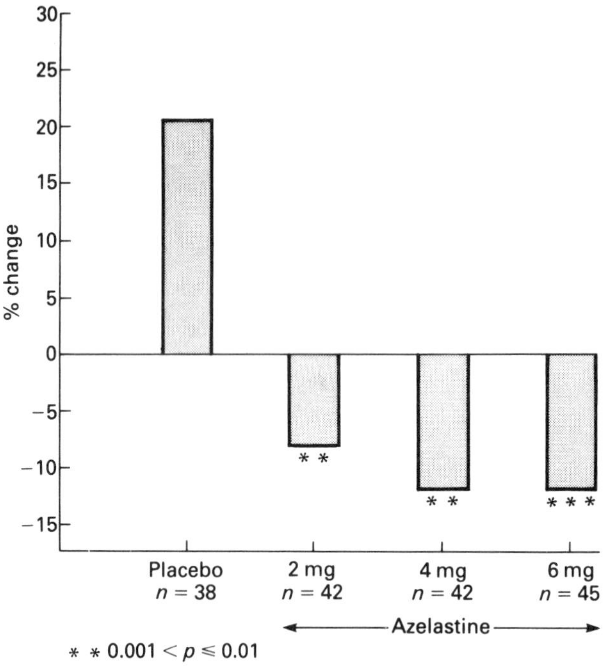

Fig. 14.4 Average daily use of albuterol aerosol during 12 weeks of treatment as per cent change from baseline.

Clinical experience in seasonal and perennial allergic rhinitis

The initial clinical evaluation of the safety and efficacy of azelastine involved multicentre clinical trials at outdoor environmental laboratories [85]. In these studies rhinitis patients were evaluated for grass or ragweed-induced symptoms of seasonal allergic rhinitis. The subjects were randomized into double-blind, positive-controlled, parallel groups. The oral treatment groups were placebo, chlorpheniramine maleate (4 mg q.i.d. × six doses) and azelastine hydrochloride (equivalent to 0.1, 0.25, 0.5, 1, 2 or 4 mg azelastine base b.i.d. × three to four doses). Symptom cards were used to assess the severity of nose blows, sneezes, runny nose, sniffles, stuffiness, postnasal drip, dry nose, watery eyes, cough, and ocular, aural and nasopharyngeal itchiness. Symptoms were analysed as total symptom complexes, including and excluding nose blows and sneezes. Statistical significance was determined from comparisons with placebo therapy [86–88].

Treatment with oral azelastine resulted in clinically and statistically significant mean improvement in the total symptom complexes including nose blows and sneezes when compared with placebo therapy (Table 14.8). Statistically significant mean improvement among the chlorpheniramine-treated subjects was delayed until after the second dose. When nose blows and sneezes were excluded from the total symptom complex, all azelastine-treated groups retained overall statistically significant improvement when compared with placebo therapy ($p < 0.026$).

Table 14.8 Per cent improvement from baseline in total symptom complex scores in short-term trial

	Total symptom complex		
Treatment	*n*	Mean	*p**
Azelastine 1 mg	44	45.07	0.023
Azelastine 2 mg	47	51.23	0.005
Azelastine 4 mg	43	45.75	0.020
Chlorpheniramine	46	26.70	0.469
Placebo	43	18.32	

* Compared with placebo.

Long-term azelastine evaluations in seasonal allergic rhinitis were conducted over a 28-day treatment period [89]. During a multicentre trial, rhinitis subjects were evaluated for symptoms of grass- or ragweed-induced seasonal allergic rhinitis. The subjects were randomized into double-blind, parallel treatment groups consisting of placebo, chlorpheniramine maleate (4 mg q.i.d.) and azelastine (0.5, 1 or 2 mg b.i.d.) for 28 days. Azelastine 1 and 2 mg b.i.d. in long-term studies generally produced statistically and clinically significant mean improvement in the total symptom complex including nose blows and sneezes. The mean improvements in the azelastine 0.5 mg- and chlorpheniramine-treated groups were not as consistent as those observed with 1 or 2 mg dosages during the 4 weeks (Table 14.9).

Azelastine is effective in a dosage range of 0.5 to 2 mg b.i.d. for the long-term treatment of symptomatic seasonal allergic rhinitis [90–92].

Table 14.9 Overall per cent improvement from baseline in total symptom complex scores over an entire season

		Total symptom complex			
		Including nose blows and sneezes		Excluding nose blows and sneezes	
Treatment	*n*	Mean	*p**	Mean	*p**
Azelastine 0.5 mg	26	27.92	0.0248	33.02	0.114
Azelastine 1.0 mg	26	35.05	0.079	41.18	0.073
Azelastine 2.0 mg	26	45.56	0.009	48.83	0.021
Chlorpheniramine	23	30.41	0.187	36.37	0.158
Placebo	27	14.37		15.36	

* Compared with placebo.

In view of azelastine's therapeutic utility in seasonal allergic rhinitis, a study was carried out to evaluate its safety and efficacy in subjects with perennial allergic rhinitis. Rhinitis patients were given either azelastine 1 or 2 mg b.i.d. or placebo for the first 4-week period, followed by a cross-over to placebo or active medication for an additional 4-week treatment period. Symptom severity scores for nose blows, sneezes, stuffiness, runny nose, and ocular, aural and nasopharyngeal itchiness (total symptom complex) were recorded. The need for supplemental decongestant tablet administration was also monitored [93].

The azelastine 2 mg treatment group achieved statistically and clinically significant mean symptom relief after 2 weeks of therapy that persisted through 4 weeks of therapy when compared with placebo therapy. The azelastine 1 mg treatment group also demonstrated clinically significant mean symptom improvement throughout the study although the statistically significant mean improvement observed after 2 weeks of therapy ($p=0.02$) did not persist through 4 weeks ($p=0.40$). The daily use of decongestant tablets during treatment with both azelastine dosages was approximately 50% less than that used during placebo treatment ($p<0.022$) [94].

These clinical responses correlate well with the suppression of skin test reactivity seen after single and multiple oral doses of azelastine [95] and with the effects on blockade of histamine nasal challenge [9, 96].

Safety

Throughout all the US clinical trials, encompassing over 1450 subjects ranging in age from 12 to 60+ years, azelastine was safe and well tolerated.

There were no serious and unexpected adverse drug experiences with a definite or probable relationship to azelastine. Altered taste and somnolence were the most frequently reported adverse drug reactions and accounted for the majority of the adverse drug reactions reported for the 2 and 4 mg b.i.d. dosage regimens. This favourable safety profile is supported by the Japanese experience in their postmarketing surveillance.

Pharmacokinetic profile

Azelastine is well absorbed after the oral route. It reaches peak plasma concentration approximately 5 h after dosing and displays linear pharmacokinetics after single and multiple dosing. No evidence has been found for accumulation of azelastine and its identified metabolite, desmethylazelastine, after steady state is attained [81, 84]. The elimination half-life for azelastine is approximately 20 h after oral administration. Excretion is primarily by the biliary route.

Concluding remarks

The clinical relevance of azelastine's pharmacological properties has been demonstrated in asthma. Azelastine is orally effective in the treatment of symptoms associated with reversible airway disease and has a duration of action sufficient to allow a twice-a-day dosage regimen. The minimum effect dosage for the treatment of symptoms related to asthma is 2.0 mg b.i.d. A dosage regimen of 4.0 mg b.i.d. provides a consistent favourable clinical response. The most important aspects of the effectiveness of azelastine are an improvement in baseline FEV_1 values and a reduction in the requirement for supplemental antiasthmatic medication during long-term therapy. Additionally, dose-related improvements in pulmonary function tests are evident in short-term trials.

Short- and long-term studies have consistently demonstrated azelastine's clinical effectiveness in the treatment of symptoms associated with seasonal and perennial allergic rhinitis. This effectiveness is manifested by a significant reduction in the symptoms associated with allergic rhinitis and by a reduction in the need for supplemental decongestant medication. The minimum effective dosage regimen is 0.5 mg b.i.d. Consistent clinical improvements are seen with the 1.0 and 2.0 mg dosage regimens.

The only available oral therapeutic agents effective for the treatment of both upper and lower reversible airway disease are corticosteroids. Azelastine is the first oral non-steroid drug that affords clinically meaningful symptomatic improvement in asthmatic and rhinitis subjects.

Acknowledgement

The authors wish to express their appreciation to Ms CA Parsons for her assistance in preparation of the manuscript.

References

1 Kaliner M. Mast cell mediators and asthma. *Chest* 1985; **87**:2S–5S.
2 Burka JF. The interaction of histamine with other bronchoconstrictor mediators. *Can. J. Physiol. Pharmacol.* 1987; **65**:442–447.
3 Baggiolini M, Dewald B. Stimulus amplification by PAF and LTB_4 in human neutrophils. *Pharmacol. Res. Commun.* 1986; **18**:51–59.
4 Raphael GD, Metcalfe DD. Mediators of airway inflammation. *Eur. J. Respir. Dis.* 1986; **69**:44–56.
5 Middleton E Jr, Atkins FM, Fanning M, Georgitis JW. Cellular mechanisms in the pathogenesis and pathophysiology of asthma. *Med. Clin. North Am.* 1981; **65**:1013–1031.
6 Austen KF. The role of arachidonic acid metabolites in local and systemic inflammatory processes. *Drugs* 1987; **33**:10–17.
7 Chand N, Altura BM. Lipoxygenase pathway and hydroperoxy acids: possible relevance to aspirin-induced asthma and hyperirritability of airways in asthmatics. *Prostaglandins Med.* 1981; **6**:249–256.
8 Perhach JL, Connell JT, Kemp JP. Treatment of upper and lower airway disease with azelastine. *NER Allergy Proc.* 1987; **8**:121–124.
9 Akagi M, Mio M, Tasaka K, Kiniwa S. Mechanism of histamine release inhibition induced by azelastine. *Pharmacometrics* 1983; **26**:191–198.
10 Chand N, Pillar J, Diamantis W, Perhach JL, Sofia RD. Inhibition of calcium ionophore (A23187)-stimulated histamine release from rat peritoneal mast cells by azelastine: implications for its mode of action. *Eur. J. Pharmacol.* 1983; **96**:227–233.
11 Chand N, Pillar J, Diamantis W, Sofia RD. Inhibition of allergic histamine release by azelastine and selected antiallergic drugs from rabbit leukocytes. *Int. Arch. Allergy Appl. Immunol.* 1985; **77**:451–455.
12 Chand N, Pillar J, Diamantis W, Sofia RD. Inhibition of allergic histamine release from rat peritoneal mast cells: interactions of azelastine with salbutamol, theophylline, DSCG and prednisolone. *Fed. Proc.* 1987; **46**:928.
13 Chand N, Pillar J, Diamantis W, Sofia RD. Inhibition of allergic histamine release from rat peritoneal mast cells by azelastine interaction with selected antiasthmatic drugs. *Int. Arch. Allergy Appl. Immunol.* 1988; **86**:240–256.
14 Chand N, Pillar J, Diamantis W, Sofia RD. Inhibition of IgE-mediated allergic histamine release from rat peritoneal mast cells by azelastine and selected antiallergic drugs. *Agents Actions* 1985; **16**:318–322.
15 Chand N, Pillar J, Diamantis W, Perhach JL, Sofia RD. Inhibition of calcium ionophore (A23187)-induced histamine release by a novel anti-allergic agent, 4-(*p*-chlorobenzyl)-2-(hexahydro-1-methyl-1H-azepine-4-yl)-1(2H)-phthalazinone hydrochloride (azelastine; A5610). *Int. J. Immunol.* 1982; **4**:342.
16 Chand N, Pillar J, Natarajan V, Diamantis W, Sofia RD. Inhibition of allergic histamine secretion from rat peritoneal mast cells by azelastine, a novel, orally acting antiallergic agent. *Pharmacologist* 1983; **25**:181.
17 Chand N, Pillar J, Natarajan V, Diamantis W, Sofia RD. Pharmacological modification of allergic and nonallergic histamine release by inhibitors of arachidonic acid (AA) metabolism and azelastine. In *Prostaglandins and Leukotrienes '84. Symposium*, Washington DC 1984, Abstract 268.
18 Chand N, Pillar J, Diamantis, W, Sofia RD. Inhibition of allergic and nonallergic histamine release by azelastine and lipoxygenase inhibitors. *J. Allergy Clin. Immunol.* 1985; **75**:194.

19 Diamantis W, Pillar J, Chand N, Sofia RD. Inhibition of IgE-mediated allergic histamine release by azelastine from rabbit basophils. *Fed. Proc.* 1984; **43**:389.

20 Diamantis W, Chand N, Pillar J, Sofia RD. Effect of azelastine and diltiazem on activation (first) and release (second) stages of allergic histamine secretion in rabbit leukocytes. *Fed. Proc.* 1987; **46**:929.

21 Fields DAS, Pillar J, Perhach JL, Sofia RD. Inhibition of secretagogue-induced histamine release from rat peritoneal mast cells *in vitro* by a novel anti-allergic agent, 4-(*p*-chlorobenzyl)-2-(hexahydro-1-methyl-1H-azepine-4-yl)-1(2H)-phthalazinone hydrochloride (azelastine). *Pharmacologist* 1981; **23**:161.

22 Fields DAS, Pillar J, Diamantis W, Perhach JL Jr, Sofia RD, Chand N. Inhibition by azelastine of nonallergic histamine release from rat peritoneal mast cells. *J. Allergy Clin. Immunol.* 1984; **73**:400–403.

23 Fischer B, Schmutzler W. Inhibition by azelastine of the immunologically induced histamine release from isolated guinea pig mast cells. *Arzneim Forsch.* 1981; **31**:1193–1195.

24 Inoue Y. Basic studies on the antiallergy drug, 4-(*p*-chlorobenzyl)-2-[*N*-methylperhydro-azepinyl-(4)]-1-(2H)-phthalazinone hydrochloride (azelastine). *Nippon Ika Daigaku Zasshi* 1983; **50**:371–378.

25 Katayama S, Akimoto N, Shionoya H, Morimoto T, Katoh Y. Antiallergic effect of azelastine hydrochloride on immediate type hypersensitivity reactions *in vivo* and *in vitro*. *Arzneim Forsch.* 1981; **31**:1196–1202.

26 Tasaka K, Akagi M. Anti-allergic properties of a new histamine antagonist, 4-(*p*-chlorobenzyl)-2-[*N*-methyl-perhydro-azepinyl-(4)]-1-(2H)-phthalazinone hydrochloride (azelastine). *Arneim Forsch.* 1979; **29**:488–493.

27 Katayama S, Tsunoda H, Sakuma Y, Kai H, Katayama K. Inhibitory effect of azelastine on the production and release of leukotriene and its antagonistic effect on leukotriene. *Prog. Med.* 1986; **6**:1173–1178.

28 Tomiuga T, Yamanishi Y, Mori N, Kobayashi M, Igarashi T. Effects of some antihistaminics on the skin reactions and histamine release induced by a nonionic-surfactant in the dog. *Jap. J. Pharmacol.* 1980; **30**:115P.

29 Sakuma Y, Shoji T. Antiallergic properties of azelastine and its principal metabolite studied *in vitro*. *Folia. Pharmacol. Jap.* 1984; **84**:55–56P.

30 Schmutzler W, Riesener KP, Radeck W. Study of antiallergic properties of DSCG, theophylline, ketotifen and azelastine in isolated human mast cells. *Naunyn Schmiedebergs Arch. Pharmacol.* 1982; **319**:R55.

31 Little M, Ecklund P, Casale T. Azelastine's therapeutic action in asthma. *J. Allergy Clin. Immunol.* 1988; **81**:278.

32 Little MM, Casale TB. Azelastine inhibits both human lung and basophil degranulation. *J. Allergy Clin. Immunol.* 1987; **79**:204.

33 Schmutzler W, Riesener KP, Radeck W. Differences in the dependence on adenosine of the histamine release inhibiting effects of theophylline, DSCG, ketotifen and azelastine in human mast cells. *Abstracts of XIth International Congress of Allergology and Clinical Immunology*, London 1982.

34 Katayama S, Hashida N, Shionoya H. Induction of experimental asthma in rats sensitized by mouse IgE and inhibitory effect of azelastine. *Prog. Med.* 1986; **6**:1429–1434.

35 Chand N, Nolan K, Diamantis W, Perhach JL Jr, Sofia RD. Inhibition of leukotriene (SRS-A)-mediated acute lung anaphylaxis by azelastine in guinea pigs. *Allergy* 1986; **41**:473–478.

36 Chand N, Nolan K, Diamantis W, Sofia RD. Inhibition of acute lung anaphylaxis by aerosolized azelastine in guinea pigs sensitized by three different procedures. *Ann. Allergy* 1987; **58**:344–349.

37 Chand N, Nolan K, Diamantis W, Perhach JL, Sofia RD. Inhibition of leukotriene (SRS-A)-mediated allergic bronchospasm by azelastine, a novel orally effective antiasthmatic drug. *J. Allergy Clin. Immunol.* 183; **71**:149.

38 Diamantis W, Chand N, Nolan K, Sofia RD. Inhibition of aeroallergen-induced pulmonary mechanics changes by oral administration of azelastine in actively sensitized guinea pig. *Pharmacologist* 1987; **29**:173.

39 Chand N, Nolan K, Diamantis W, Sofia RD. Effect of aerosolized azelastine on acute lung anaphylaxis in guinea pigs sensitized by two different procedures. *Pharmacologist* 1985; **27**:162.

40 Chand N, Nolan K, Diamantis W, Sofia RD. Inhibition of acute bronchial anaphylaxis by azelastine in aerosol-sensitized guinea pig asthma model. *Ann. Allergy* 1985; **55**:393.

41 Zechel HJ, Brock N, Lenke D, Achterrath-Tuckermann U. Pharmacological and toxicological properties of azelastine, a novel antiallergic agent. *Arzneim Forsch.* 1981; **31**:1184–1193.

42 Chand N, Harrison JE, Rooney SM, Sofia RD, Diamantis W. Inhibition of passive cutaneous anaphylaxis (PCA) by azelastine: dissociation of its antiallergic activities from antihistaminic and antiserotonin properties. *Int. J. Immunopharmacol.* 1985; **7**:833–838.

43 Diamantis W, Chand N, Harrison JE, Rooney SM, Sofia RD. Dissociation of antiallergic (anti-PCA) activity from the antihistaminic and antiserotonin activities of azelastine. *Pharmacologist* 1984; **26**:151.

44 Kaji R, Kamijo T, Kojima S. Inhibitory effects of a new antiallergic agent azelastine, on passive cutaneous anaphylaxis and expulsion of *Nippostrongylus brasiliensis. Immunopharmacol.* 1981; **3**:49–52.

45 Diamantis W, Chand N, Harrison JE, Rooney SM, Sofia RD. Inhibition of IgE-mediated passive cutaneous anaphylaxis (PCA) by azelastine in rat. *J. Allergy Clin. Immunol.* 1984; **73**:184.

46 Katayama S, Hashida N, Shionoya H. Inhibitory effect of azelastine on skin reactions and peritoneal anaphylaxis in rats. *Med. Cons. New Rem.* 1986; **23**:103–109.

47 Chand N, Diamantis W, Nolan K, Pillar J, Sofia RD. Inhibition of allergic leukotriene C_4 (LTC_4) formation by azelastine in actively sensitized chopped guinea pig lung. *Pharmacologist* 1987; **29**:173.

48 Chand N, Pillar J, Nolan K, Diamantis W, Sofia RD. Inhibition of 5-HETE, LTB_4 and LTC_4 synthesis by azelastine and its D- and L-isomers in rat mixed peritoneal cells. *Am. Rev. Respir. Dis.* 1987; **135**:A381.

49 Chand N. Possible role for the products of arachidonate lipoxygenase pathway as regulators of airway reactivity and release of chemical mediators. *Kongr. Ber. Wiss. Tag. Norddtsch. Ges. Lungen Bronchialheilk.* 1983; **18**:93–103.

50 Takanaka K. Effects of azelastine on polymorphonuclear leukocytes: arachidonate cascade inhibition mechanism. *Prog. Med.* 1987; **7**:275–278.

51 Katayama S, Tsunoda H, Sakuma Y, Kai H, Tanaka I, Katayama K. Effect of azelastine on the release and action of leukotriene C_4 and D_4. *Int. Arch. Allergy Appl. Immunol.* 1987; **83**:284–289.

52 Diamantis W, Chand N, Harrison JE, Pillar J, Perhach JL, Sofia RD. Inhibition of release of SRS-A and its antagonism by azelastine an H_1 antagonist-antiallergic agent. *Pharmacologist* 1982; **24**:200.

53 Yamanaka T, Shoji T, Murakami M, Igarashi T. Effects of azelastine hydrochloride, a new anti-allergic drug, on the gastrointestinal tract. *Arzneim Forsch.* 1981; **31**:1203–1206.

54 Diamantis W, Harrison JE, Melton J, Perhach JL, Sofia RD. *In vivo* and *vitro* H_1 antagonist properties of azelastine. *Pharmacologist* 1981; **23**:149.

55 Chand N, Diamantis W, Sofia RD. Antagonism of leukotrienes, calcium and histamine by azelastine. *Pharmacologist* 1984; **26**:152.

56 Chand N, Diamantis W, Sofia RD. Antagonism of histamine and leukotrienes by azelastine in isolated guinea pig ileum. *Agents Actions* 1986; **19**:164–168.

57 Fujimori T, Harada K. Inhibitory effect of azelastine against platelet activating factor (PAF) on human platelet. *Med. Cons. New Rem.* 1986; **23**:2251–2253.

58 Achterrath-Tuckermann U, Weischer CH, Szelenyi I. Azelastine, a new antiallergic/antiasthmatic agent, inhibits PAF-acether induced platelet aggregation, paw edema and bronchoconstriction. *Pharmacology* 1988; **36**:265–271.

59 Chand N, Mahoney TP, Diamantis W, Sofia RD. Inhibition of antigen-induced airway hyperreactivity to cold provocation by azelastine and selected antiasthmatic drugs in rat isolated tracheal segments. *J. Allergy Clin. Immunol.* 1987; **79**:157.

60 Chand N, Mahoney TP, Diamantis W, Sofia RD. Phospholipase A_2 (PLA_2) causes airway hyperreactivity to cold provocation: effect of azelastine, Ca^{++} channel blockers and selected drugs. *Am. Rev. Respir. Dis.* 1987; **135**:A271.
61 Chand N, Diamantis W, Mahoney TP, Sofia RD. Allergic responses and subsequent development of airway hyperreactivity to cold in the rat trachea: pharmacological modulation. *Eur. J. Pharmacol.* 1988; **150**:95–101.
62 Chand N, Diamantis W, Sofia RD. Pharmacological modulation of *in vitro* anaphylactic responses of isolated guinea pig tracheal rings by azelastine and inhibitors of arachidonic acid metabolism. *Fed. Proc.* 1984; **43**:387.
63 Chand N, Diamantis W, Sofia RD. Modulation of *in vitro* anaphylaxis of guinea-pig isolated tracheal segments by azelastine, inhibitors of arachidonic acid metabolism and selected antiallergic drugs. *Br. J. Pharmacol.* 1986; **87**:443–448.
64 Honda M, Kazunori M, Tanigawa T. Effect of azelastine hydrochloride on macrophage chemotaxis and phagocytosis *in vitro*. *Allergy* 1982; **37**:41–47.
65 Taniguchi K, Takanaka K. Inhibitory effects of various drugs on phorbol myristate acetate and N-formyl methionyl leucyl phenylalanine induced $\cdot O_2^-$ production in polymorphonuclear leukocytes. *Biochem. Pharmacol.* 1984; **33**:3165–3169.
66 Busse W, Randlev B, Sedgwick J, Sofia RD. The effect of azelastine on neutrophil and eosinophil generation of superoxide ($\cdot O_2^-$). *J. Allergy Clin. Immunol.* 1988; **81**:212.
67 Tanigawa T, Honda M, Miura K. Effect of azelastine hydrochloride on vascular permeability in hypersensitivity reaction skin site in guinea pig. *Arzneim Forsch.* 1981; **31**:1212–1215.
68 Gigl G, Hartweg D, Sanchez-Delgado E, Metz G, Gietzen K. Calmodulin antagonism: a pharmacological approach for the inhibition of mediator release from mast cells. *Cell Calcium* 1987; **8**:327–344.
69 Chand N, Perhach JL Jr, Diamantis W, Sofia RD. Heterogeneity of calcium channels in mast cells and basophils and the possible relevance to pathophysiology of lung diseases: a review. *Agents Actions* 1985; **17**:407–417.
70 Chand N, Diamantis W, Sofia RD. The obligatory role of calcium in the development of antigen-induced airway hyperreactivity to cold provocation in the rat isolated trachea. *Br. J. Pharmacol.* 1987; **91**:17–22.
71 Sofia RD, Middleton E Jr, Ferriola P, Drzewiecki G. Some biochemical properties of azelastine. *J. Allergy Clin. Immunol.* 1988; **81**:277.
72 Motojima S, Ohashi Y, Otsuka T, Fukuda T, Makino S. Effects of azelastine on allergen- and exercise-induced asthma. *Asian Pacific J. Allergy Immunol.* 1985; **3**:174–178.
73 Gould CAL, Ollier S, Davies RJ. The effect of single and multiple dose therapy with azelastine on the immediate asthmatic response to allergen provocation testing. *Ann. Allergy* 1985; **55**:232.
74 Ollier S, Gould CAL, Davies RJ. The effect of single and multiple dose therapy with azelastine on the immediate asthmatic response to allergen provocation testing. *J. Allergy Clin. Immunol.* 1986; **78**:358–364.
75 Kamburoff P. Recent advances on the use of antihistaminics in the treatment of bronchial obstruction. *Bull. Eur. Physiopathol. Respir.* 1981; **17**:8P.
76 Magnussen H. The inhibitory effect of azelastine and ketotifen on histamine-induced bronchoconstriction in asthmatic patients. *Chest* 1987; **91**:855–858.
77 Mandi A, Galgoczy G, Galambos E, Aurich R. Histamine protection and bronchodilation with azelastine, a new antiallergic compound. *Bull. Eur. Physiopathol. Respir.* 1981; **17**:5P.
78 Albazzaz MK, Patel KR. Effect of single and multiple dose therapy with azelastine on leukotriene (LTC_4) and histamine-induced bronchoconstriction in patients with asthma. *Thorax* 1987; **42**:724.
79 Kemp JP, Meltzer EO, Orgel HA, Welch MJ, Bucholtz GA, Middleton E Jr, Spector SL, Newton JJ, Perhach JL Jr. A dose-response study of the bronchodilator action of azelastine in asthma. *J. Allergy Clin. Immunol.* 1987: **79**:893–899.
80 Spector SL, Perhach JL, Rohr AS, Rachelefsky GS, Katz RM, Siegel SC. Pharmacodynamic evaluation of azelastine in subjects with asthma. *J. Allergy Clin. Immunol.* 1987; **80**:75–80.

81 Spector S, Rohr A, Rachelefsky G, Katz R, Siegel S, Perhach J, Newton J. Pharmacodynamic evaluation of azelastine in asthmatics. *J. Allergy Clin. Immunol.* 1986; **77**:249.

82 Kemp JP, Meltzer EO, Orgel HA, Welch MJ, Lockey RF, Middleton E Jr, Spector SL, Perhach JL, Newton JJ. A dose-response study of the bronchodilator action of azelastine in asthma. *J. Allergy. Clin. Immunol.* 1986; **77**:249.

83 Storms W, Middleton E, Dvorin D, Kemp J, Spector S, Newton J, Perhach JL. Azelastine in the treatment of asthma. *J. Allergy Clin. Immunol.* 1985; **75**:167.

84 Tinkelman D, Bucholtz G, Kemp J, Repsher L, Spector S, Storms W, Van As A, Koepke J. Evaluation of the safety and efficacy of long term use of multiple doses of azelastine in asthmatics. *J. Allergy Clin. Immunol.* 1988; **81**:280.

85 Connell JT, Howard JC Jr, Dressler W, Perhach JL. Antihistamines: findings in clinical trials relevant to therapeutics. *Am. J. Rhinol.* 1987; **1**:3–16.

86 Connell JT, Perhach JL, Weiler JM, Rosenthal R, Hamilton L, Diamond L, Newton JJ. Azelastine, a new antiallergy agent: efficacy in ragweed hay fever. *Ann. Allergy* 1985; **55**:392.

87 Perhach J, Connell J, Hamilton L, Diamond L, Weiler J, Melvin J. Multicenter trial of azelastine in allergic rhinitis. *J. Allergy Clin. Immunol.* 1984; **73**:144.

88 Rhodes BJ, Iwamoto P, Donnelly AL, Perhach JL, Weiler JM. Azelastine in seasonal allergic rhinitis: the Iowa experience. *Ann. Allergy* 1986; **56**:528.

89 Weiler JM, Rhodes BJ, Iwamoto PKL, Donnelly AL, Perhach JL. Effectiveness and safety of azelastine in a chronic study in patients with allergic rhinitis. *J. Allergy Clin. Immunol.* 1986; **77**:180.

90 Fukutake T, Kusumoto T. Studies of the clinical effect of E-0659 on so-called vasomotor rhinitis. *Otolarynology* 1983; **29**:496–504.

91 Okuda M. Effect of E-0659 for nasal allergy. *Oto-Rhino-Laryngol.* 1980; **23**:8.

92 Emoto K, Sunagane H, Azuma E, Asakura M. Experiences in the use of E-0659 for treatment of allergic rhinitis. *Jap. J. Clin. Expt. Med.* 1983; **60**:308–316.

93 Meltzer EO, Storms WW, Pierson WE, Cummins LH, Perhach JL, Hemsworth GR. Efficacy of azelastine in perennial allergic rhinitis. *J. Allergy Clin. Immunol.* 1987; **79**:205.

94 Meltzer EO, Storms WW, Pierson WE, Cummins LH, Orgel AA, Perhach JL, Hemsworth GR. Efficacy of azelastine in perennial allergic rhinitis: clinical and rhinomanometric evaluation. *J. Allergy Clin. Immunol.* 1988; **82**:447–455.

95 Atkins P, Merton H, Karpink P, Weliky I, Zweiman B. Azelastine inhibition of skin test reactivity in humans. *J. Allergy Clin. Immunol.* 1985; **75**:167.

96 Shapiro GG, Bierman CW, Pierson WE, Furukawa CT, Altman LC. Azelastine differential induced blockade of histamine nasal challenge test. *J. Allergy Clin. Immunol.* 1987; **79**:255.

97 Tatsumi K, Toshio OU, Yamada H, Yoshimura H. Studies on metabolic fate of a new antiallergic agent, azelastine (4-(*p*-chlorobenzyl)-2-[*N*-methydroazepinyl-(4)]-1-(2H)-phthalazinone hydrochloride. *Jap. J. Pharmacol.* 1980; **30**:37–48.

98 Bernstein IL (Ed.). Report of the AAAI task force on guidelines for clinical investigation of nonbronchodilator anti-asthma drugs. *J. Allergy Clin. Immunol.* 1987; **78**:489–546.

General discussion (Chapters 10–14)

BARNES: inhaled and intranasal corticosteroids are remarkably effective in controlling asthma and rhinitis and I do not believe that we have heard of any drug today that goes any way near them in clinical effectiveness. My question for the panel is what direction would they proceed in if they were looking for something that was going to be as good or better than steroids.

HOLGATE: if a little bit more was known about how corticosteroids work in clinical situations we would be able to partly answer your question. The amount of available data on the cellular changes in the upper and lower airways following administration of corticosteroids is minimal. There are also no detailed studies of the effects of this class of drugs on individual cells *ex vivo*. One is tempted to speculate on various sites of action that Professor Flower referred to in his chapter. Does Professor Flower think it is possible to be able to divorce the anti-inflammatory effects of corticosteroids from their endocrine side effects using the sort of molecules he is working with?

FLOWER: just to crystal gaze for a little bit, personally I doubt it. I think you might improve it. The problem is that what we call the side effects are not really the side effects as far as the cell is concerned. The cell probably employs a variety of second messenger molecules like lipocortin to perform all sorts of jobs like switching off eicosanoid synthesis, immunosuppression and so on. We happen to think that immunosuppression, generally speaking, is a bad side effect but this is not so for the cell. Obviously it simply uses the same second messengers to perform the same task. I think, therefore, that we will be able to achieve a significant advance but we can never expect to get a complete split of activities for the very simple reason that the cells you are trying to hit in inflammation are exactly the same cells which are providing your protection against infection. If you interfere at all with their activity then you are going to lay yourself open to some degree of increased chance of infection and I think we are never going to get around that. Back to your original point, I think you are absolutely right, we simply do not know enough about how these drugs act, how long they take to work and which are the principal cell types, to be able to do anything sensible.

KAY: a proportion of asthmatics, albeit small, do not seem to respond very effectively to treatment with corticosteroids, even oral prednisolone, and these are a very important group of patients. They represent the hard core of the asthmatic problem. It has been estimated by Ian Grant and his colleagues in Edinburgh to be about 20% of all chronic asthmatics. Our recent data suggest that these patients have chronically activated T-cells. Maybe these subjects are telling us something about the really important mode of action of steroids in asthma.

WASSERMAN: I do not know that I am as pessimistic as Professor Flower. Part of the problem is the way glucocorticoid drugs have been selected in the past. This has been to look at either their anti-inflammatory potency in the skin or at their ability to replace an adrenalectomized animal. We have never looked at the most appropriate assay to see if we cannot design a drug that would be more effective. Secondly, what we have learned about molecular biology and the mechanism whereby steroids induce or suppress induction of genes suggests that they bind to selected segments of DNA. It may be possible, when we learn about the selected segments of DNA, to engineer selected materials and transport them to the appropriate cells. I do not know that it is impossible. I think it is impossible for our patients who now have asthma given that it takes 10–12 years between thinking of an idea, having a drug and marketing it.

PERHACH: I share Dr Wasserman's more optimistic view that once we understand the process from a molecular point of view then we will attempt to attack it and change it.

BACH: I would like to take the reverse side of your point. Talking of the cromolyn-like drugs there has been recent evidence suggesting that these compounds may be inhibiting protein kinase C. This may well be an underlying general mechanism to explain all the phenomenology that we have been exposed to. If this is how these drugs act we would expect them to do a lot of other things. Do they do this?

HOLGATE: to me, the interesting thing about cell signals which couple activation to secretion, or activation to contraction, is not the biochemical mechanism that undertakes the transduction from outside the cell to within the cell, but the receptor recognition of the agonists. That seems to be where nature has directed its specificity. There may be common mechanisms whereby these enzymes are stimulated. In the case of sodium cromoglycate and more recently with nedocromil sodium binding characteristics on the surface of cells have been identified to suggest that there are specific protein or glycoprotein binding sites. Since these are still poorly defined it is not justified to use the term 'receptor'. The binding sites might influence the behaviour of the cell with respect to protein kinase activity, particularly the activation state of protein kinase C and its relationship to the phosphatidyl inositol pathway. For example, I am aware that Professor Barnes' group are doing some good work looking at the effect of phorbol esters on the expression of β-receptor function. This demonstrates that by modifying this group of enzymes you can produce large changes in the way the cells respond to other agonist stimuli. With cromoglycate and nedocromil sodium it is tantalizing to think that there may be effects on these important enzyme pathways, but until they are looked at in great detail one cannot be sure.

BARNES: I think it is an attractive idea but it seems to me that any drug which interferes with such a fundamental enzyme which regulates many cellular functions would have a lot of problems and side effects. That is one thing that cromoglycate does not appear to have, i.e. side effects.

EADY: can I just come back on that. We have to assume that cromoglycate does not go into the cell. There is a lot of evidence to show that the compound cannot get into the cell, so Professor Holgate's point is important, the compound is going to bind to something on the outside of the cell which in turn could affect protein kinase C. That is where the selectivity could come in terms of the binding. Therefore you might not expect it to have a generalized toxic effect.

PERHACH: I want to respond with respect to azelastine. At least in some preliminary work in Elliott Middleton's laboratory, it does not appear to have any influence on protein kinase C, for what that is worth at this point in time.

VARSANO (Israel): what are the side effects of azelastine?

PERHACH: the most frequently observed side effect is one that I was personally intrigued to see with nedocromil—altered taste sensation with liquids. This phenomenon does not occur in all subjects, it does appear to be dose related. The best way I can describe it to you is that there is not a persistent bitter taste in the mouth but certain subjects experience a bitter aftertaste upon ingestion of certain liquids. Colas are quite popular in the USA and that is what is frequently reported. Some people who ingest reasonable quantities of milk report it. Based on knowledge of the reporting of this experience in the USA versus Japan and the European trials, the USA population seems to be more sensitive to this aftertaste than in other countries.

AUTY: could I correct the impression about nedocromil sodium. Professor Holgate correctly reported taste perversion with nedocromil sodium. That is the expression used by the WHO classification of symptoms needed in correlating all the adverse experience data. I think it is an unfortunate expression. What people in fact describe with nedocromil sodium is a taste—about 15% of people reported a taste. The data that Professor Holgate showed were correct, and about 3% found it an unacceptable taste, but it is a taste and not a perversion of the ability to taste other things, which is what is being reported.

HOLGATE: in our trials on azelastine the symptom of an unpleasant 'taste' is one of a persisting metallic taste, but with the large doses being administered the major side effect was drowsiness.

PERHACH: we have just completed the overall tabulation of side effects from our rhinitis studies and those studies ranged from 0.10 mg up to 4 mg b.i.d. and the incidence of drowsiness or somnolence reported by placebo in our rhinitis controls is approximately 6%, with 0.5 mg this increases to 8%, 1–2 mg approximately 10% but then very interestingly with no explanation at 4 mg it drops back to 4%.

ANDRESCICOSA (Italy): my question regards the late reaction. In occupational asthma we have roughly about 10% of immediate reaction, 45% of dual reaction and 45% of late reaction without any immediate reaction. I would like to know if there is some idea about the possible pathogenesis of the late reaction that manifests itself without any immediate reaction?

KAY: my guess would be that isolated late reactions probably have an element of disordered cell-mediated immunity and that immediate reactions are due to mast-cell activation. The smaller chemicals tend to be the ones which give the isolated late reactions, whereas the more organic and larger molecules seem to be associated with the immediate and dual responses.

FORD-HUTCHINSON: you did not say what the incidence of this taste experience with azelastine was.

PERHACH: it is about 17%. In the rhinitis study doses above 4 mg give a higher incidence but our current anticipation is that we will not exceed 4 mg b.i.d. as the dosage regimen.

FORD-HUTCHINSON: another thing I was not clear about. I noticed a plethora of effects of this compound around about micromolar range. In a comparable system, what is the ratio of activity against histamine compared with all these other properties?

PERHACH: if one were to categorize the effects on a scale with most potent effects, the H_1-blocking property of azelastine is the single most potent property. That is at approximately 10^{-9} M concentration.

FORD-HUTCHINSON: so effectively, it is about 1000-fold more potent as an antihistamine than any of the other properties you describe. It seems to me to be a highly selective antihistamine and all the other effects are occurring at higher doses. What are the blood levels?

PERHACH: the effects of azelastine on neutrophils and eosinophils with respect to superoxide formation occur at concentrations that have been found in the lungs of animals given radiolabelled azelastine. Furthermore, this effect on superoxide formulation is not seen with classical H_1-receptor blocking agents (Busse WW *et al.*, *J. Allergy*, in press). The fact that the drug is delivered to the target site in concentrations sufficient to produce the desired effects allows for the possibility that factors other than H_1-receptor antagonism are contributing to the response observed in man.

BARNES: I think Dr Patel in Glasgow has shown that azelastine does not have any inhibitory effect against inhaled leukotriene.

PERHACH: that is correct and not surprising since the more predominant effect of azelastine is on synthesis and release than on receptor antagonism. In an acute challenge situation inhibition of an inhaled leukotriene may not be possible.

MORLEY: are *in vitro* tests of cell function relevant? Perhaps more effort should be given to using *in vivo* tests. We certainly find that looking at eosinophil accumulation. Substances either work or do not work if you use *in vitro* tests with quite high concentrations of the drug. With *in vitro* tests, increasing the concentration, often produces an effect eventually.

WASSERMAN: the danger is that you move from one irrelevant model to a more expensive irrelevant model.

IV

Concluding Remarks

15

Summing up and general discussion

KAY: many subjects have been covered and many issues have been raised. We must now get the views as to the areas which are felt to be important if we are to make progress in improving treatment in the general area of allergy and asthma. What, for instance, does the future hold for antagonists of lipid formation and action? Agents with a wide range of action have very similar effects in the various experimental models. None of them seem to be exceptional in the clinic. Is this telling us something? On the other hand, is bronchial asthma one disease? Could different therapeutic approaches suit some patients better than others? What are the variants of the disease? The pharmaceutical industry has tremendous courage in tackling many of these problems. It has to think over large timeframes in order to achieve its goals. At the same time it has to be prepared for a change in direction in the light of new knowledge. Thus, at one point in time a particular agent may only be providing a stopgap in the treatment of allergic disease. In other words there have to be both short-term and long-term strategies. Many people feel that corticosteroids remain positive 'gold standard' for prevention of allergy and asthma. Inhaled corticosteroids have been a tremendous advance in the treatment of moderate to moderately severe asthma. They are surprisingly free of side effects. They do not seem to interfere with local immunity at recommended therapeutic concentrations. The main target of corticosteroids in asthma is inflammation and in particular the eosinophil. It should be borne in mind nevertheless that eosinophils, sometimes in large numbers, are often found at mucosal surfaces in health. Their function in this context remains unclear. One of the questions I would like to put is—can there be chronic asthma without eosinophils?

BARNES: I agree that asthma is an inflammatory disease (although this information was known over a hundred years ago). At one time it was thought inflammation was an aspect of fatal asthma not present in patients who come to the clinic. I think the recent appreciation that inflammation is crucial to asthma has been very important. A lot of emphasis has been placed on bronchial hyperresponsiveness as a key abnormality. However, steroids are not very good in reducing bronchial hyperreactivity and yet they have remarkable effects in relieving the symptoms. If you screen for drugs solely in terms of their effects on bronchial hyperreactivity you may miss those which are good at controlling asthma symptoms (that is all the patient is interested in!).

In my opinion, animal systems can give extremely misleading results. What the guinea pig cannot do is predict things you are *not* expecting, because what you are *not* expecting is by definition something unexpected. There are problems in extrapolating from animals to humans. I think that a lot of money has been wasted on inappropriate models and building up whole series of drugs which turn out in the end to be quite useless. The important thing in research is to try and test (wherever possible) the actions of drugs not only in humans but in the disease which the drug is meant to be dealing with. This is a lesson which has not been learnt very well. Even today, people are introducing drugs without doing appropriate background tests. They are not designing studies well. One of the reasons for this is the commercial pressure to get drugs into the marketplace quickly. There are great dangers in that approach in contrast to a strategy involving a logical series of experiments. Maybe that is because drug companies do not take good advice, at least not all of them. I feel very strongly that inhaled steroids are very much better than any other drug we have for asthma at present. If you want to look for new drugs you should be attempting to understand the precise mode of action of steroids. The goal is to develop a drug with actions of steroids but without the unwanted side effects. The unwanted effects with inhaled steroids are actually very small. Nevertheless we need a safe oral 'steroid' because of the problem of compliance. Inhaled steroids do not produce immediate relief of symptoms and so one important goal in the next few years is to encourage patients to take these drugs regularly since undoubtedly their asthma will get better and it will reduce the mortality (which is still an important problem). We need to be able to educate people to take inhaled steroids and tell them why they are taking them. A recent study at Guy's Hospital looking at compliance with antiasthma therapy was very revealing. They found that patients took more β-agonists than they admitted to. β-Agonists relieve symptoms immediately and the patients like this treatment because it makes them feel better. As a consequence they were taking insufficient inhaled steroids. The ideal treatment for asthma should be one where bronchodilators are no longer necessary (because there would be nothing to bronchodilate). A further point I will bring up now is specificity in drug design. For example, we have looked at two contrasting drugs, which in fact are very similar in that they are both antihistamines. With one, the company were at pains to demonstrate how selective it was, and in the other case the company were at pains to demonstrate that it had other actions. You can view these two approaches in different ways. On the one hand, specific drugs are extremely useful for understanding mechanisms of disease and at the present time they are probably the only way of dissecting a complex inflammatory disease like asthma. If you happen to get the right antagonist you could have a major clinical effect. For example, antihistamines are very useful in allergic rhinitis, in that particular disease histamine appears to be the most important mediator. This does not seem to be the case with asthma. On the other hand, perhaps one mediator that has got a lot going for it is platelet-activating factor (PAF-acether). Its actions rather closely mimic all the features of asthma that we know of. PAF-acether antagonists are now being tested clinically and some very potent and selective drugs will shortly be going into clinical trials. We will know within the next 2–3 years whether PAF-acether really is important in asthma. Thromboxane antagonists are another promising area. The compounds are specific and may turn out to have major therapeutic benefit. If I were betting I would actually go for drugs that lack specificity, given what we know now. Also, it would seem more prudent to develop drugs that are non-specific. For example, in the bronchodilator field, β-agonists are the best because they will block constriction of airway smooth muscle irrespective of the constricting agent. Anticholinergic drugs on the other hand only possibly block constriction due to acetylcholine. They have no other known mode of action and therefore it is not particularly surprising that such drugs are far less effective in asthma than β-agonists. That gives you an example of how specificity can be a disadvantage. That is possibly why steroids are so effective because they have so many different actions. It may be difficult to isolate a single action which can be mimicked by another drug. We should really be grateful to steroids because they inhibit virtually every inflammatory mechanism that has been described in asthma (apart from mast-cell degranulation which again emphasizes the fact that mast cells are probably not that important in chronic asthma).

HOLGATE: the evidence that inflammation is causally related to clinical asthma has a long way to go before we are certain about it. Can an attack of asthma occur in the absence of airway inflammation? The answer seems to be 'do not know' and therefore how can you be so certain that inflammation is so important? Unlike so many other diseases, it should be borne in mind that very little is known about the histopathology of asthma. Furthermore, is there more than one form of asthma, i.e. diisocyanate asthma, platinum asthma, allergic asthma, non-allergic asthma, asthma related to bronchitis and on and on, and do they all have the same final common pathway or are they totally different conditions? Just as one can have many different causes of hypertension why could not the same apply to asthma? I agree, of course that corticosteroids are effective in disease control, but this should not stop one thinking about developing new chemical entities and new ways of interrupting the disease process, possibly from early childhood.

Another subject of current interest are the newer bronchodilators. Soon bronchodilators such as salmeterol will be available which maintain airway calibre for up to 12–24 h after a single inhalation. How are we going to use drugs like this? Patients are going to find them very acceptable, but how are we going to advise our patients to use them? Since many inflammatory leucocytes express inhibitory β_2-receptors, what effect will constant β-agonist stimulation have on airway inflammation? At present there is little data on this and even some speculation that such therapy may enhance inflammatory processes.

We now come to the very difficult question of the role of individual mediators. Are drugs needed with broad specificities or is it preferable to develop agents which target on individual mediators such as PAF-acether? At the present time I do not think our current state of knowledge tells us in which direction effort should be concentrated. However, I do disagree with the view that the *only* way to answer this question is to improve the currently available drugs.

Finally, is asthma an immunological disease and should we not be targeting our efforts more towards the processes relating to the induction of the disease process? The drugs that are most effective in treating chronic human diseases tend to be those that promote healing and yet the amount of information on corticosteroids or sodium cromoglycate (and nedocromil sodium) on the process of healing of the airway is negligible. I believe that it is our job as clinical investigators and basic scientists to try to answer some of these crucial questions before extrapolating our limited knowledge too far.

WASSERMAN: the link between inflammation and asthma is not proven, and between the late-phase reaction and inflammation is only assumed. Are we really missing the boat by putting all of our attention on the late phase? Do we have better models to study asthma in the clinic and in the clinical laboratory at the present time?

HOLGATE: to be able to answer that I think we need a better understanding of the relationship between late-phase reactions, the cellular correlates and the acquisition of hyperresponsiveness. The only way I can see of getting these answers is from quantitative pathology and its relation to pharmacology and physiology.

BARNES: I agree with what Professor Holgate says. We do not know what models are good until we have tested them. A drug may fail to have an effect in a model but is effective clinically and vice versa. There has been a lot of interest in drugs that affect bronchial hyperreactivity but the relationship between bronchial hyperreactivity and asthma is not at all straightforward.

WASSERMAN: certainly, cough, cold air and exercise-induced bronchospasm are associated with bronchial hyperreactivity. However, one of the problems is that, since there may be different 'asthmas', no single model can be used to devise all drug therapies. I think that better clinical descriptions are needed. Classification needs to take into account variables such as inborn hyperreactivity, differences in genetic background, IgE levels and aspirin sensitivity. Antigen provocation model is an unsatisfactory model because of the variations in responsiveness. We often give more antigen in a bronchoprovocation reaction than a person would inhale in a year.

KATZ: I would like to throw open for discussion the possibility that the best model of human asthma is perhaps the disease at a later stage, i.e. chronic obstructive pulmonary disease. I have believed for a long time that this is the end stage of some of the inflammatory processes and that these patients are functionally compensated asthmatics.

KAY: inflammation in general and inflammation in the lung in bronchial asthma in particular result from a continuous stimulus and seem to be perpetuated if there is persistent release of chemotactic agents. Once the mechanism for recruiting cells has stopped then the healing process can start. What is the fundamental inflammatory focus in asthma? Is it an abnormal epithelial cell, a defect in T-cells controlling local immunity, leaky mast cells or, most likely, a quite unknown lesion? I would not be surprised if it turns out that there is something very fundamentally wrong with the bronchial mucosa in asthma and that a quite new approach is needed to address this issue.

MENZ (Switzerland): I was very interested to hear about the findings of Lichtenstein's group concerning the heterogeneity of IgE and the ability to bind histamine-releasing factor. What is the opinion of the panel and other experts here about this?

WASSERMAN: I think it is a very interesting observation which at this point is totally phenomenology. For those of you who have not kept up with the field, Lichtenstein's group has demonstrated that it is now possible to take basophils and to remove IgE from the IgE receptors and then have a 'virgin' basophil. Then one can take IgE from a variety of donors and resensitize this basophil and then ask the basophil to release mediators. It is then possible to show that IgE from some donors only will mediate the release of histamine from these basophils when interacting with a factor. The factor is called an 'IgE-binding factor' or 'histamine-releasing factor' and is generated from monocytes or lymphocytes. Other IgE donors do not do this. My understanding is that individuals do not have mixtures of the two forms of

IgE, at least it has not been possible to separate the two forms on any kind of column. What the nature of the heterogeneity is, is totally unclear. From my understanding of the genome there is only one β-gene and it may have to do with glycosylation rather than other aspects of the IgE molecule, or it may have to do with other factors that we have no understanding of.

FORD-HUTCHINSON: allergic rhinitis and allergic conjunctivitis seem to be quite important clinical entities. I am concerned about the way in which these diseases are evaluated in drug trials. They are essentially symptom scores. Has anyone any ideas about how we could be more objective about evaluating these diseases?

KAY: although symptom scores are subjective they often seem to be the most useful measurement in many of these studies.

AUTY: a well-selected, well-trained patient group who diligently complete symptom scores on diary cards provide as good a measure of efficacy of a drug as anybody has managed to come up with. Rhinomanometry had a vogue for some years but turned out not to be reproducible and extremely difficult to manage with large numbers of patients. I do not know of any other objective measures which have been tried in the nose.

PERHACH: I think the priming phenomenon demonstrated by John Connell (*J. Allergy* 1969; **43**:33–44) with insufflation of allergen is important. It shows that symptoms are related to the native condition of the nasal mucosa. You can have high allergen exposure with low symptoms but after challenge (priming) you can get the reciprocal with very low amounts of allergen insufflation giving high degree of symptoms. I agree that neither anterior nor posterior rhinomanometry correlate with symptoms to any reliable degree, especially when one tries to do a multicentre trial. In the end, the patients telling you how they feel is really the best we have (Connell JT *et al., Am. J. Rhinology* 1987; **1**:3–16).

BARNES: if you have a drug that relieves symptoms then that is good news. If it makes anterior rhinomanometry better that is irrelevant if it does not make the patient feel better. I think people feel that measuring symptoms is not scientific because it is easy to do. That is not the case because symptom scores need to be carefully evaluated. Visual analogue scales have been used to measure dyspnoea. There is no other way of measuring dyspnoea because it is a subjective sensation. If you apply such a technique incorporating all the controls that are used in any objective measurements (i.e. of lung function tests) you will find that it is extremely reproducible. I do not think we should denigrate the measurement of symptoms especially when that is what we should be interested in.

WASSERMAN: we do not know enough about these diseases to know whether we should be treating the asymptomatic phases. As we learn more about allergic rhinitis and asthma, the question arises as to whether we should intervene early to reverse changes such as inflammation. We should be able to measure, in some objective fashion, changes in the eye and the nose so that we can attempt to reverse changes. We only treat symptoms, but asymptomatic allergic rhinitis may progress to rhinosinusitis and asthma. I do not think it is a trivial point. We should try to determine by some objective criteria what is going on in these patients with measurements other than IgE and RAST.

KAY: having said that, is there in fact any way in which we can improve the design of clinical trials?

HOLGATE: I think the major problem in clinical-trial design is maintaining homogeneity for the population of patients involved. Returning to asthma, one of the disturbing features about this group of disorders is the tremendous discordance between symptoms and objective measurements of airway obstruction. There cannot be many diseases of the respiratory tract where discordance between objective and subjective assessments is so evident. Symptoms are the only thing we have hard data on in our clinical trials. In clinical trials in which peak expiratory flow measurements have been used there is frequently too great a scatter to make use of the results. With some of the newer drugs that are coming along, which we believe modify an underlying process, I would like to suggest that we obtain more pathological information. Take, for example, the histopathological data which has been generated recently in the nose with inhaled steroids. Two or three years ago who would have thought that one could greatly reduce the mast-cell population in the nose with corticosteroids? At present we frequently try to answer too many questions with our clinical trials rather than focusing on just one or two.

SCHEINMANN (Paris): as a paediatrician I do not think that subjective assessment of symptoms is sufficient to assess the value of a drug in children. You have children who feel perfectly well with any drug and yet they have persistent airway obstruction.

DREBORG: I think that diary-card scores are as reliable in children as in adults. It depends on how the scoring system is designed. With simple scoring systems it is possible to show efficacy of a drug in children as in adults. I have supervised a lot of trials and I am quite confident about this.

SCHEINMANN: I think it is much more dangerous to have bronchial asthma than allergic rhinitis and objective measurements are needed in asthma. Perhaps they are not needed in rhinitis. No child will die of allergic rhinitis but some children die of undetected asthma.

MALOLEPSZY (Poland): we have two main types of asthma, atopic and non-atopic, and we are still looking for the triggering factor in non-atopic asthma. Do we have any provocation test for non-atopic asthma?

BARNES: in intrinsic asthma there is by definition no specific allergen to use in inhalational challenge tests. However, I do not think there is anything which differentiates between intrinsic and extrinsic asthma other than the fact that extrinsic asthmatics respond to allergens. Everything we know about intrinsic asthma tends to suggest that it has a very similar pathogenesis and pathology to extrinsic asthma. Both have eosinophils, both respond to steroids and you cannot clinically differentiate extrinsic and intrinsic asthma nor is there any point in doing so.

WASSERMAN: is not all asthma essentially extrinsic? Air pollution, cold air, exercise; they all come from outside. All we are saying is that there are many triggers.

BARNES: they are non-allergic. Intrinsic equals non-allergic.

WASSERMAN: how do we know that it is non-allergic?

HOLGATE: you are talking about a clinical entity not about pathology. We know nothing about the pathology of non-allergic (intrinsic) asthma as classically described in women over 40 years.

WASSERMAN: in our country very few clinical trials include substantial numbers of women. A high proportion of patients with asthma are obviously female but we are systematically excluding them from our studies. I think we may be missing a major factor.

KAY: this is a very difficult area. What do we really mean by extrinsic and intrinsic? We can have an older person who has severe chronic asthma, where from the clinical history allergy triggers do not seem to be important as, say infection, and yet he is atopic in the sense of having positive skin tests to common allergens. Several years previously he may have had a typical history of allergen-induced asthma but the pattern of the disease often changes and allergy seems to fade into the background. We also recognize the traditional, classical, so-called intrinsic or cryptogenic type of asthmatic, who is skin test negative and where infection also seems to be a very important trigger. Are these two types of individual the same? Many of us believe that it would be very useful to have a model of asthma provoked by an infectious agent. There would be quite a lot of ethical problems with this approach but it may be something worth considering. Personally I think it would be very useful to have a model of viral-induced asthma.

VIRCHOW (Switzerland): there is a model of asthma for a subgroup of both intrinsic or non-extrinsic asthma, and this is asthmatics who are sensitive to analgesics (non-steroidal anti-inflammatory drugs). You can perform tests very easily, either orally or by inhalation, as we do in Davos. I think this could be an interesting model to study non-extrinsic asthma.

VOLOVITCH (Israel): to answer your question about the virus. In Buffalo, New York, we did studies in which we showed that viral infection induced the production of leukotrienes in nasal secretion. The viruses were RSV or para-influenza. We also have been able to show that IgE-mediated immunity is a part of this problem since children with RSV-induced production of leukotrienes had higher levels of IgE in their nasal secretions. We know that this is not the only mechanism because other people without IgE also had some leukotrienes. Nevertheless we consider leukotrienes to be one of the mediators worth considering in asthma and viral infection.

SHEINMANN: will the natural history of asthma be modified by antiviral drugs given to bronchiolitis in infancy?

KAY: that is an interesting thought. As you say, every paediatrician has the experience of bronchiolitis in childhood progressing to asthma.

DREBORG: we need better epidemiological studies, and longer follow-up studies in asthma so that we can learn more about the natural development of the disease.

HOLGATE: I endorse entirely your views on longitudinal studies. Last year we completed a longitudinal study of 20 asthmatics, looking at their symptoms, peak flows and methacholine responsiveness every 2 weeks for a year (Josephs *et al. Am. Rev. Respir. Dis.* 1989; In press). In some of these subjects, methacholine responsiveness varied up to 1000-fold within a period of a few weeks, while in others it did not vary at all. Attacks of asthma which were severe enough for children or adults to consult their family doctor were not related to changes in hyperresponsiveness. They were, however, frequently correlated with the patient recognition of an 'influenza-like illness', coryza or upper respiratory tract viral infections. Finally, what was most interesting to us was that baseline changes correlated most closely with what patients perceived as symptoms. Factors which cause changes in reactivity and factors which produce asthma attacks and make the patient consult the doctor may be different things.

FORD-HUTCHINSON: one of the problems in industry is that we are limited by our preclinical models. We started with the rat PCA test and this was obviously a disaster. Apart from anything else, rat mast cells are not like human mast cells. *Ascaris* challenges in monkeys were believed to be more like the human situation. Five years ago clinicians were saying that the late phase was important. People developed models that affected the late phase as well. Now you say that it is hyperresponsiveness and eosinophil migration and companies are busy developing models of those and finding drugs. I predict that in 5 years time you will say provocation tests are useless and that asthma is a chronic disease and you have to have a chronic model. To forestall you I have decided to start working with horses with chronic obstructive pulmonary diseases (heaves)!

KAY: that is very interesting, but I would like to correct one thing. I think it was the pharmaceutical industry who were responsible for the 'blind alley' of using rat passive cutaneous anaphylaxis as a model of human asthma. Clinicians have been telling industry for years that mast-cell stabilizers are not helpful in the treatment of chronic asthma of any severity. Also, do not dismiss the late-phase reaction. It still has a lot going for it. The trouble is that inhibition of allergen-induced late reactions in humans is so time consuming. It requires considerable patient cooperation and you need formidable resources.

SMITH (Beecham Pharmaceuticals): I think that compounds which stabilized rat mast cells often stabilized the mast cell in humans. I and others in the pharmaceutical industry produced such compounds. In the clinic they protected against the immediate allergic reaction as well as the late reaction and yet in clinical trials they were not effective. Thus, one begins to doubt whether we should be working on the late reaction. We ourselves are working on the involvement of the eosinophil in airway hyperreactivity.

WASSERMAN: the mast cell does a lot more than release histamine. There are mechanisms for dissociating histamine release from leukotriene generation and other membrane remodelling events. Before we abandon the idea of mast-cell stabilizing drugs we should look at more than the release of histamine as an index of mast-cell activation.

DREBORG: there are always some patients who respond to drugs and some who do not, for example, Professor Kay's study on sodium cromoglycate (SCG). Patients are not homogeneous. Have you ever tried to study the subpopulations, those who respond, those who do not respond, their characteristics and how they behave?

KAY: very little is known about SCG non-responders. Quite a lot is known about corticosteroid-resistant patients. They may have features in common such as chronically activated T-cells. We are actively involved in this area at present.

HOLGATE: at present we do not have any information on the relationship between acute challenges and clinical day-to-day asthma. Maybe we could learn a lot from a detailed longitudinal study of classical allergic asthmatic patients. This might involve some invasive procedures, mediator measurements, drug studies and so forth, but would provide considerable information about the character of the disease. Historically, a detailed and careful study of a single patient in many diseases has led to a key observation and a major step forward in understanding. In most of our work, i.e. clinical trials, we are compelled to use statistics and generate group data. We are possibly deluded into thinking that a p value of <0.05 tells us about the nature of a disease process! It clearly does not.

PERHACH: in those subsets of patients who exhibit a late-phase reaction do you then find that they respond predictably to routine clinical treatments?

KAY: that is a difficult question to answer because the late-phase reaction is itself highly artificial. If you pressed hard enough with your allergen you will also get a late response but you are limited by the magnitude of the early response. Generally speaking, patients who get late responses elicited by a reasonable allergen concentration are the ones who tend to more severe asthma but this is certainly not a hard and fast rule or association.

SMITH: the late-phase reaction is a rather loose term. Late-phase reactions in animals are often defined just as cellular infiltration. Do patients who respond with an early reaction without a clinical late phase have a cellular infiltration?

KAY: when we talk about the allergen-induced late-phase reaction in the context of asthma or rhinitis or late-phase reactions in the skin we are talking about a clinical phenomenon. We do not mean a pathological response. Nevertheless there is always some cellular infiltration in so-called single early responders and I am sure this is important. Do products from the infiltrating cells actually cause the clinical symptoms in the late reactions? I do not think we know.

HOLGATE: I would like to come back to the bronchodilators. I wonder if Peter Barnes might like to comment on the new β-agonists with their new profile of activities and how we should use them?

BARNES: the drug that looks best in that respect is salmeterol which appears to cause bronchodilation for more than 12 h. This drug will probably be extremely popular with patients because it will relieve their symptoms for a long time. It should also be useful for nocturnal asthma which is still quite difficult to treat in some patients. I do think, however, that there is a danger associated with its effectiveness. It will reduce the demand of the patients for some other treatments and they might be even less inclined to take their inhaled steroids or other prophylactic drugs such as cromoglycate because symptom relief will be so dramatic. There is data from Laitinen's group where he has actually biopsied asthmatics after two types of treatment. In one instance patients were treated with β-agonists for about 3 months and in the other instance they were treated with inhaled steroids. What he found was that after inhaled steroids, the airway epithelium completely reformed and all the inflammatory cells disappeared. In those patients maintained on β-agonists, the inflammation was exactly the same as in people having no treatment at all. I am concerned that patients are going to be taking effective treatments but this will mask their underlying disease process and may lead to long-term deterioration in lung function, particularly in children.

VARLEY (Glaxo): what effects do β-stimulants have on inflammatory cells?

BARNES: we do have the evidence that they do not clear inflammation from the airway.

VARLEY: that is with salbutamol, presumably.

BARNES: yes, but it was taken regularly. Salbutamol taken regularly would seem to me to be identical pharmacologically with salmeterol administered less frequently.

LEE: I wonder whether anyone would like to comment on the data suggesting that inhaled β_2-agonists, maybe terbutaline, actually increase bronchial hyperresponsiveness over a period of time. Does this have physiological relevance? How does it alter our use of other β_2-agonists other than terbutaline?

BARNES: I think the studies you are referring to are from Holland where they have documented a very small increase in reactivity. It was very small and unlikely to be of clinical significance. It is a small change and goes away with continued use. The important thing that those studies highlighted was the fact that the reactivity did not improve (which people had expected).

HOLGATE: I am aware of a similar study that Anne Tattersfield's group has undertaken. They have formally looked at this and has shown with salbutamol that the same phenomenon occurs. I believe the increase in responsiveness was no greater than a 1.5–2-fold doubling dilution of histamine and, therefore, a fairly small effect. The patients themselves did not detect any symptomatic deterioration.

COLEMAN (Ware): Peter Barnes anticipated that one dose of salmeterol will be essentially the same as repeated doses of salbutamol. This may or may not be the case. Salmeterol and salbutamol do have a dramatically different profile in terms of their duration of action. It appears that after treatment with salmeterol the effects are very long lasting and it may be that twice-a-day treatment will give you a true 24-h cover, whereas with four-times-a-day salbutamol, by the time the next dose is due the effect of the salbutamol is effectively no longer there. I think it may be that the compound will prove itself one way or the other, but I think it will be very interesting to see whether in fact salmeterol does behave just like a long-acting salbutamol or whether it has a little something extra.

BARNES: I think that is the very danger. It does give 24-h cover and the patients are pleased but if you biopsied them you might find the airways were very inflamed.

COLEMAN: yes, I agree it could be a problem but on the other hand it could mean that the compound does have a truly different profile. It may have a chance to do something at the cellular level that salbutamol is not doing. I think it will have to be checked. Hopefully we will be pleasantly surprised but that remains to be seen.

BARNES: regarding the effects of drugs on inflammatory cells, we have been looking at the eosinophil. β-agonists have no effect on that cell, nor on the neutrophil, nor on the macrophage. Yet steroids are dramatically effective against eosinophils and macrophages, which may be important.

WASSERMAN: it is of some concern that the only drugs (i.e. β-agonists) that act rapidly might be lost if used chronically. Suppose it is exerting maximal effect, continuously, in the face of oedema, mucus secretion, mucus plugging and inflammatory cell exudate. When the patient finally does become truly symptomatic and 'breaks through' the β-agonist there is nothing else that will work acutely.

SMITH: whilst β-stimulants may have no direct effect upon the inflammatory cells they are very effective indeed at inhibiting cellular infiltrations. In some models they are very effective at inhibiting the release of leukotrienes. So one may ask why are we looking for an inhibitor of leukotriene release when the β-stimulants already do this. They also prevent extravasation of plasma proteins so perhaps the newer β-agonists will reduce inflammation in the lung.

QUESTIONER (unidentified): is azelastine related to the flavonoid compounds?

PERHACH: structurally not. Elliott Middleton in Buffalo (who is an expert on the flavonoids) and I have had a number of conversations about the similarities of the actions as opposed to the structure. There are similarities between azelastine and flavonoids but not of a structural nature.

KAY: I am very grateful to the contributors for giving such excellent presentations and to those who entered into the lively discussion and debate. In the circumstances I think we should have another of these meetings in the near future. I hope that by then some of the issues raised here will be resolved.

Index